Efficient Use of Energy

(The APS Studies on the Technical Aspects of the More Efficient Use of Energy)

AIP Conference Proceedings
Series Editor: Hugh C. Wolfe
No. 25

Efficient Use of Energy

(The APS Studies on the Technical Aspects of the More Efficient Use of Energy)

Editors

Part I - A Physics Perspective

K.W. Ford, G.I. Rochlin, R.H. Socolow

Principal Investigators: Marc Ross, R.H. Socolow

Part II- The Role of Physics in Combustion

D.L. Hartley, D.R. Hardesty, M. Lapp, J. Dooher, F. Dryer

Part III- Energy Conservation and Window Systems

S.M. Berman, S.D. Silverstein

American Institute of Physics

New York 1975

L.C. Catalog Card No. 75-18227
ISBN 0-88318-124-X
ERDA CONF-740752

American Institute of Physics
335 East 45th Street
New York, N.Y. 10017

Printed in the United States of America

PREFACE

The APS Studies on the Technical Aspects of the More Efficient Utilization of Energy

The traditional role of the American Physical Society has been to serve the physics community through the organization of technical meetings and publication of journals. In 1972, the Society established an Ad Hoc Committee to consider its future direction and this committee's report strongly recommended that the Society assume broader responsibilities for non-traditional activities. About this same time, government, scientific, and educational society representatives began to explore new roles for the societies whereby they could constructively serve emerging public needs.

Representatives of the American Physical Society met with the Science Advisor to the President on 2 March 1973 and identified a wide spectrum of potential Society activities in the public arena. The APS Council, on 22 April 1973, empowered the President of the American Physical Society to appoint an ad hoc committee to examine the technical and administrative problems which would have to be resolved if the Society were to undertake specialized short-term studies. This committee, chaired by Professor Jack Sandweiss of Yale University, met in Los Alamos in the summer of 1973 and then recommended that the Society explore the possibility of undertaking summer studies on a number of specialized topics related to the energy problem. After further investigation, three summer study topics were chosen: (1) The Technical Aspects of the More Efficient Utilization of Energy, (2) Technical Aspects of Nuclear Reactor Safety, and (3) Radiation Effects on Materials.

The Summer Study Group on the Technical Aspects of the More Efficient Utilization of Energy met at Princeton University from 8 July to 2 August 1974. The objective of the group was to identify areas of physics research and development which promise to improve the technologies governing energy efficiency at points of use. Three parallel substudies were conducted which are included in this volume. They are:

I. *A Physics Perspective*—which develops a conceptual framework to evaluate research opportunities in the use of energy indoors, by the automobile, and in industrial processes.

II. *The Role of Physics in Combustion*—which defines physics research possibilities in Experimental Diagnostics, Combustion Modeling, and Emulsified Fuels.

III. *Energy Conservation and Window Systems*—which relates fundamental physics research on windows to reduction of energy consumption for residential and commercial climate control.

The Council of the American Physical Society established an advisory panel of three distinguished physicists to review the study reports. This panel was chaired by Dr. James Comly, Manager of the Thermal Branch of the General Electric Company Corporate Research and Development, with Professor Luis Alvarez, Nobel Laureate, of the University of California, and Dr. Alan Chynoweth, Director of Materials Science Research at Bell Telephone Laboratories as its members. Their critical review has contributed substantially to both the quality and clarity of these reports and we are grateful to them for their efforts and patience. The support of this summer study group by the National Science Foundation, the Federal Energy Administration and the Electric Power Research Institute is gratefully acknowledged.

W. W. Havens, Jr.
Executive Secretary
The American Physical Society

PARTICIPANTS AND BRIEFERS

Part I **A Physics Perspective**

Participants:

Walter Carnahan, Hammondsport, NY

Barry M. Casper, Carleton College and Stanford Linear Accelerator Center

Kenneth Ford, Physics Department, University of Massachusetts, Boston

Andrea Prosperetti, Engineering Science Department, California Institute of Technology

Gene Rochlin, Energy and Resources Program, University of California, Berkeley

Arthur Rosenfeld, Lawrence Berkeley Laboratory, University of California

Marc Ross, Physics Department, University of Michigan, Ann Arbor

Joseph Rothberg, Physics Department, University of Washington, Seattle

Thomas Schrader, Department of Aerospace and Mechanical Sciences, Princeton University

Michael Schwartz, Department of Chemical Engineering, Princeton University

George Seidel, Physics Department, Brown University

Robert Socolow, Center for Environmental Studies, Princeton University

Gary Thomas, Electrical Sciences Department, State University of New York, Stony Brook

Myron Uman, National Academy of Sciences

Richard Werthamer, Bell Telephone Laboratories

Briefers:

Shephard Bartnoff, Jersey Central Power and Light

Frank Beldecos, Westinghouse Electric Corporation

Charles Berg, Federal Power Commission

Manfred Breiter, General Electric Company

Richard Garwin, IBM Watson Center

John Gibbons, Federal Energy Administration

Del Hoover, Westinghouse Electric Corporation

Heinz Jaster, General Electric Company

J. T. Kummer, Ford Motor Company

Norman Kurtz, Flack and Kurtz, New York, NY

Stephen Mallard, Public Service Electric and Gas Company, New Jersey

Herbert Nash, Pennsylvania Power and Light

Frank Powell, U. S. National Bureau of Standards

William Rollwage, Dow Chemical Company

Craig Smith, Applied Nucleonics, Los Angeles, CA

Chauncy Starr, Electric Power Research Institute

Robert Williams, Energy Policy Project, Ford Foundation

Part II **The Role of Physics in Combustion**

Participants:

Russell Barnes, Battelle Memorial Institute

Frediano V. Bracco, Department of Aerospace and Mechanical Sciences, Princeton University

Julius Chang, Lawrence Livermore Laboratory

Paul Chung, Department of Energy Engineering, University of Illinois, Chicago Circle

John Dooher, Physics Department, Adelphi University

Frederick Dryer, Department of Aerospace and Mechanical Sciences, Princeton University

Harry Dwyer, Mechanical Engineering Department, University of California, Davis

Robert Gelinas, Theoretical Physics Division, Lawrence Livermore Laboratory

Frederick L. Gouldin, Department of Mechanical and Aeronautical Engineering, Cornell University

Donald Hardesty, Combustion Research Division, Sandia Laboratories

Danny Hartley, Combustion Research Division, Sandia Laboratories

Phillip Hooker, Union Oil Company and Department of Petroleum Engineering, Stanford University

Marshall Lapp, Corporate Research and Development, General Electric Company

Sullivan Marsden, Department of Petroleum Engineering, Stanford University

Andrea Prosperetti, Engineering Science Department, California Institute of Technology

Marvin Ross, Hydrodynamics Division, Lawrence Livermore Laboratory

Sydney Self, Mechanical Engineering Department, Stanford University

Briefers:

John Brogan, U. S. Environmental Protection Agency

Wen Chiu, Cummins Diesel

Coleman Donaldson, ARAP, Princeton, NJ

W. J. Ewbank, Department of Mechanical Engineering, University of Oklahoma, Norman

Irvin Glassman, Center for Environmental Studies, Princeton University

Part III **Energy Conservation and Window Systems**

Participants:

Stanley Barker, Bell Telephone Laboratories

Philip Baumeister, Optical Sciences Institute, University of Rochester

Samuel Berman, Stanford Linear Accelerator Center

David Claridge, Physics Department, Stanford University

Howard Gillery, Glass Research Center, Pittsburgh Plate Glass

Mortimer Labes, Chemistry Department, Temple University

Robert Langley, Engelhard Mineral and Chemical Corporation

Henry Mar, Systems Research Center, Honeywell Corporation

Leonard Muldawer, Physics Department, Temple University

Steven Schnatterly, Physics Department, Princeton University

Seth Silverstein, Corporate Research and Development, General Electric Company

CONTENTS

PART I – A PHYSICS PERSPECTIVE

PART II – THE ROLE OF PHYSICS IN COMBUSTION

PART III – ENERGY CONSERVATION AND WINDOW SYSTEMS

EFFICIENT USE OF ENERGY

PART I

A PHYSICS PERSPECTIVE

Edited by

Walter Carnahan

K. W. Ford

Andrea Prosperetti

G. I. Rochlin

Arthur Rosenfeld

Marc Ross

Joseph Rothberg

George Seidel

R. H. Socolow

PART I

PREFACE AND ACKNOWLEDGMENTS

The main body of this report is intended for persons with scientific training who are interested in energy but are not yet energy experts. It is unabashedly tutorial in character. Most of the time, we show not only the numbers but how to get the numbers. For readers interested primarily in conclusions and recommendations, a summary is provided at the beginning (pp. 1–11). This summary should also be a useful guide for readers who intend to dig more deeply into the report. For those who wish to go beyond the summary yet still bypass many of the details, we recommend the following route through the text:

All of Chapter 1;
Chapter 2, Sections A, B, F;
Chapter 3, Sections A, D, F (plus selections of special interest in Sections C and E);
Chapter 4, Section G (Sections D, E, and F provide a more complete discussion, and Section C provides the background for the recommendations);
Chapter 5, the first few pages for an overview, and selections thereafter of special interest to the reader

A cautionary word for readers who are experts in specific areas: you will not find this report to be competitive in depth or precision with the best that is available in the technical literature of specific devices or systems. Our goal is not to provide "state-of-the-art" information, but to present a unified overview of energy efficiency from the perspective of physics. This means bringing the laws of thermodynamics to stage center and it means adopting the typical strategies of a physicist: simplifying, idealizing, and seeking common elements in different problems. We hope thereby to reveal new research opportunities and attract new talent to the field.

We have endeavored, as a rule, to quote numerical results in both British (now, increasingly, exclusively American) engineering units and in international standard (SI) units. When calculations were done in one system (usually, engineering units), there will often be discrepancies in the last significant figure in the other. The watt (W) is frequently used as a unit of power, whether or not the energy is electric, and subscripts e and t are used for electric and thermal power where there might be confusion. Throughout, we have assumed that one unit of electric energy can be created in the present economy with the consumption of three units of fuel energy; thus $1\ W_e \longleftrightarrow 3W_t$.

This summer study was conceived at an ad-hoc week-long meeting in August 1973 at Los Alamos, New Mexico, held under the auspices of the American Physical Society, chaired by Jack Sandweiss, and organized by Frank von Hippel and Marvin Goldberger. In the period of proposal development, the wisdom and encouragement of William Havens, APS Executive Secretary, and W. K. H. Panofsky, APS President, were indispensable.

Paul Craig, Kurt Riegel, and Craig Smith, representatives of our sponsoring agencies and Sidney Millman at APS headquarters were most attentive and helpful as we planned the study. For careful readings of various parts of the report, we are also indebted to Samuel Berman, Stanley Blois, John Brogan, David Claridge, Margaret Fels, Henry Kolm, Kenneth Kreider, J. T. Kummer, Dillard Murrell, Lawrence Sweet, Robert Williams, and Princeton undergraduates enrolled in AMS 213. (Short-comings in the report are, of course, the responsibility only of the authors.)

Two Princeton graduate students, Thomas Schrader and Michael Schwartz, played a critical role as research assistants. Caryl Wiggs ably headed a very busy secretarial staff. The preparation of the final drafts of this report at the University of Massachusetts, Boston, was made possible by the capable work of Joyce Kauffman, who did the typing, and Roderick Macdonald, who drew the figures.

PART I
A PHYSICS PERSPECTIVE

INTRODUCTION

SUMMARY

Framework of the report

As every student of energy knows, there is a large and complex set of energy problems having economic, social, political, and technical components. In this report we focus on the *technical* aspects of energy use, with emphasis on energy efficiency. We also focus primarily on *end use* of energy. Our perspective is that of physics.

In limiting ourselves largely to the technical components of energy use, we do not derogate the *regulatory, economic,* or *persuasive* tools that may be brought to bear on energy consumption, nor do we assess the relative importance of such measures either among themselves or with respect to technical improvement or innovation. Our report serves to introduce scientists and engineers to problems of energy efficiency and it points out areas where they may contribute to invention or improvement in the technical structure of the energy economy.

The first part of this introduction (pp. 1–7) is a summary of the main ideas in the report. The second part (pp. 7–11) is a listing in outline form of many of the specific research opportunities identified in the report.

Task efficiency

Specifying the service and defining the task

The first objective of any technical study of energy use is to establish a norm, a standard of performance against which present use of energy can be evaluated and a goal towards which technical innovation can strive. This presented us with the immediate problem of determining the purpose for which energy is consumed. Since this purpose is often psychologically or socially determined, it was necessary to separate the problem further to distinguish matters of individual or social choice from those of technical efficiency. If a discussion were to be organized according to the services performed by energy use—keeping warm in the winter, for instance, or obtaining structural materials, or transporting goods—perspectives other than that of physics would be required. Our physics perspective leads us to focus on specific energy-using tasks—providing heat to a home, or making aluminum, or moving a vehicle. By so limiting our discussion of energy use, we avoid the problems of such nontechnical tradeoffs as sweaters for heat, steel for aluminum, or mass transit for automobiles. However, we take care not to limit ourselves to analyzing the specific devices used to perform these tasks—furnaces, the Hall process, the internal combustion engine—even though these devices dominate our present energy-use patterns. Their efficiency as devices must be compared to a more general physical performance standard defined not by the device, but by the task. Such analysis delineates the limitations and inefficiencies of the devices we now have, and indicates where they should either be improved or be replaced with new devices or integrated to form new systems which perform joint tasks more efficiently than either could separately accomplish.

Thermodynamics and available work

The focus on task rather than device emphasizes the need for the definition of a minimum energy path for performing a given task by means of *any* possible device or system. [The idea of minimum task energy has been emphasized in some recent technical literature (references appear in the body of the report); the idea has its roots long ago in the work of J. W. Gibbs]. This makes it possible to estimate the potential for interchanging one device or system for another, and provides guidelines for seeking scientific and technological innovations which more closely conform to the path of least energy use.

The use of the first law of thermodynamics, the conservation of energy, is inadequate for considering minimum task energy. We know that energy is not lost; from the viewpoint of this law, the task of minimizing energy consumption appears to be primarily one of hoarding. Yet we also know that, in any process involving heat, the constraints of the second law of thermodynamics, of the inexorable increase of entropy, usually guarantees that not all of the energy can be made available in useful form. The second law makes it fruitful to define a quantity that has the dimension of energy, yet is actually consumed in a process, *available work*. There is an upper limit to the amount of ordered work energy that can be produced from thermal energy at a temperature T_1 in a setting of ambient temperature T_0. This upper limit is, roughly speaking, the available work, the maximum work that can be provided by a system as it proceeds to its final state in equilibrium with the atmosphere. Work is the highest quality (lowest entropy) form of energy, and consequently the most valuable.

Second-law efficiency

The "available work" concept makes possible a useful general definition of efficiency, which we choose to call *second-law efficiency:* For any device or system, the second-law efficiency ϵ is the ratio of the least available work that could perform the task to the available work actually consumed in doing the job with a given device or system. This second-law efficiency, whose maximum value is always $\epsilon = 1$ by definition, provides rapid insight into the performance of a specific device executing a specified task in terms of how efficiently that task might be performed were an ideal device available. It shows how much room there is for improvement. Maximizing ϵ necessarily minimizes energy consumption, and the maximization of ϵ for any given task provides a realistic technical goal for energy use, to be ranked alongside economic, environmental, and other goals to help form a coherent energy policy.

The evaluation of energy use in terms of second-law efficiency is a point of major emphasis in our study, and a unifying thread in our report. We strongly recommend that this formulation, or a similar one, be widely adopted by the scientific and technical community as a standard from which all tasks should be measured, and against which all devices should be evaluated.

A few simple examples will suffice to show the great power of this method: (1) A certain type of hydrogen-oxygen fuel cell is said by its inventor to have an efficiency (electric energy produced per unit heat of combustion) of 70 percent. The available work for this process, however, is only 82 percent of the heat of combustion, so that this device is operating at 85 percent of its maximum second-law efficiency. The distinction may be crucial for determining allocation of further research and development funding. (2) An air-conditioner has a quoted coefficient of performance (COP) of 2. Although under specific conditions it extracts two units of heat from the room for every unit of electricity consumed, the COP changes rapidly as ambient and room temperatures change, and is not restricted in value. The maximum COP might be considerably greater than 2. The calculated second-law efficiency for this device might be only $\epsilon = 0.1$, and depend only weakly on temperature. This is a relatively stable figure of merit for comparing different units or alternate methods of cooling. (3) The quoted first-law efficiency of a Carnot engine operating between flame temperature and ambient temperature might be 0.87. For internal combustion, however, the flame is a finite, not an infinite, reservoir, and 30 percent of the available work is lost in the combustion process. Carnot efficiency is seen to be an unachievable goal for engines of this type.

Focal points of this report

Energy use can be treated in this manner at all points along its path from resource to ultimate dispersal as heat at near-ambient temperature. The bulk of our report, however, concentrates not on energy production, nor on aggregated sector use, but on the end-use, the task ultimately responsible for the energy consumed. We have chosen three categories of end use—the house, the automobile, and industrial processes based on chemical and physical changes of state—as major energy-consuming sectors in which our analysis appears to offer constructive new insights into present inefficiencies and the potential for further improvements from scientific and technical research and development. These three cases form a nicely complementary set for evaluating the ultimate technical potential for reduction in energy use.

The house

The house is in itself a device for providing a service: an environment for its occupants. The energy consuming aspects of the service the house is asked to provide are, in the most general terms, warmth in the winter, coolness in the summer, hot water, light, food storage, and a means for cooking. To the extent that a unit loses heat in the winter or gains it in the summer, or fails to take advantage of such natural aids as sunlight, solar heating, or temperature averaging, it executes the task of delivering these services inefficiently. We review in our report a number of areas where heat loss, or gain, could be reduced. These include better insulation, increased heat capacity for thermal averaging, and better use of house aerodynamics and convective heat flow to minimize losses due to air infiltration and at windows.

Energy-using tasks in the house. Within the house, various devices perform the tasks of heating or cooling air or water, providing light, etc. A second-law analysis shows that many of these tasks are being performed at very low efficiency, and that the fault often lies more with the choice of device than with device efficiency *per se.* A typical home furnace, for instance, may be delivering 60 percent of the fuel energy as heat at 110 °F (43 °C) from the register. But as the high flame temperature is "wasted" in delivering low-temperature heat, the second-law efficiency is only about 7 percent ($\epsilon = 0.074$) for this task, assuming an ambient temperature of 40 °F (4 °C). A gas hot water heater which raises a finite reservoir of 55-°F (13-°C) water to 120 °F (49°C) with an energy efficiency ("first-law efficiency") of 50 percent has a second-law efficiency of only 3 percent. ($\epsilon = 0.03$). What the second-law efficiency tells us is that even a "perfect" furnace falls far short of being ideal for these tasks. (At 100-percent energy efficiency, these devices would have second-law efficiencies of only 12 and 6 percent, respectively.) The use of a high-quality flame, at site or power plant, to provide low-quality, low-temperature heat is a fundamental abuse, and there is a much larger potential for reduction in energy use by using other devices, such as heat pumps, or by using combined systems in which both electricity and low-temperature heat are produced simultaneously by a fuel-consuming device. In both cases, the second-law efficiency provides a basic parameter for evaluating the system.

Low-quality waste heat. Another example is the waste heat from electric power plants. As over one-half the input fuel energy is discharged to the cooling water, the opportunities for use of this waste heat seem large. Yet, for a typical temperature of 100 °F (40 °C), the cooling stream contains only 1.5 percent of the fuel available work. Inefficiencies where the temperature is high account for much larger losses of available work. This waste heat may, however, be put to good use if it replaces heat generated by devices operating at low second-law efficiencies; alternatively, the output stream temperature might be raised to deliberately convey more of the fuel available work for another application. By using second-law efficiencies, the values of these tradeoffs can be quickly and accurately estimated, and intelligent choices made.

The automobile

The automobile is somewhat more difficult to evaluate, because much of the energy waste is attributable to social choice of transportation. Nevertheless, any increase in technical efficiency will reduce energy consumption for any combination of transportation choices. In physical terms, the net difference in energy between any initial and final position at rest at the same gravitational potential is zero. The minimum amount of available work consumed is just the ambient temperature times the amount of entropy produced during the motion. To minimize involvement with nontechnical aspects of the automobile, we have focused on the driving wheels—separating the system into the task of converting fuel available work to energy delivered at the wheels and the complementary task of using that energy to do work against dissipative forces—without considering whether the service of providing transportation might be made more energy efficient by other means, such as improving passenger load factors or shifting to mass transit.

The power train; matching engine to load. We lead the reader through an analysis, pioneered by Kummer, which shows that the second-law efficiency for delivering energy to the rear wheels is only about 10 percent. More interesting is the further decomposition of the power train into stages. Even neglecting friction and heat loss, an ideal fuel-air cycle internal combustion engine can convert less than half of the fuel available work into useful mechanical work. Modern engines, particularly diesels, could be made to operate fairly close to this maximum efficiency (less another 20 percent or so for heat loss and friction). But, in addition to energy wasted in inefficient powering of accessories, automobile efficiency is reduced considerably by poor matching of the engine and transmission to the load. Very overpowered vehicles can be operating at such low efficiency that any reduction in load, by reducing drag or lightening the vehicle, is nullified by further decreases in engine efficiency. An enormous potential exists for decreasing fuel consumption by such strategies as using a more efficient power source, or designing a combined system using a small engine running at high efficiency and high load with a booster to meet peak power demand.

Strategies for reducing fuel consumption. Once the energy is delivered to the rear wheels, it is used to give the vehicle kinetic or gravitational potential energy and to overcome air and rolling resistance. We note that such quasi-social strategies as reducing speed limits (which reduces air resistance) and reducing mass (which gives proportional reductions in all but air resistance) can have large effects on fuel consumption. A number of more purely technical options are available which could also result in large fuel savings. The kinetic and gravitational potential energy dissipated in the brakes need not be dissipated; through the use of flywheels or some other regeneration system this energy could, in principle, be recovered and stored. Air drag coefficients are much larger than they have to be. The rolling resistance dissipates between one-third and one-half of the total energy delivered to the rear wheels in urban driving, and is almost entirely due to viscoelastic flexing of the tires. The design of new tires, or tires using new materials, or entirely new methods of mating the car to the road, could yield very large reductions in vehicle power requirements. The reduction of energy needed at the rear wheels for any combination of these strategies has a proportional effect on reducing fuel consumption, but the scale factor depends on the load matching. A combined strategy that includes concurrent reduction in engine size must be used to realize the full potential savings from reductions in vehicle weight or drag.

Industrial processes

Chemical change. The energy efficiency of industry presented us with a different problem. The use of energy in industry is complex and varied, ranging through refining, smelting, weaving, fabrication, assembly, and so on. In some industries the flow of energy is internally audited and the energy cost of each component tabulated. Other industries have been analyzed for aggregate energy flows, which accomplished roughly the same goal. Such industry-specific studies are very useful for seeking ways to reduce energy use for a given product. More useful, perhaps, for discussing research opportunities is an organization focused not on the product itself, but on the chemical/physical nature of the process. The industrial processes are then classified as belonging to one of three general types: change in chemical state, change in physical state, and fabrication. The first of these can be discussed in terms of available work and second-law efficiency with some facility. Examples of areas where research provides great opportunities are electrochemical processes, such as fuel cells, photolysis of water, and electrochemical energy storage. In all three of these areas there is not only critical technical work to be done to make economically viable and competitive devices with greater second-law efficiencies, there is also a considerable amount of basic research to be done on the microscopic mechanisms, particularly at surfaces.

Physical change and fabrication. Changes in physical state, such as in the desalination of water, involve separative work and typically operate at very low second-law efficiencies at present. Important reductions in energy use could be obtained through a better understanding of the mechanisms of separative processes, such as solute mobility or transport through membranes, and through the development of materials with better selectivity for different molecular species.

Opportunities in fabrication are more difficult to specify; it is quite difficult to estimate the minimum energy path for tasks such as machining a part. There are, however, a wide variety of physical processes, such as beneficiation of ores, alloying, or paint or paper drying, for which the second-law efficiency can easily be specified. Such strategies as selective energy absorption to remove solvents or new ways to distribute dopant materials in the alloy phase could greatly reduce the amount of energy used. Research into other areas such as heat transfer or convection would have implications not only for industry, but for all sectors of the energy economy.

Concluding remarks

Future scientists attacking problems of energy efficiency will require facility in subjects of classical physics—such as fluid mechanics and thermodynamics—that are all too often neglected in our educational enterprise. These subjects need renewed emphasis; they can offer challenge and satisfaction to students and teachers alike.

We believe that the use of second-law efficiencies and available work, and the resulting systematic organization by process and task rather than by industry and device, offers new and powerful tools for seeking areas where research payoffs can be large. At present, our energy resources are being consumed with an overall second-law efficiency of only 10 to 15 percent. This is not only wasteful, but inelegant. There is much room for improvement, for designing our machines to be technically excellent and to reflect the powerful insights of modern science and technology.

SELECTED RESEARCH OPPORTUNITIES

We have not arrogated to ourselves the authority to establish priorities for research. The list of research opportunities below contains problems and areas that we found interesting, but is surely incomplete. Those within the scientific and technical community who are concerned, individually or collectively, with energy efficiency will add additional ideas and will judge the ripeness of the various opportunities.

SECOND-LAW EFFICIENCY (CHAPTER 2)

Systematic investigation of points of entropy increase in energy-using systems

Calculations of second-law efficiencies for industrial processes with a view to identifying the most fruitful areas for technical improvement

Systems studies of the cascading of heat energy in specific industries

Classification of processes according to the minimum required energy quality with a view to better matching of energy source and energy use

Interdisciplinary investigation of more sophisticated procedures of energy and entropy accounting

ENERGY INDOORS (CHAPTER 3)

Energy management

Studies of heat transport over long distances (including heat transported as chemical potential energy)

Systems analysis of district systems to supply both heat and electricity to a building or a group of houses

Systems studies of house-scale heat and electricity systems based on fuel cells (possibly ones deliberately chosen to have low electrical efficiency)

Extensive data gathering and modeling of energy-use patterns of existing houses and buildings of various types

Research on heat storage for better load averaging of both solar and electric power

Thermal properties of buildings

More flexible computer modeling to include the effect of wind

Measurement and modeling of thermal response functions of buildings

Investigation of building materials with large specific heats

Theoretical and experimental studies of external features, including sun control and surfaces of variable reflectivity

Insulation

Theoretical modeling together with experimental studies to select new materials and filler gases, to control cell size, and to enhance fire resistance

Investigation of a "thermal diode"

Development of layered material with variable, controllable conductivity

Development of window coatings to control heat loss (see APS 1974a)

Optimization of insulation for the distribution of space heat

Aerodynamics

Outside: studies of micrometeorology, wind effects on heat transfer and infiltration, control of local air flow patterns

Inside: development of methods to control air flow, especially near walls, windows, and ducts

General: microscopic studies, experimental and theoretical, of heat transfer at solid surfaces bounded by moving air

Air conditioning

Broad search for new techniques using heat to power air conditioners; experimental studies of absorption and adsorption cycles

Search for new desiccants, especially with a view to lowering the regeneration temperature

Systems studies of ground-water cooling

Interdisciplinary studies of ground-water hydrology for heat-transfer applications

Space heating

Continued research on heat-pump cycles and technology to extend the useful temperature range

Systems studies of solar-assisted heat pumps

Systems studies (and hydrology) of using ground water as the low-temperature reservoir of a heat-pump system

Basic heat-transfer studies at solid-fluid interfaces, with a view to decreasing ΔT across heat exchangers working not far from ambient temperatures

Hot water

Detailed measurements and modeling of patterns of hot-water heating and use in houses and buildings

Systems studies of heating or "boosting" water at the point of use

Continued development of solar hot-water heating systems

Lighting

Basic research on gas discharges and on fluorescence with a view to increasing the efficacy of light sources and controlling the color

Development of small and/or screw-in fluorescent lamps (mostly engineering)

Development of efficient kHz power supplies and further studies of frequency-dependence of fluorescent-lamp efficacy

Interdisciplinary studies of lighting needs for various tasks—intensity, uniformity, spectral distribution

Exploration of more effective ways to use sunlight indoors

Instrumentation

Development of easy-to-use local heat flux meter

Development of cheap home instruments for monitoring and control, such as a clock-programmed thermostat, a degree-days-per-gallon meter (with suitable time averaging) and an on-line monitor of furnace burner efficiency

Packaging of a meter to measure air exchange rate in a room or a building

The development for larger buildings of on-line monitoring of several features of air quality coupled with control of reconditioned air vs. fresh air; the development of devices to recondition air

Development of infrared equipment for diagnostic studies of houses

THE AUTOMOBILE (CHAPTER 4)

The power plant

Basic studies of the combustion process; development of advanced combustion diagnostic techniques (see APS 1974b)

Continued research on external combustion engines (Rankine, Stirling)

Continued research on small diesel engines, to reduce their weight, noise, smoke, and odor

Studies of a variety of hybrid power plants (small internal combustion engines plus a source of boost power)

Extended studies of fuel emulsions and novel fuels (see APS 1974b)

Energy storage

Further systems studies of harnessing braking energy with a dynamotor

Basic research on rechargeable batteries (see Electrochemical processes, below)

Experimental and theoretical studies of flywheel storage

Weight

Exploration of lighter automobile structural materials

Studies of the crashworthiness of lighter (not necessarily smaller) cars

Tires and suspension

Basic studies of viscoelastic materials and exploration of radically new tire designs

Exploration of tradeoffs in function among tires, wheels, and suspension, with a view to increasing the role of the suspension and decreasing that of the tire, to preserve ride and safety while decreasing rolling resistance

Air drag

More elaborate studies, experimental and theoretical, of automobile (also truck) air drag, including the effect of relative motion of the vehicle belly and the road

Instrumentation

Development of a sophisticated yet inexpensive miles-per-gallon meter capable of integrating over various times or distances

Implementation of other monitoring meters of engine efficiency, fuel flow, automatic-transmission gear

Accessories

Development and refinement of heat-powered air conditioners for automobiles and trucks

INDUSTRIAL PROCESSES (CHAPTER 5)

Electrochemical processes

Basic research on the physics of charge transfer at electrolyte-electrode interfaces

Basic physical studies of the principles of solid electrolytes

Continued search for rechargeable batteries of greater energy density

Basic research on all aspects of surface phenomena, especially as applied to fuel-cell performance

Photochemical processes

Exploration of catalyzed solar photolysis of water

Continued studies of the semiconductor physics of photovoltaic cells

Physical processes

Studies of adsorption techniques for molecular separation

Basic research on transport in membranes

Careful studies of the energy inefficiencies in ore beneficiation and water desalination to seek ways to approach more closely the minimum separative work

Heat transfer

Basic studies of heat transfer at interfaces—the role of convection, the role of surface irregularities

Basic studies of boiling

Fundamental investigation of two-phase flow

1. COMPLEMENTARY APPROACHES TO ENERGY CONSERVATION

The subject of efficient energy use ("energy conservation") lacks any single organizing principle, and for good reason. We have found at least six complementary perspectives to be of interest in sorting out the subject. In seeking to give the intellectual context of this report (and of the companion reports of the other two working groups of this APS Summer Study), we begin by reviewing these approaches and pointing out, where appropriate, the ways in which our report consciously circumscribes its subject matter. The six approaches organize the subject of energy conservation by:

A. the choice of service and means of providing the service (task definition);

B. the policy tools available for changing the patterns of consumption;

C. the economic sector and the energy source or fuel;

D. the perspective of thermodynamics (how energy is being transformed—high-temperature heat to low-temperature heat, heat to light, etc.);

E. the stage in the chain of transformations from fuel to service;

F. the time period in which a given change can occur.

The single precursor of our study which most effectively incorporates such a range of perspectives is *Technology of Efficient Energy Utilization,* the report of a NATO Science Committee Conference held in France in October, 1973 (NATO, 1973). Like our report, the NATO report emphasizes technological rather than social change. It makes considerable use of economic-sector analysis. It takes adequate account of the second law of thermodynamics. Our report differs in being more tutorial. We have tried to write for persons (especially among our fellow physicists) who are not already energy specialists. This makes our report considerably longer, yet at the same time less comprehensive, than the NATO report.

In general, we try to bring the physicist's style and approach to bear on the problems we discuss. This does not mean ignoring real-world concerns. It does mean sometimes simplifying and idealizing in order to search for the essence of a problem; it means trying to reveal common scientific threads in disparate processes; and it means placing all energy efficiency considerations in the common context of thermodynamics.

A. Choice of service and means of providing it

Suppose the amount of energy to perform a given service is being investigated. Keeping people warm in winter is an example of a service; providing ways to exchange information is another. It is possible to formulate the investigation in such a way that the aspects of individual choice and the aspects of technological efficiency are nearly disjoint. For the example of winter comfort, one can separate the choices of where to live, what to wear, and how to set the thermostat from the energy efficiency of the alternative ways of achieving a given indoor temperature in a given climate. For the example of information exchange, one can separate the choice of whether to telephone, write a letter, or visit (and if to visit, then whether by bus or car, train or plane) from the energy efficiency of alternative ways of distributing the mail or alternative ways of powering planes. The aspects of individual choice depend on attitudes and economic conditions that are largely outside the sphere of the natural scientist and engineer, and they are largely excluded from this study. The factoring is not perfect, however, for there are technical components even to these "non-technical" choices: more versatile controls on space-conditioning systems in the home, for example, might alter perceptions of comfort; the availability of more sophisticated telecommunications might alter travel decisions. We would encourage examination of these technical components in future studies.[1]

Our emphasis in this study is on the efficiency with which a given service is provided. This requires that the service—including underlying assumptions about individual choice—be translated into a specific energy-using task. The task definition is crucial for efficiency considerations.

B. Tools of policy and the role of technology

New technology is not always included among the tools available for changing the patterns of consumption in the direction of energy conservation. There is a considerable body of opinion which presumes that energy conservation is primarily a matter of economics, with higher prices leading to reduced consumption and new products. Others grant that government regulation is an appropriate complement to (or substitute for) higher prices, and still others insist that questions of culture and personal choice are deeply imbedded in the problem. We do not deny the applicability of these concerns. New technology does play a role, however, by restricting and expanding options. Our studies have limited themselves to identifying some technological constraints on achievable energy conservation, and to assessing the possibilities for modifying these constraints.

New technology can take two forms, new systems and new devices. Cooling a house in summer by pumping ground water through coils inside a forced air furnace is a systems idea, whereas improving the separation of oxygen from nitrogen

---

[1] *Exploring Energy Choices*, the preliminary report of the Energy Policy Project of the Ford Foundation (1974), presents a thoughtful discussion of the role of new services and lifestyle changes in reducing energy consumption, as well as the role of technical modification of existing subsystems.

Table 1.1. Sample policy options to reduce energy use in residential heating and cooling.

Area	Sample options
Research and development	Basic research on insulation, on house aerodynamics (see pp. 78–80)
	R & D on absorption and adsorption air conditioning (see p. 63)
	Development and analysis of solar-assisted and ground-water-based heat-pump systems (see pp. 66–68)
	Systems analysis of control and equipment aspects of combined heating and electric generating systems (see pp. 83–93)
Regulation	Requirements for better insulation
	Requirements for the display of life-cycle cost information on heating/cooling systems
Economic tools	Addition of environmental costs to the costs of energy generation and transmission
	Peak pricing for electricity at summer/winter demand peaks
	Provision of attractive financing and/or tax benefits for retrofitting of insulation
Government demonstration	Construction of solar heated and cooled government office buildings
Persuasion and information	Campaign to promote lower indoor temperature in winter and higher indoor temperature and humidity in summer
	Diffusion of information on air infiltration and how to reduce it (see pp. 51, 54–55)
	Promotion of interest in a "degree days per gallon" meter (see p. 51)

by a new molecular sieve is a device idea. Most ideas discussed in this report combine device aspects and systems aspects, with perhaps a preponderance of systems. Our suggestions about where it would pay to look for new devices are based in large measure on modeling of some of the major energy-consuming activities of the society as simplified thermodynamic systems. The two companion reports pursue device questions considerably further than they are pursued here; the discussions of window coatings and combustion in those reports give some indication of the level of detail to which many ideas and questions lightly touched upon here must be carried in order to make an adequate technical assessment of their practicality. Important questions of price, regulation, and acceptability are not emphasized in our reports. Additional analysis by social scientists would be expected to winnow further the ideas we have selected to present.[2]

The breadth of policy options, partially influenced by technology, is illustrated by example in Table 1.1 for residential heating and cooling.

C. Economic sector and energy source

When the subject of energy conservation is organized according to sectors of the economy, one must rely on the categories used by the relevant data gathering organizations; in the United States this gives primacy to the categories of the Bureau of the Census, in its Population Census and its Census of Manufactures, and the billing categories of the gas and electric utilities. One report organized in this fashion has dominated the recent literature, *Patterns of Energy Consumption in the United States*, prepared in 1970 by the Stanford Research Institute, based on what were then the latest available complete data (1968), and released by the Office of Science and Technology in January 1972. We are aware, as are others who worked with the SRI report, that it is a crude cut at the problem, that its methodology is imperfectly documented and often arguable where it *is* documented. The need for carefully documented data in this field is so substantial that a new effort to appraise how much energy goes where is clearly in order immediately, perhaps dealing with available data for 1973. The development of a program of more systematic and consistent acquisition of data is perhaps even more critically needed, so that energy accounts of substantially greater sophistication and reliability can be prepared, perhaps for a year as early as 1975. Such a program might well require the monitoring of a broad sample of actual facilities in order to learn how to reduce the ambiguities inherent in any reporting procedure.

A few numbers from the SRI tables appear in Table 1.2. Somewhat more detail is provided in Appendix I. The unit for measuring energy (or fuel) consumption in the SRI tables is quadrillion (10^{15}) Btu per year[3]; total U. S. consumption was 60.5×10^{15} Btu in 1968.[4] A more congenial number is power consumption per capita, measured in kilowatts thermal (kW_t). The 1968 value was 10 kW_t and the value is about 12 kW_t now. This may be compared with the average rate at which a human being metabolizes the energy in his food. The consumption (and complete metabolism) of 3300 kilocalories (kcal) of food per day delivers 150 watts thermal

[2]The reader is referred to *An Agenda for Research and Development on End Use Energy Conservation* (Mitre Corporation, 1973) for a perspective which emphasizes research in the social sciences. The "Agenda" is replete with imaginative suggestions for systems studies, studies of consumer motivation, and demonstration projects using existing technologies.

[3]Some conversion factors (a must in energy studies!) appear in Appendix II.

[4]Apart from inevitable uncertainties, the figure for total energy consumption is influenced by the way in which hydroelectric and nuclear power plants are treated. Only the *electric* energy generated by these plants enters the SRI accounts, whereas most other studies assign a multiplicative factor of approximately 3 to this electric energy—based on the assumption that if these plants did not exist, fossil fuel combustion would have been required to generate the electricity. (The factor reflects the average efficiency of conversion of chemical energy to electric energy in fossil-fuel generating stations.) A system of accounting along these lines gives 62.4 quadrillion Btu for the U. S. consumption in 1968 and an estimated 75.6 quadrillion Btu in 1973 (Bureau of Mines, 1969, 1974).

Table 1.2. Energy equivalent of fuel consumed in the United States in 1968. [a]

Classification by location of fuel consumption			Classification by end use		
Sector	10^{15} Btu	Percent of total	Sector	10^{15} Btu	Percent of total
Residential and commercial (non-electric)	13.6	22.5	Residential	11.6	19.2
Industrial (non-electric) (including feedstock [b])	19.3	32.0	Commercial	8.8	14.4
Transportation (non-electric)	15.1	25.0	Industrial (including feedstock [b])	25.0	41.2
Electric generating stations	12.4	20.5	Transportation	15.2	25.2
	60.5	100.0		60.5	100.0

[a] Source: SRI (1972). See Appendix I for more detailed tables.
[b] Feedstock is fuel used as a raw material rather than as an energy source.

(0.15 kW_t) to the body, and eventually to the environment. [The per capita *electric* consumption in the U. S. is about 1.0 kilowatts electric (kW_e), half of the installed electric capacity.] Per capita energy consumption in the U. S. is about six times the world average and twice the European average.

One common way of grouping energy-use data divides the economy into four sectors of roughly equal fuel consumption (see Table 1.2): residential and commercial non-electric consumption (dominated by space heating); transportation (dominated by the automobile); industrial non-electric consumption (dominated by process heat below 500 °F); and the consumption of fuels by electric generating stations. The latter represented 20.5 percent of total fuel consumption in 1968, but that sector has been growing at about 7 percent per year, versus 4 percent per year for the other three sectors, so that it is about 25 percent by now (see Table A4 in Appendix I).

In only rare instances is the byproduct heat at power plants actually utilized. This "waste heat" (13 percent of the national energy consumption in 1968, according to the SRI report) is a favorite target for energy conservation strategists. Its size is deceptive, however, since the heat is at temperatures not far above ambient on an absolute scale. Indeed, as we develop the institutional mechanism to use that heat, we will find ourselves choosing to generate electricity less efficiently, to maximize the joint value of the heat and the electric energy. A recurrent theme of this report is that descriptions of the magnitudes of energy uses in terms of Btu's alone, without reference to the thermodynamic *quality* of that energy, are seriously deficient. The SRI tables, and much of the other recent literature on energy conservation, must be used carefully.

The SRI report aggregates energy consumption over all locations in the country and all times of the year. Thus, the consequences of regional variations in relative fuel availability and price and in climate are lost, as are the fluctuations in consumption over time of day and season of year. Several regional studies have been carried out since the SRI report,[5] and attempts to display the results of these studies in ways which help visualize the similarities and differences over regions seem to us to be badly needed. An important direction for improvements in the acquisition of data on energy consumption is toward greater detail about temporal fluctuations. These temporal fluctuations usually imply the under-utilization of capacity in "troughs" and the utilization of inefficient energy conversion devices during "peaks." The energy costs of time-varying load are particularly well known in the context of the generation of electricity, but the principle is broadly generalizable, as our discussions of the hot water heater, the combined heating and electric system for a district, and the automobile engine in later parts of the report are intended to make clear. A number of the strategies discussed in this report either exploit or are constrained by fluctuations.

The SRI report, and much of the more recent literature, strives to treat all fuels on an equal footing. Thus, the Btu's released from the complete combustion of coal (whether strip mined or surface mined), of oil (whether imported or domestic), or of natural gas (whether provided on an interruptible or a non-interruptible basis) are treated as completely equivalent. These Btu's are also treated as equivalent to the Btu's carried in the electric distribution system which originated from nuclear or hydroelectric generating stations.

Analyses which deliberately weight the various energy sources unequally can easily be imagined. They might be carried out to assess more complex objective functions than energy conservation. The attractiveness of a given energy conservation strategy might appear quite different if the mix of sources were weighted to emphasize the impact on capital or balance of payments, or the relative environmental despoliation, or the relative risks of depletion. We are dubious, however, about whether the needs of present-day America would depart very far from an equal weighting of all energy sources. For each source, the advocates and the antagonists are both well supplied with good arguments. In any event, we have abstained from such refinements here.[6]

---

[5]See, for example, *California's Electric Quandary* (Rand Corporation, 1972); *Energy*, A Report to the Governor of New Jersey from the Task Force on Energy (New Jersey, 1974); and *Patterns of Energy Consumption in the Greater New York City Area* (Regional Plan Association and Resources for the Future, 1973). For a significant recent study from another country, see NEDO (1974).

[6]One of the basic references in the field, *The Potential for Energy Conservation*, a staff study by the Office of Emergency Preparedness, appeared in two parts. The first study, October 1972, treated all fuels alike, and the second, January 1973, emphasized fuel substitution (OEP, 1972, 1973).

D. Thermodynamics

We pay considerable attention to another aspect of energy accounting in this report, one which concerns the second law of thermodynamics, and the critical concepts of temperature and entropy. The discussion in most of the energy-use literature falls within the purview of the first law of thermodynamics, which is simply the conservation of energy, where conservation is now used in its physics sense—a quantity is conserved when its total value in a closed system cannot change. The second law of thermodynamics permits one to identify a quantity that *can* be consumed, *available work*. According to the second law, there is an upper limit to the amount of ordered work energy that can be produced from disordered thermal energy at a temperature T_1 in a setting (typically the atmosphere) of fixed ambient temperature T_0. This upper limit, roughly, is the available work.

The further is the temperature T_1 from the ambient temperature, the greater is the amount of potentially available work. From the perspective of the second law, organized coherent motion is most precious, very high (and very low) temperature heat is next most precious, and heat at a temperature near ambient (lukewarm, cool) is degraded energy.[7] Good energy strategies aim to harness high temperatures, now often unusable because of materials limitations; they also aim to avoid degradation, by reducing friction, by reducing thermal gradients, and by reducing the mixing of energy streams having substantially different temperatures. One of the most significant uses of the second law is in providing a yardstick, the minimum available work that *must* be consumed to perform a given task, against which the actual consumption may be laid. Thermodynamics gives such minima both for chemical reactions (like the reduction of aluminum oxide to aluminum by the simultaneous oxidation of carbon, as in the Hall process) and for physical separation of substances, as in the desalination of water. We discuss the concept of available work at length in Chapter 2, and we search throughout the report for ways to make the second law of thermodynamics more vivid and more prominent in energy accounting.

In Table 1.3 we present an abbreviated version of the SRI use data, rearranged according to second-law concerns. This gives some indication of the division of energy by quality. It is, of course, far from a detailed comparison, for given tasks, of minimum available work needed vs. available work presently consumed. In our study we begin to provide that information. Much more thorough data gathering and analysis will be needed.

E. Stage in the transformation from fuel to service

Energy may be husbanded at each stage, from its removal from the ground (atmosphere in the case of sunlight, seawater in the case of deuterium fusion) to its final dispersal as heat at ambient temperature. Until recently, energy conservation meant conservation at the point of extraction; energy conservation in this sense motivated the organization of natural-gas collection and the huge capital

---

[7] Choosing ambient temperatures at or near the earth's surface as reference temperatures for defining *high* and *low* grafts the specificity of the environmental sciences onto the generality of thermodynamics; the meaning of low and high temperature for an industrial process on the moon or in outer space would be quite different.

Table 1.3. End-use of energy in the United States in 1968: Classification by thermodynamic quality of the energy. [a]

	Energy equivalent of fuel consumed (10^{15} Btu)	Percent of total fuel	Percent of total electricity allocated to sector
Residential/commercial space conditioning and other low-temperature energy (lowest-quality energy)	16.4	27.0	26
Industrial process steam and residential/commercial cooking (low-quality energy)	10.9	18.0	3.5 [b]
Industrial direct heat (higher-quality energy)	6.9	11.5	~2 [c]
Miscellaneous, mostly very high-quality energy, such as electric drive, lighting, and electrochemical processes	8.0	13.2	68
Transportation (highest-quality energy: work = "infinite-temperature" energy)	15.0	24.8	0.4
Feedstock [d]	3.3	5.5	...
	60.5	100.0	100

[a] Source: SRI (1972).
[b] This is all cooking; we assume zero use of electricity for generating process steam.
[c] The SRI tables show that less than 2.6 percent of electricity was used for industrial heat in 1968, but do not give an exact figure.
[d] Feedstock is defined in Table 1.2.

investment in pipelines, as well as regulations on the maximum rate of withdrawal of oil from underground reservoirs.[8] It motivates the insistence on appraising "the energy to make energy," so that we can avoid concealed treadmills of energy conversion which involve a larger input of some energy resource than the output of that same resource. Energy conservation also properly includes the avoidance of losses in energy distribution, and motivates the thrust toward the transmission of electricity at ever higher voltages, and, perhaps, someday, in superconducting transmission lines.

This report, however, attempts to isolate the "point of use," the final stage

[8] A discussion of these "conservation" opportunities may be found in the excellent overview by Perry (1972).

where bulk quantities of energy no longer appear and further control over the energy is not expected. The new literature on energy conservation, accordingly, focuses on farms, factories, stores, homes, cars, and buses, a highly diverse set of systems, some of them not ordinarily conceptualized as technical systems with inflows and outflows of materials and energy. When one focuses on such systems, one is led to the subsystems finally responsible for the amount of energy consumed: the viscoelastic flexing of rubber tires, the pyrolysis of fuel droplets, the gas discharge in fluorescent bulbs, the behavior of boundary layers at the surfaces of windows. This is where the physics, chemistry, and engineering of energy use lie, and it is here that we and our colleagues must work if the current constraints within which tradeoffs must be made are to be transformed to give a more advantageous set of tradeoffs. The tradeoffs in question include tradeoffs between energy conservation and safety, or performance, or comfort, or reduction in the emission of pollutants, or conservation of materials, including water.

A possible pitfall in emphasizing the point of use is neglect of systems integration, which offers additional rich opportunities for savings. As energy passes from a high-quality form to its impotent final form as ambient-temperature heat, it may pass through a number of stages, with the quality dropping at each stage. Energy efficiency is improved if processes requiring energy can cascade from one to the next. Where cascading is not feasible, or where the cascade includes a large temperature gap, there is the potential for running heat engines and extracting work.

Another potential pitfall of an undue emphasis on the point of use is the neglect of the energy required to build an energy-using system (or energy-saving component) and, sometimes, the energy to take it apart again after use. Quantitatively, the energy consumed to make, install, and eventually dispose of a pound of insulation is usually a small part of the energy saving that can be provided by the insulation. The energy cost of an automobile or an oven is much less than the energy used by the automobile or the oven in its lifetime. This is typical of items whose use significantly influences energy consumption. Nevertheless, it would be prudent to test all proposed conservation strategies in terms of life-cycle energy costs. Estimates of this kind are gradually becoming tractable, following the painstaking and pioneering work of Berry and Fels (1973), Hannon (1973), and Herendeen (1973a, 1973b).

F. Timing

A final classification which is indispensable in approaching the subject of energy conservation concerns timing. Timing questions run from what can be done now with available equipment and knowledge to what may be feasible after research and development. This range is illustrated by example in Table 1.4. For R & D, the question is: What is the time scale? There is no doubt that our study's sponsors would have liked us to sort through the areas where we recommend further research, and to say which ones are ripe and which ones are far from maturation. We have not done this. The NATO report, with a much longer shopping list, did

Table 1.4. Examples of current and future technological opportunities for energy conservation.[a]

Use sector	Timing of technology: Current	Timing of technology: Future
Residential/commercial space conditioning	Clock-programmed thermal control to replace thermostat Solar assisted heat pump	Selective window coatings Fuel cell total energy system
Industrial process steam	Integrated electric power and steam system	Organic solvents with low heat of vaporization to replace water in some processes
Industrial direct heat	Improved high-temperature insulation Recovery and use of exhaust heat "Dry" paint that requires no heat	Heating by radiation of wavelength specific to a process New catalysts to lower the temperature required for chemical processes Solar furnaces
Transportation	Radial tires Diesel engines	Hybrid electric and internal-combustion engines Flywheel storage of braking energy

[a] Following Table 1.3, the examples are arranged in order of increasing quality of energy.

not either. It is not clear what kind of meeting is required to eke such judgements out of people.[9]

G. General remarks

Many areas of physics are not mentioned in this report. Many subfields that we neglected (including, for example, the study of superconductivity and of high-temperature ceramic materials) have applicability to a wide variety of social needs; this has been demonstrated in several comprehensive reports (NAS, 1972; NAS, 1974; Hein, 1974). It was beyond our powers to add anything.

One realization we shared during our study is the importance of classical physics, notably areas such as fluid mechanics and classical physical and chemical

[9] In the summer of 1974, the National Science Foundation conducted more than 20 workshops to recommend research priorities related to energy.

thermodynamics. Despite the power and scope of these subjects (and despite their practical importance), they have been neglected in our educational enterprise. Academic physicists, while witnessing with satisfaction the adoption and adaptation of these subjects within other disciplines, have counseled their students, openly or subtly, to put their efforts elsewhere. This lack of breadth and diversity in physics education is a disservice to our science. Teachers who seek to remedy this situation will, we believe, find challenge and satisfaction in the classical subjects.

The research areas we identify in subsequent chapters, but do not order, all appear interesting to us. It is possible that some, after only a small amount of investigation, will turn out to be dry holes. But research, like wildcat drilling, has the promise of unusual payoff. The energy conservation enterprise at times seems a bit dreary, when posed entirely in terms of working within the existing technological reality. Research in science and engineering offers at least glimpses into a more appealing reality, in which it becomes commonplace to expect the machines around us to be technically excellent, and to reflect, far more than is the case today, the comprehensive and elegant insights of the past hundred and fifty years of physical science.

2. SECOND-LAW EFFICIENCY:

THE ROLE OF THE SECOND LAW OF THERMODYNAMICS IN ASSESSING THE EFFICIENCY OF ENERGY USE

A. First-law efficiency; the need for a more general measure of efficiency

Consider the following examples, which illustrate typical measures of efficiency of energy use.

(1) A certain household furnace is described as being 60-percent efficient; this means that the ratio of the heat usefully delivered within the house to the heat of combustion[10] of the fuel burned is 0.6. This measure suggests that a 100-percent-efficient furnace would be "perfect," which is incorrect. One could do better in various ways—for instance, by having the fuel power an engine to drive a heat pump (possibly with the intermediary of electricity) to provide more heat to the house than the fuel's heat of combustion.

(2) A certain air conditioner has a coefficient of performance (COP) of 2. This means that the ratio of the heat extracted to the input electric work is 2. To assess this number as a figure of merit, one needs to compare it with a maximum COP. The maximum COP depends on temperature (actually on the inside-outside temperature ratio), and will be considerably greater than 2.

(3) A modern coal-fired power plant has an efficiency of 40 percent. This means that the ratio of its output electric energy to the input heat of combustion of coal is 0.40. As is well known, this efficiency is constrained by the second law of thermodynamics and has a theoretical upper limit that is less than 1.

(4) A large electric motor is 90-percent efficient; the ratio of its output mechanical work to its input electric work is 0.9. Its ideal maximum efficiency is 1.[11]

The figure of merit applied in each of these examples—and in many more—is basically the same. We may call it the *first-law efficiency*. It is

---

[10] Heat of combustion = maximum heat provided to environment in constant-pressure combustion = decrease of enthalpy of a fuel reacting ideally at constant temperature and pressure in standard air.

[11] In the context of concern for fuel conservation, it is often preferable to think of the motor as an extension of the thermal power plant that provides the electricity. The motor's "input" is then power-plant fuel and its efficiency is about 0.3, not 0.9.

$$\eta = \frac{\text{energy transfer (of desired kind) achieved by a device or system}}{\text{energy input to the device or system}} . \quad (2.1)$$

[first-law efficiency]

When the theoretical maximum value of this ratio is greater than 1, it is usually called a coefficient of performance. When $\eta_{max} \leqslant 1$, it is usually called an efficiency. Table 2.1 shows numerators and denominators that define η for various classes of devices and also gives η_{max} and standard nomenclature. (Although η itself is defined without reference to the second law of thermodynamics, its ideal maximum value is—for all but work-in-work-out devices—limited by the second law. The temperatures that specify η_{max} are, of course, absolute temperatures.)

As a general figure of merit, the first-law efficiency has several drawbacks: (a) Its maximum value depends on the system and on temperatures and may be greater

TABLE 2.1. First-law efficiency. [*Note*: T_1(hot) > T_2(warm) > T_0(ambient) > T_3(cool).]

Type of device or system[a]	Numerator in ratio defining η	Denominator in ratio defining η	η_{max}	Standard nomenclature
Electric motor (W/W)	Mechanical work output	Electric work input	1	Efficiency
Heat pump, electric (Q/W)	Heat Q_2 added to warm reservoir at T_2	Electric work input	$\frac{1}{1-(T_0/T_2)} > 1$	Coefficient of performance (COP)
Air conditioner or refrigerator, electric (Q/W)	Heat Q_3 removed from cool reservoir at T_3	Electric work input	$\frac{1}{(T_0/T_3)-1}$ (not restricted in value)	COP
Heat engine (W/Q)	Mechanical or electric work output	Heat Q_1 from hot reservoir at T_1	$1-\frac{T_0}{T_1} < 1$	Efficiency (thermal efficiency)
Heat-powered heating device[b] (Q/Q)	Heat Q_2 added to warm reservoir at T_2	Heat Q_1 from hot reservoir at T_1	$\frac{1-(T_0/T_1)}{1-(T_0/T_2)} > 1$	COP or efficiency
Absorption refrigerator[c] (Q/Q)	Heat Q_3 removed from cool reservoir at T_3	Heat Q_1 from hot reservoir at T_1	$\frac{1-(T_0/T_1)}{(T_0/T_3)-1}$ (not restricted in value)	COP

[a] The symbols W and Q refer to work and heat, respectively.

[b] A furnace is a special case; for it, $\eta_{max}=1$. More generally, the device could include a heat engine and heat pump; then $\eta_{max} > 1$.

[c] "Absorption refrigerator" means any heat-powered device for cooling.

than, less than, or equal to 1. (b) It does not adequately emphasize the central role of the *second law* in governing the possible efficiency of energy use. (c) It cannot readily be generalized to complex systems in which the desired output is some combination of work and heat.

B. Second-law efficiency; the concept of available work

Various other special and general figures of merit have been proposed.[12] The one that we recommend for general adoption is a quantity that we shall call the second-law efficiency.[13] It is a measure of performance relative to the optimal performance permitted by both the first and second laws of thermodynamics. It is equally applicable to a simple device with a single output or a complex system with multiple outputs.

We define second-law efficiency first for a device or system whose output is the useful transfer of work or heat (not both)—for instance, a motor, refrigerator, heat pump, or power plant. It is

$$\epsilon = \frac{\text{heat or work usefully transferred by a given device or system}}{\text{maximum possible heat or work usefully transferable for the same function by any device or system using the same energy input as the given device or system}} \quad . \tag{2.2}$$

[second-law efficiency for single-output system]

As is obvious from this definition, the maximum value of ϵ is 1 in all cases. The numerator in the defining ratio is the same as that in the first-law-efficiency ratio (Eq. 2.1). The new denominator represents a simple but powerful change because it brings the laws of thermodynamics directly into the definition of efficiency. The "maximum" in the denominator means the theoretical maximum permitted by the first and second laws. Note that this is a *task* maximum, not a *device* maximum. To maximize the heat delivered to a house by fuel, for instance, a furnace should be replaced by an ideal fuel cell and an ideal heat pump.

The second-law efficiency provides immediate insight into the quality of performance of any device relative to what it could ideally be.[14] It shows how much room there is for improvement in principle. It measures the "waste" of fuel (some of it inevitable in practice, of course). For any specified task requiring heat or work, maximizing ϵ is equivalent to minimizing fuel consumption. In case no fuel consumption is involved—such as with hydroelectric plants, wind mills, geothermal sites, or solar collectors—maximizing ϵ has an impact on capital in-

---

[12]Recently, for example, D. O. Lee and W. H. McCulloch (1973) have introduced the concept of "utility," defined as the ratio (work plus heat usefully transferred by a system)/(available work consumed by the system). Utility can have a maximum value greater than 1.

[13]This quantity has previously been termed "effectiveness" (Keenan, 1948). Besides changing its name, we generalize its meaning.

[14]We are indebted to Charles Berg for stimulating our thinking along these lines (see Federal Power Commission, 1974; also Berg, 1974a). Also influential was a report by Keenan, Gyftopoulos, and Hatsopoulos (1973).

vestment in power-producing units. Inevitably, therefore, the maximization of ϵ becomes a matter for *policy* consideration. It is a technical goal to be placed alongside economic, environmental, and conservation goals.

Under certain circumstances, the first-law and second-law efficiencies can differ dramatically. For example, a furnace providing hot air at 110 °F (43 °C) to a house when the outside air temperature is 32 °F (0 °C) has a second-law efficiency $\epsilon = 0.082$ if its first-law efficiency is $\eta = 0.60$.[15] For a power plant, on the other hand, the first-law and second-law efficiencies are nearly the same.

In order to understand and calculate second-law efficiencies, an important concept is *available work*.[16] Its definition in terms of thermodynamic quantities is given in Section D. Here we state its meaning in words:

B = maximum work that can be provided by a system (or by fuel) as it proceeds (by any path) to a specified final state in thermodynamic equilibrium with the atmosphere; interaction with the atmosphere is permitted but work done on the atmosphere is not counted.[17]

[available work]

This quantity specifies only harnessable work, not work done on the atmosphere. The reference is to *work* (rather than heat) because work is the highest "quality" form of energy—equivalent to heat at infinite temperature. Work is the best overall measure of capacity for doing *any* task. For a raised weight (to take an elementary example), the available work is simply mgh, the work that can be done by lowering the weight (ground level must then be specified as part of the "ambient" conditions). For heat Q_1 extracted from a hot reservoir at temperature T_1, the available work is $B = Q_1[1 - (T_0/T_1)]$. For fuel with heat of combustion $|\Delta H|$, B turns out to be roughly equal to $|\Delta H|$. (The difference between a fuel's heat of combustion and its available work is discussed in Section E.) For a chemical cell, the available work is $B = -\Delta G_0$, the change of Gibbs free energy in the reaction carried out at ambient temperature and pressure.[18]

In terms of the available-work concept, the definition of second-law efficiency can be re-stated much more simply and more generally. It is[19]

$$\epsilon = \frac{B_{\min}}{B_{\text{actual}}} \tag{2.3}$$

[second-law efficiency in general].

---

[15]The calculation of ϵ appears as an example in Section C.

[16]This concept appears in most engineering thermodynamics texts (see, for example, Keenan, 1948). It is probably unfamiliar to most physicists. It is often called "availability." Symbols for it include A, Φ, and (for molar availability) b. We use B (for bereitstehender brauchbarer Arbeitsbestand).

[17]In some cases, some base reservoir other than the atmosphere might be appropriate.

[18]The equation $B = |\Delta G_0|$ is not quite precise because of a small additional contribution to B from a term that measures the available work resulting from the diffusion into the atmosphere of the products of the chemical reaction. This is discussed in Section D.

[19]For a device where output is heat or work (not both), it is easy to show the equivalence of Eqs. 2.2 and 2.3.

TABLE 2.2. Available work provided by sources and needed by end uses. [*Note*: T_1(hot) > T_2(warm) > T_0(ambient) > T_3(cool).]

	Work W_{in}	Fuel with heat of Combustion $\lvert \Delta H \rvert$	Heat Q_1 from hot reservoir at T_1
Sources	(e.g., water power, wind power, raised weight) [or electricity if the wall socket is treated as the source]	(e.g., coal, oil, gas)	(e.g., geothermal source, solar collector source) [also fission reactors and fossil-fuel plants if alternatives to thermal operation are excluded]
	$B = W_{in}$	$B \cong \lvert \Delta H \rvert$[a] (usually to within 10%)	$B = Q_1\left(1 - \frac{T_0}{T_1}\right)$
End Uses	Work W_{out}	Heat Q_2 added to warm reservoir at T_2	Heat Q_3 extracted from cool reservoir at T_3
	(e.g., turning shafts, pumping fluids, propelling vehicles)	(e.g., space heating, cooking, baking, drying)	(e.g., refrigerating, air conditioning)
	$B_{min} = W_{out}$	$B_{min} = Q_2\left(1 - \frac{T_0}{T_2}\right)$	$B_{min} = Q_3\left(\frac{T_0}{T_3} = 1\right)$

[a]An exact consideration is required for each fuel. See Table 2.6.

This can be applied to anything from a cuckoo clock to a nation's total energy economy. It states that the efficiency[20] is equal to the ratio of the least available work that could have done the job to the actual available work used to do the job. Note that available work (unlike energy) *is* consumed. (The consumption of available work is related to total entropy change; see Section D.)

Table 2.2 displays available work provided by three kinds of source and minimum available work needed by three kinds of end use. To find the second-law efficiencies ϵ for single-source–single-output devices, it is only necessary to divide one of the end-use entries in Table 2.2 by one of the source entries. This gives the set of 9 efficiency formulas shown in Table 2.3. For comparison, first-law efficiencies (or coefficients of performance) are also shown in this table.

To help emphasize the central role of temperature in efficiency considerations, Figures 2.1 and 2.2 show the minimum available work required to move a unit of heat from one temperature to another. Figure 2.1 is explained in its caption. Note that over a wide summer to winter range, $B_{min}/Q < 0.15$, a very small value attributable to the small fractional changes of absolute temperature that are involved.

[20]When no confusion is likely to result, we shall often use "efficiency" or "efficiency ϵ" to mean "second-law efficiency."

TABLE 2.3. First-law and second-law efficiencies for single-source–single-output devices.

End use \ Source	Work W_{in}	Fuel: Heat of combustion $\|\Delta H\|$ Available work B	Heat Q_1 from hot reservoir at T_1
Work W_{out}	1. $\eta = W_{out}/W_{in}$ $\epsilon = \eta$ (e.g., electric motor)	2. $\eta = W_{out}/\|\Delta H\|$ $\epsilon = \frac{W_{out}}{B}$ $(\cong \eta)$ (e.g., power plant)	3. $\eta = W_{out}/Q_1$ $\epsilon = \frac{\eta}{1-(T_0/T_1)}$ (e.g., geothermal plant)
Heat Q_2 added to warm reservoir at T_2	4. $\eta(\mathrm{COP}) = Q_2/W_{in}$ $\epsilon = \eta\left(1-\frac{T_0}{T_2}\right)$ (e.g., electrically driven heat pump)	5. $\eta(\mathrm{COP}) = Q_2/\|\Delta H\|$ $\epsilon = \frac{Q_2}{B}\left(1-\frac{T_0}{T_2}\right)$ (e.g., engine-driven heat pump)	6. $\eta(\mathrm{COP}) = Q_2/Q_1$ $\epsilon = \eta\frac{1-(T_0/T_2)}{1-(T_0/T_1)}$ (e.g., solar hot water heater)
Heat Q_3 extracted from cool reservoir at T_3	7. $\eta(\mathrm{COP}) = Q_3/W_{in}$ $\epsilon = \eta\left(\frac{T_0}{T_3}-1\right)$ (e.g., electric refrigerator)	8. $\eta(\mathrm{COP}) = Q_3/\|\Delta H\|$ $\epsilon = \frac{Q_3}{B}\left(\frac{T_0}{T_3}-1\right)$ (e.g., gas powered air conditioner)	9. $\eta(\mathrm{COP}) = Q_3/Q_1$ $\epsilon = \eta\frac{(T_0/T_3)-1}{1-(T_0/T_1)}$ (e.g., absorption refrigerator)

The *actual* expenditure of available work in a typical furnace is $B_{actual}/Q \cong 1.7$, more than 10 times the minimum required. Note that the vertical scale in Figure 2.1 is *not* efficiency. To find the second-law efficiency of any actual furnace, heat pump, or air conditioner, it would be necessary to divide the value B_{min}/Q shown in the figure by the quantity B_{actual}/Q for the device in question. The two different curves in Figure 2.1 are calculated for different temperatures at which heat is transferred to or from a room. If the transfer could be accomplished at the actual room temperature, the required expenditure of available work would be less.

Figure 2.2 shows the minimum available work required to provide heat at high temperature. In the range of temperature relevant to process-steam production, $B_{min}/Q \approx \frac{1}{3}$. In current practice, $B_{actual}/Q > 1$. As the quality of the heat increases (higher T_2), the available work required to provide it increases. Thus, the inefficiency of current practice is more pronounced at lower temperatures.

C. Examples of second-law efficiencies

A second-law efficiency is calculated for a specific *task*, defined by the quantities of work and/or heat to be transferred for stated purposes, the temperature(s) at which heat transfer is accomplished, and the ambient temperature of the atmosphere (or other large reservoir). In general, the *task* (providing heat to the

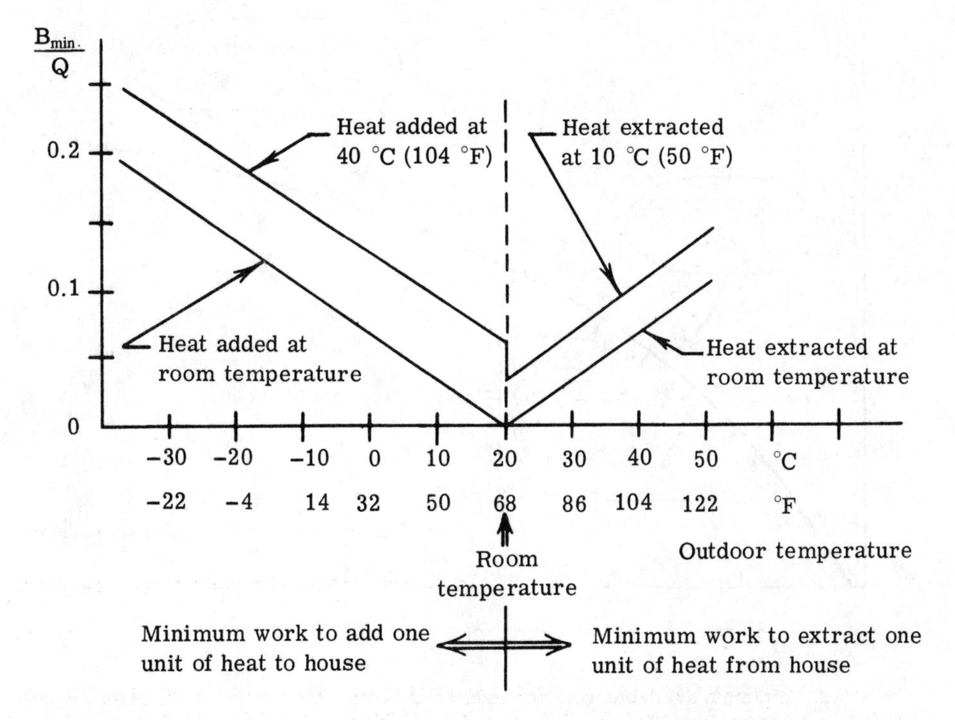

Figure 2.1. Minimum available work for space conditioning (for a room maintained at 68 °F year-round). The vertical scale gives B_{min}/Q, the least available work required to add 1 unit of heat to a room in the winter (to the left of the dashed line) or to extract 1 unit of heat in the summer (to the right of the dashed line), as a function of the outdoor temperature. The upper curve refers to heat addition and extraction at typical practical temperatures. The lower curve shows the even smaller expenditure of available work that would be required if heat could be added and extracted at room temperature. [The straight lines are graphs of $B_{min}/Q = 1 - (T_0/T_2)$ and $B_{min}/Q = (T_0/T_3) - 1$ (using absolute temperatures).]

interior of a house, for example) must be more narrowly defined than the final *goal* (human comfort in winter). Minimum available work for the task is an essential ingredient of an efficiency calculation (Eq. 2.3). For heating a house—to continue the same example—the minimum available work could be that required for a well-insulated reference house of the same dimensions, or it could be that required for the actual house in question (the latter is our choice). It could be calculated for heat supplied at room temperature or for heat supplied above room temperature (in deference to "realism," we recommend the latter choice). Conventional choices of this kind must be made before second-law efficiencies can be calculated and compared. (See also Chapter 3, Section D.)

1. Power generation from fossil fuels

If the available work in fuel is approximated as being equal to the heat of combustion, then $\epsilon = \eta$ (square 2 in Table 2.3); this is in the range 0.3 to 0.4 for most

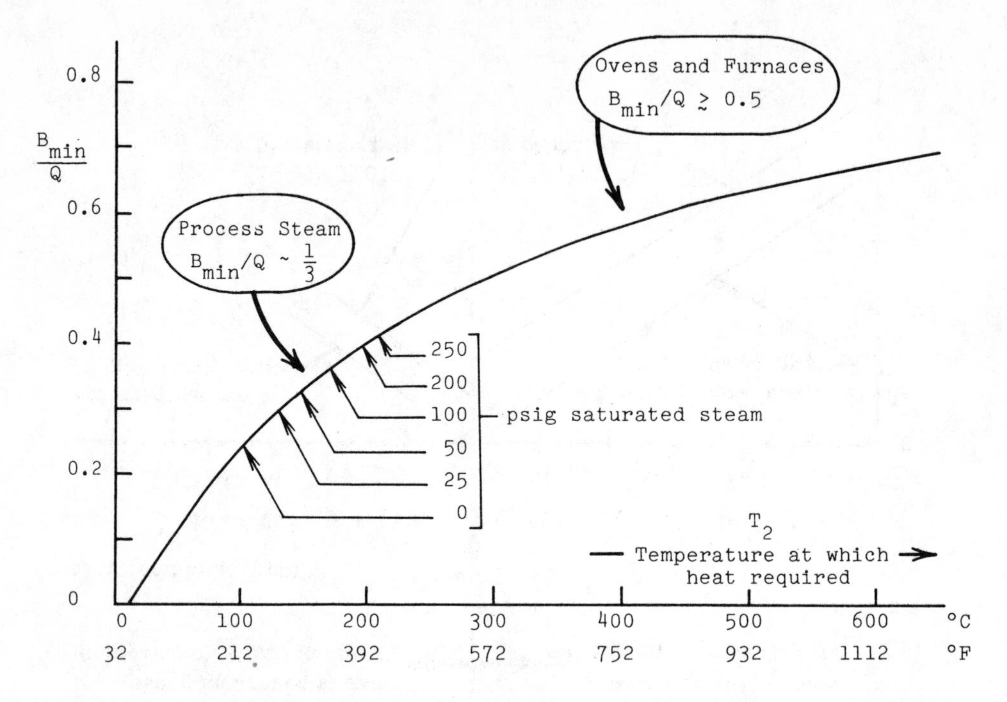

Figure 2.2. Minimum available work for industrial heat. The vertical scale gives B_{min}/Q, the least available work required to provide 1 unit of heat at the temperature shown on the horizontal scale. Ambient temperature is T_0=55 °F=13 °C. [The curve is a graph of $B_{min}/Q = 1-(T_0/T_2)$ as a function of T_2. It is insensitive to changes in the assumed ambient temperature T_0 within reasonable limits.]

operating power plants. Usually η is said to have a theoretical maximum value (the Carnot efficiency) that is less than 1. For a flame temperature of 2000 °C, the Carnot efficiency is 0.87; for a peak steam temperature of 550 °C, it is 0.64. What the second-law efficiency helps to emphasize is that for the conversion of chemical energy to electricity, the Carnot efficiency is *not* an upper limit. The ideal fuel cell is a realization of a device that does not suffer the irreversibility of combustion and therefore has the higher upper limit $\epsilon_{max} = 1$.

2. Oil or gas fired furnace

It is in providing relatively low-temperature heat—in space heating and cooling, and also in industrial process steam—that the most wasteful consumption of available work occurs (as suggested by Figures 2.1 and 2.2). Consider the same furnace cited earlier as an example: Its first-law efficiency is $\eta = 0.6$; it provides air at 110 °F ($T_2 = 316$ K); the outside air is at 32 °F ($T_0 = 273$ K). Using the approximate equality between available work and heat of combustion, we get, from square 5 in Table 2.3,

$$\epsilon = \frac{Q_2}{B}\left(1 - \frac{T_0}{T_2}\right) \cong \eta\left(1 - \frac{T_0}{T_2}\right) = 0.082 \ . \qquad (2.4)$$

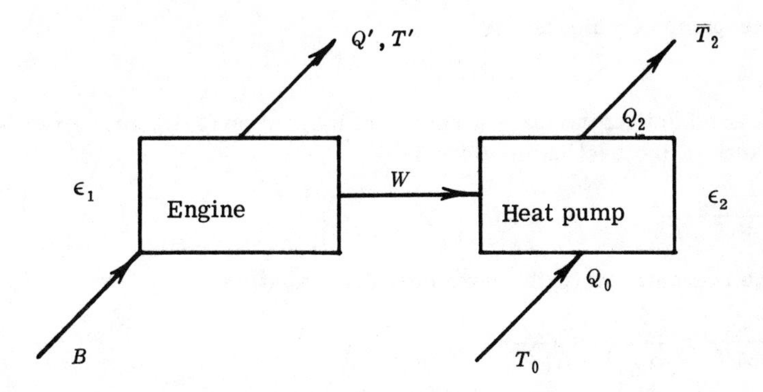

Figure 2.3. Schematic presentation of an engine-driven heat pump. The engine uses fuel with available work B, provides work W to the heat pump, and exhausts waste heat Q' at temperature T'. Its (second-law) efficiency is ϵ_1. The heat pump, with efficiency ϵ_2, takes heat Q_0 from a reservoir at temperature T_0 and supplies heat Q_2 at temperature T_2. If the engine's exhaust temperature T' is adjusted to satisfy $T' \geq T_2$, the waste heat Q' may be added to the useful heat Q_2.

If we postulate the possibility of heat transfer at room temperature and set $T_2 = 70\ ^\circ\text{F} = 294\ \text{K}$, the calculated efficiency is even less, $\epsilon = 0.043$. Space conditioning of buildings is discussed further in Chapter 3.

3. *Heat pumps*

A heat pump is any device that transfers heat "uphill" from a cooler to a warmer place. Refrigerators and air conditioners use heat pumps. More often, however, the name "heat pump" is used when the goal is heating, not cooling. Consider a heat pump whose input energy is work W (electricity or mechanical drive) and whose output energy is heat Q_2 supplied at temperature T_2 when the cooler reservoir (e.g., the atmosphere) is at temperature T_0. Its coefficient of performance is $\eta(\text{COP}) = Q_2/W$ and its second-law efficiency is $\epsilon = (Q_2/W)[1 - (T_0/T_2)]$ (square 4 in Table 2.3). Typical existing heat pumps have efficiencies ϵ of about 0.3 and COP's that depend on temperature[21]:

$$\eta(\text{COP}) = \frac{\epsilon}{1 - (T_0/T_2)} \ . \tag{2.5}$$

A heat pump is most effective over small temperature steps. *Example*: Let $\epsilon = 0.3$ and $T_2/T_0 = 1.1$, a 10 percent "boost" in temperature. Then $\eta(\text{COP}) = 3.3$. This heat pump would lose its "magnification" (i.e., $\eta \leqslant 1$) if the step up in absolute temperature exceeded 43 percent (if $T_2/T_0 \geqslant 1.43$).

Engine-driven heat pump. Figure 2.3 shows an engine-driven heat pump in simplest schematic form. The (second-law) efficiency of the engine is $\epsilon_1 = W/B$(see the figure for notation); that of the heat pump is $\epsilon_2 = (Q_2/W)[1 - (T_0/T_2)]$. It is easy to show that if the waste heat Q' is thrown away, the overall second-law efficiency of

---

[21]The efficiency may also depend (weakly) on temperature.

the engine-pump combination is

$$\epsilon = \epsilon_1 \epsilon_2 , \tag{2.6}$$

the same as if the engine were a remotely located power plant. The coefficient of performance of the heat pump alone is

$$\eta = \frac{\epsilon_2}{1-(T_0/T_2)} ; \tag{2.7}$$

that of the combination (with waste heat discarded) is

$$\eta_t = \frac{Q_2}{|\Delta H|} \cong \frac{Q_2}{B} = \frac{\epsilon_1 \epsilon_2}{1-(T_0/T_2)} , \tag{2.8}$$

where $|\Delta H|$ is the heat of combustion of the fuel consumed by the engine.

Suppose now that the waste heat Q' is added to Q_2 as useful heat (the exhaust temperature must satisfy $T' \geqslant T_2$). From Table 2.2, the minimum available work to supply this heat is

$$B_{\min} = (Q_2 + Q')[1-(T_0/T_2)] .$$

From Eq. 2.3, the second-law efficiency is therefore

$$\epsilon = \frac{B_{\min}}{B_{\text{actual}}} = \frac{(Q_2 + Q')[1-(T_0/T_2)]}{B} . \tag{2.9}$$

To express this in terms of the engine efficiency ϵ_1 and the heat-pump efficiency ϵ_2, we can substitute the following expressions for Q_2, Q', and B: from square 4 in Table 2.3,

$$Q_2 = \frac{\epsilon_2 W}{1-(T_0/T_2)} ;$$

from square 2, $B = W/\epsilon_1$; and, from energy conservation, $Q' = |\Delta H| - W$. These substitutions in Eq. 2.9 give

$$\begin{aligned} \epsilon &= \epsilon_1 \epsilon_2 + \left[\frac{|\Delta H|}{B} - \epsilon_1\right][1-(T_0/T_2)] \\ &\cong \epsilon_1 \epsilon_2 + (1-\epsilon_1)[1-(T_0/T_2)] . \end{aligned} \tag{2.10}$$

Compare this with Eq. 2.6. The new terms are positive, so $\epsilon > \epsilon_1 \epsilon_2$, as one would expect. (The exhaust temperature T' does not appear explicitly. However, it influences the efficiency ϵ_1.)

Numerical results for two models appear in Table 2.4. Waste-heat recovery nearly doubles the efficiency in the space-heating model and more than doubles it in the steam-production model. The substantial gains are not surprising in view of the fact that the engine puts out twice as much heat as work. The greater gain in the higher temperature application is a reflection of the fact that the waste heat is being used for a higher quality application in the steam-production model. Note also that the heat pump itself is relatively ineffective in the steam-production model because of the large temperature step. Without waste-heat recovery, the steam-producing heat pump is inferior to a standard boiler. (We must emphasize that these are simplified *model* calculations. Other factors—such as cost of the local

Table 2.4. Efficiency of an engine-driven heat pump with and without waste heat recovery.

Models (For both, $\epsilon_1 = \epsilon_2 = \frac{1}{3}$)	COP of heat pump for the model (first-law efficiency) (Eq. 2.7)	Second-Law Efficiencies: Waste heat thrown away (Eq. 2.6)	Waste heat used (Eq. 2.10)	Comparison: direct combustion, no engine or heat pump (Eq. 2.4)
Space-heating model				
T_2 = 110 °F = 43 °C = 316 K				Furnace, $\eta = 0.6$
T_0 = 32 °F = 0 °C = 273 K	$\eta_e = 2.45$	$\epsilon = 0.111$	$\epsilon = 0.202$	$\epsilon = 0.082$
Steam-production model				
T_2 = 250 °F = 121 °C = 394 K				Boiler, $\eta = 0.8$
T_0 = 50 °F = 10 °C = 283 K	$\eta = 1.18$	$\epsilon = 0.111$	$\epsilon = 0.299$	$\epsilon = 0.225$

installation and only partial recovery of its waste heat—could erode the apparent advantage of the engine-driven heat pump.)

D. The thermodynamics of available work

Consider a system characterized by energy E, entropy S, and volume V (see Figure 2.4). Its energy may include, in addition to its internal energy, also gravitational potential energy, kinetic energy of bulk motion, etc. (Temperature, pressure, and other intensive variables may vary from one part of the system to another.) Now let the system proceed via chemical reactions and/or other changes until it comes into thermodynamic equilibrium with the atmosphere. (By "atmosphere" we mean an appropriate large reservoir comprising the environment of the system; usually it will in fact be the earth's atmosphere.) The atmosphere has temperatures T_0, pressure P_0, and a specified composition. The system, as it changes, may exchange heat and work, but not matter,[22] with the atmosphere, and it may also do work on other systems. The latter is called "useful work." Upon reaching equilibrium with the atmosphere, the system has energy E_f, entropy S_f, and volume V_f. The maximum useful work that can be transferred by the system is called its *available work*.[23] The available work is determined by the condition that the total entropy of the system plus the environment not change (i.e., that all

[22] We generalize to matter exchange (diffusion) later.

[23] As noted earlier, this quantity is frequently called "availability."

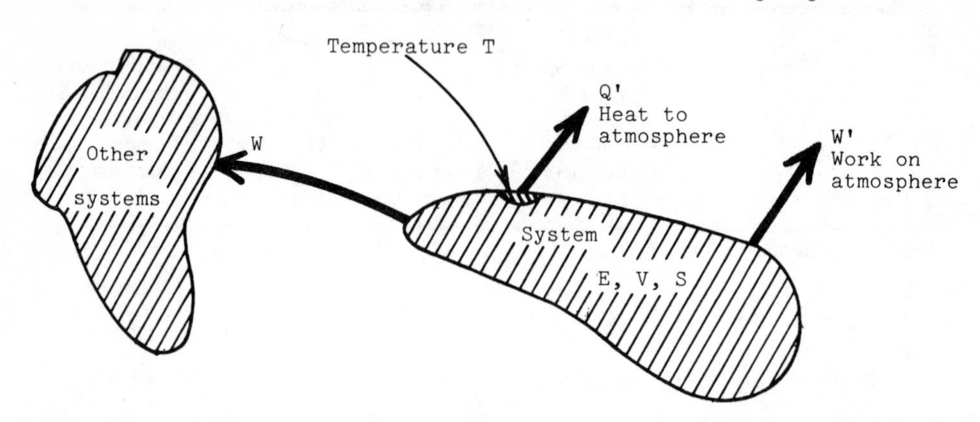

Figure 2.4. Schematic diagram of system interacting with the atmosphere and transferring work to other systems. The available work of the system is the maximum useful work it can transfer to other systems as it comes to thermodynamic equilibrium with the atmosphere. For reversible change, the temperature T at which heat leaves the system must be equal to the atmospheric temperature T_0. This can be achieved, if necessary, by including one or more ideal heat engines in the system. In the absence of irreversibilities, $W=B$. Otherwise, $W<B$.

processes are carried out reversibly[24]). In the subsection below we show that for the conditions stated, the maximum useful work is

$$B = (E - E_f) + P_0(V - V_f) - T_0(S - S_f) \ . \tag{2.11}$$

[available work without diffusion]

The second and third terms on the right represent energy "bootlegged" from the atmosphere. If the system's volume decreases ($V - V_f > 0$), the atmosphere does work on the system, which can be passed on as useful work (the second term is positive). If the system's entropy increases ($S - S_f < 0$), and the increase is accomplished reversibly, the atmosphere transfers heat to the system, which can be passed on as useful work (the third term is then positive).

The available work is closely related to the change of the Gibbs function, or Gibbs free energy ($G = E + PV - TS$), but B differs from $-\Delta G$ in that the definition of B includes the temperature and pressure of the atmosphere only, not the system. This means that B can be defined for a nonequilibrium state. In one important special case, that in which the "system" consists of a quantity of fuel at atmospheric pressure and room temperature, the available work and the Gibbs free energy change are the same (except for the small additional diffusion term).

---

[24]Reversible change may require that one or more Carnot engines be incorporated in the system; see the caption of Figure 2.4.

Some formalism: Loss of available work related to entropy change and proof that $W_{max} = B$ [25]

Consider a finite change of the system in Figure 2.4, characterized as follows:

the energy, volume, and entropy of the system change by ΔE, ΔV, and ΔS, respectively;

heat Q' flows from the system to the atmosphere;

the system does useful work W on other systems;

the system does nonuseful work $W' = P_0 \Delta V$ on the atmosphere.

From the first law of thermodynamics

$$W = -\Delta E - Q' - W' . \tag{2.12}$$

From the second law

$$\Delta S_{sa} = \frac{Q'}{T_0} + \Delta S . \tag{2.13}$$

This is the entropy change of the system plus the atmosphere. It does not include the possible entropy change of the other systems on which work is done. In Eq. 2.12, set $W' = P_0 \Delta V$. Also solve Eq. 2.13 for Q' and substitute in Eq. 2.12. The result is

$$W = -T_0 \Delta S_{sa} - (\Delta E + P_0 \Delta V - T_0 \Delta S) . \tag{2.14}$$

Let Eq. 2.11 *define* the quantity B (later to be identified as available work) for an arbitrary state of the system (relative to a fixed end-state f). For a change from state i to state j, Eq. 2.14 then takes the form

$$B_i - B_j = W(i \to j) + T_0 \Delta S_{sa}(i \to j) . \tag{2.15}$$

This is a simple and powerful equation. To apply it we need one more assumption, namely that the "other systems" in Figure 2.4 are passive receivers of work. They do not otherwise interact with the system under study or the atmosphere. Accordingly, the entropy of the system of interest plus the atmosphere cannot decrease: $\Delta S_{sa} \geqslant 0$. This means that the useful work is constrained by the change of B:

$$W(i \to j) \leqslant B_i - B_j . \tag{2.16}$$

This at once identifies B as the available work. In particular, if i is any state of the system (we drop the subscript and write $B_i = B$) and j is the state f in equilibrium with the atmosphere (from the definition of B, $B_f = 0$), Eq. 2.16 shows the maximum useful work to be

$$W_{max} = B . \tag{2.17}$$

This corresponds to the limiting case $\Delta S_{sa} = 0$ in Eq. 2.15. Note the corollary to Eq. 2.17: For a given task requiring a certain work W, the minimum available work required to perform the task is

$$B_{min} = W . \tag{2.18}$$

---

[25]The treatment in this subsection is a modified version of the presentation in Keenan's text (1948).

Another interesting limiting case is one in which the system changes while doing *no* useful work (adiabatic combustion is an example). Then Eq. 2.15 gives, for the loss of available work,

$$B_i - B_j = T_0 \Delta S_{sa} (i \rightarrow j) \; . \qquad (2.19)$$

[no useful work]

The loss of available work is proportional to the "irreversibility," measured by ΔS_{sa}. More generally, Eq. 2.15 governs processes in which some useful work is performed with some irreversibility.

Entropy change and efficiency

Taking advantage of Eq. 2.15, we can write for the second-law efficiency of a process requiring work W,

$$\epsilon = \frac{B_{min}}{B_{actual}} = \frac{W}{W + T_0 \Delta S_{sa}} \, ,$$

or

$$\epsilon = \frac{1}{1 + (T_0 \Delta S_{sa} / W)} \; . \qquad (2.20)$$

For certain processes, this could provide a practical means to calculate ϵ.

In some applications (charging a battery, for example, or pumping water to a higher reservoir), it is useful to think of transferring available work from one system to another. Second-law efficiencies are then multiplicative. Let the efficiency of storage be $\epsilon_1 = B_{stored}/B_{source}$ and the efficiency of secondary use be $\epsilon_2 = W/B_{stored}$. The overall efficiency, $\epsilon_2 = W/B_{source}$, is then

$$\epsilon = \epsilon_1 \epsilon_2 \; . \qquad (2.21)$$

This is a rather obvious point. We make it only to emphasize that in an energy-storage device it is the stored available work, not the stored internal energy, that is the relevant quantity. Efficiencies of storage and re-use should be measured in terms of available work.

Most end-uses, it should be noted, strongly increase entropy and do not store available work. Imagine a vehicle powered by a hypothetically perfect fuel cell and electric motor. This supercar operates with an efficiency $\epsilon = 1$ and $\Delta S_{sa} = 0$; that is, no entropy increase is associated with getting the fuel's energy to the drive wheels of the car. Cruising along the freeway, however, the supercar is dissipating all of this hard-won energy to the environment, so the entropy change of the universe is

$$\Delta S_T = B/T_0 \; . \qquad (2.22)$$

This again is a rather obvious point. We make it as a reminder that available work (unlike energy) is a quantity that is consumed. Minimizing its consumption minimizes total entropy change.

An example: Compressed-air energy storage

Compressed air in large underground caverns is under serious consideration for energy storage at power plants (Harboe, 1971). As a simple example to illustrate the link between entropy and available work, we consider air that is compressed adiabatically and then loses some heat while it is in storage awaiting decompression. Initially the air is at pressure P_0, temperature T_0, and volume V_1. It is compressed adiabatically to P_2, T_2, and V_2. It then loses heat Q and its pressure and temperature fall to P_3, T_3 while its volume remains constant ($V_3 = V_2$). We set $P_2/P_0 = r$, the maximum compression ratio.

The air in its initial state has zero available work:

$$B_1 = 0 .$$

After adiabatic compression, its available work is

$$B_2 = -\int_1^2 P\,dV ,$$

which can be evaluated in terms of temperatures,

$$B_2 = \frac{R}{\gamma - 1}(T_2 - T_0) , \tag{2.23}$$

or in terms of the compression ratio,

$$B_2 = \frac{RT_0}{\gamma - 1}\left[r^{(\gamma-1)/\gamma} - 1\right] , \tag{2.24}$$

where $\gamma = C_p/C_v = 1.4$ and R is the gas constant. In the subsequent loss of heat, the entropy change of the compressed air plus its environment is

$$\Delta S_{sa} = -\int_{T_2}^{T_3}\left(\frac{1}{T_0} - \frac{1}{T}\right) C_v\,dT . \tag{2.25}$$

The loss of available work (see Eq. 2.19) is therefore

$$|\Delta B| = T_0 \Delta S_{sa} = C_v(T_2 - T_3) - C_v T_0 \ln(T_2/T_3) . \tag{2.26}$$

If we set $C_v = \frac{5}{2}R$, we get for the fractional loss of available work during storage

$$\frac{|\Delta B|}{B} = \frac{(T_2/T_0) - (T_3/T_0) - \ln(T_2/T_3)}{(T_2/T_0) - 1} . \tag{2.27}$$

Graphs of this quantity as a function of compression ratio r are shown in Figure 2.5 for complete cooling during storage ($T_3 = T_0$) and for partial cooling. (Note that $T_2/T_0 = r^{(\gamma-1)/\gamma}$). For complete cooling, losses are large—more than 20 percent for a compression ratio of 5 and about 30 percent for a compression ratio of 10. If the stored gas loses only 25 percent of its sensible heat (the lowest curve in the figure), its loss of available work is 9 percent at $r = 5$ and 11 percent at $r = 10$. For compression ratios greater than $r = 10$, the loss of available work increases rather slowly with increasing r. In addition to the losses shown here, there are, of course, pumping losses. For comparison with these numbers, one may note that the *total* loss of available work in the storage of water in elevated reservoirs (electricity to electricity) is about 30 percent.

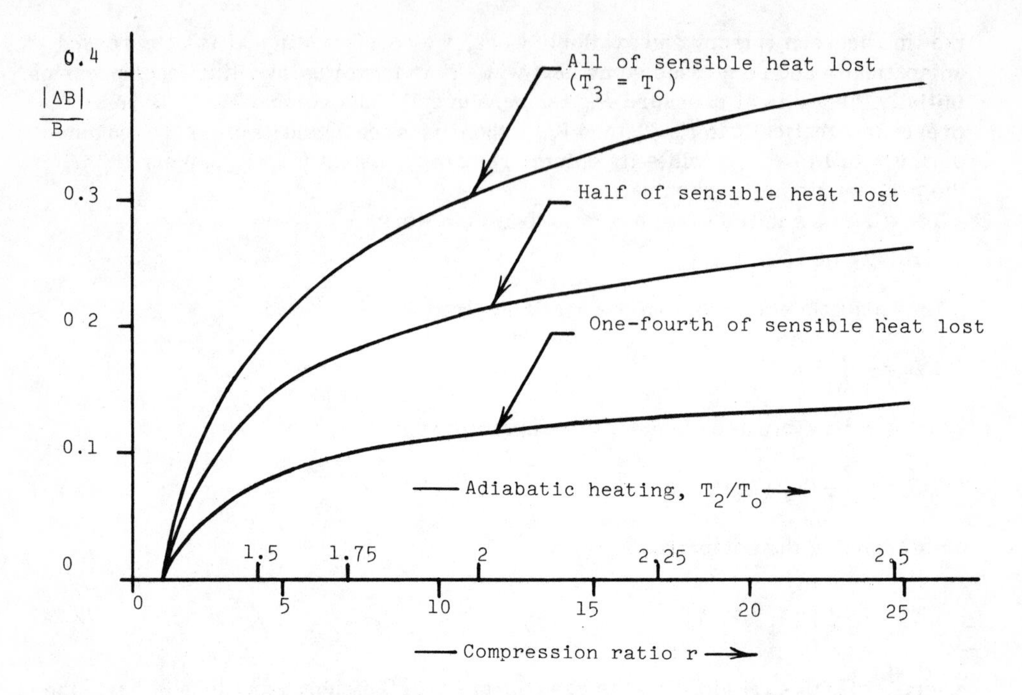

Figure 2.5. Fractional loss of available work in compressed air as a function of initial compression ratio for various losses of sensible heat during storage (at constant volume).

Diffusion of products; an additional contributor to available work

When J. Willard Gibbs introduced the concept of available work in 1878 (Gibbs, 1948), he envisioned the additional possibility of gaining work by diffusing the components of a system into the atmosphere.[26] Associated with this process is an entropy of mixing that adds another term to B.[27] With the extra mixing term we have

$$B = (E - E_f) + P_0(V - V_f) - T_0(S - S_f) - T_0(S_f - S_d), \tag{2.28}$$

[available work including diffusion]

where S_d is the entropy of the diffused products at specified concentrations in the atmosphere.[28] For the products treated as ideal gases, for example,

---

[26]Gibbs actually considered the inverse process, constructing a system from components in the atmosphere (or other large reservoir).

[27]Entropies of mixing *within* the system, before it intermingles with the atmosphere, are understood to be included in the quantities S and S_f in Eq. 2.11.

[28]We continue to use f to designate the "final" state of the system before it diffuses into the atmosphere. The diffused state d is subsequent to f.

$$-T_0(S_f - S_d) = +RT_0 \sum_i n_i \ln(x_{fi}/x_{di}), \qquad (2.29)$$

[ideal gas]

where n_i is the number of moles of the *i*th product, x_{di} is the final mole fraction of this product in the atmosphere, x_{fi} is its mole fraction in the equilibrium state f before diffusion, and R is the gas constant. The diffusion contribution to available work can be written more generally (Gibbs, 1948; Keenan *et al.*, 1973) as

$$-T_0(S_f - S_d) = \sum_i n_i (\mu_{fi} - \mu_{di}), \qquad (2.30)$$

where μ_{fi} and μ_{di} are the chemical potentials of the *i*th product in state f and in the diffused state, respectively.

Example. What is the available work of a bottle containing 1 mole of CO_2 and 1 mole of H_2O (vapor) at atmospheric temperature and pressure? An answer requires specifying the mole fractions of the diffused substances; we may choose $x_{d1}(CO_2) = 3 \times 10^{-4}$ (= 300 ppm) and $x_{d2}(H_2O) = 0.005$. The quantities $E - E_f$, $V - V_f$, and $S - S_f$ in Eq. 2.28 are all zero, because the given system is already in equilibrium with the atmosphere. Therefore (in the ideal-gas approximation, Eq. 2.29)

$$B = RT_0 \left[1 \times \ln\left(\frac{0.5}{3 \times 10^{-4}}\right) + 1 \times \ln\left(\frac{0.5}{0.005}\right) \right]$$

$$= 12.0RT_0, \text{ or } 6.0RT_0 \text{ per mole.} \qquad (2.31)$$

This is several times larger than thermal kinetic energy ($1.5RT_0$ per mole) but small compared with typical heats of combustion. Since readers may be familiar with chemical energies in different units, we convert the answer to several other units (using $T_0 = 298$ K):

$$6.0RT_0 \text{ per mole} = 3.55 \text{ kcal/mole} = 14.9 \text{ kJ/mole}$$

$$= 0.154 \text{ eV/molecule}$$

$$= 6400 \text{ Btu/1b-mole.} \qquad (2.32)$$

The diffusion contribution to available work is evidently small (even for diffusion to 300 ppm). Tapping any reasonable part of this available work hardly appears feasible. In any example involving combustion, it would be of little importance even if it were possible because of the small relative magnitude of the diffusion term.

High-energy physicists will be pleased to notice a logarithmic divergence in Eq. 2.29! If a given product is not found as a normal constituent of air, its diffusion into the atmosphere can in principle contribute an infinite term to the available work. As the CO_2 example above illustrates, however, even very great dilution does not produce a very large addition to B. The contribution to B from a 1 ppm dilution is of order $10RT_0$ ($\ln 10^6 = 13.8$), a few percent of typical heats of combustion. Even with such a cutoff, it would be quite unrealistic to add such a diffusion term to calculated available work because of the infeasibility of actually recovering the work.

E. Available work and power generation

We begin this section with a position of advocacy. Keenan *et al.* (1973) and Berg[29] have persuasively argued that available work be used as a basis for assessing the efficiency of energy use at all levels. We strongly endorse this recommendation. As we have emphasized earlier in this chapter, the concept of second-law efficiency seems to be an especially attractive way to bring the available-work concept (and the second law of thermodynamics) to bear in a unified way on fuel conservation efforts.

For power generation in particular, it seems very clear that available work is the appropriate yardstick. Useful work is exactly the goal of power generation. We recommend that available work, not heat of combustion, be generally adopted as the reference energy for computing the efficiency of power generating units. (Then the efficiency would indeed be the second-law efficiency.) It is sometimes said, for instance, that the "maximum" efficiency of a hydrogen-oxygen fuel cell is 0.83 (see, for example, Vielstich, 1970, pp. 19–20). This is only an apparent limitation imposed by adopting the wrong yardstick, heat of combustion (enthalpy change), rather than the correct yardstick, available work.[30] In an ideal fuel cell, *all* of the available work is converted to electric energy.

As part of our position of advocacy, however, we recommend that for practical reasons, the available work be calculated using Eq. 2.11 (i.e., omitting the diffusion term). There is some ambiguity in the diffusion term (what, for instance, should be adopted for the mole fraction of water in the atmosphere?); there is the problem of the logarithmic divergence; and there is the practical impossibility of harnessing much, if any, of the diffusion contribution to available work. For these reasons, we recommend that the restricted definition of available work (Eq. 2.11) be used and that the efficiencies of all power-producing devices be expressed in the form $\epsilon = W_{out}/B$.

The combustion of methane: An idealized calculation

Combustion is an irreversible process. As such, it necessarily involves loss of available work. In this subsection we seek to illuminate this fact by treating one example of combustion in an idealized way—specifically, by approximating the fuel, air, and combustion products as ideal gases with constant specific heats. The simple connection between entropy change and loss of available work developed in the preceding section will prove useful.

The stoichiometric combustion of methane in air can be represented by

$$CH_4 + 2O_2 + 7.5N_2 \rightarrow CO_2 + 2H_2 + 7.5N_2. \qquad (2.33)$$

---

[29] Charles Berg, private communication. See also Federal Power Commission (1974).

[30] We do not mean to suggest that authors who define efficiency by $\eta = W_{out}/|\Delta H|$ are in any sense wrong; it is a matter of arbitrary definition. We suggest only that this time-honored way of defining efficiency is not the best way and should be changed.

The heat of combustion is 192 kcal per mole of CH_4 ("low heating value," water in gaseous form). The adiabatic flame temperature is 2240 K (see, for example, Van Wylen, 1959).

Adiabatic combustion carries the initial fuel-air mixture in state 1 to the products in state c (see Figure 2.6). Subsequent processes (not considered here) carry the products to a final state f in equilibrium with the atmosphere. The energy that goes into heating and expanding the products in the flame is

$$-\Delta E = -\Delta H + P_0(V_f - V_1), \tag{2.34}$$

where $-\Delta H$ is the heat of combustion. In this particular example, $V_f = V_1$ because there are 10.5 moles of gas in both state 1 and state f. Therefore, in an ideal-gas approximation, the temperature of state c is determined by the energy equation

$$-\Delta H = n\overline{C_p}\,(T_c - T_0), \tag{2.35}$$

where n is the number of moles (10.5) and $\overline{C_p}$ is an approximate average specific heat at constant pressure for the products. It is helpful to express $\overline{C_p}$ as a multiple of the gas constant R and to express $-\Delta H$ as a multiple of the thermal

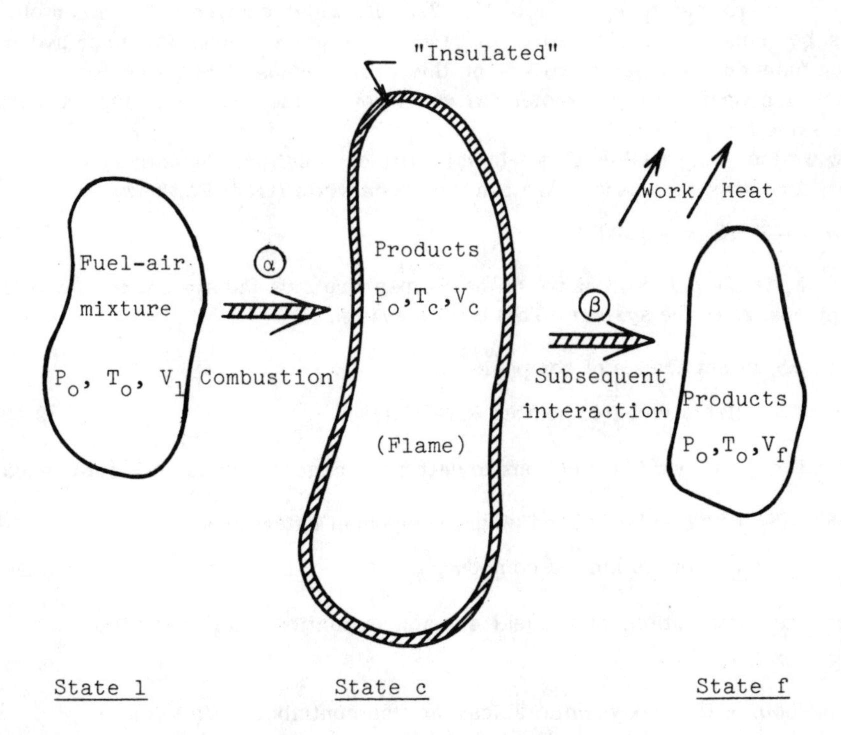

Figure 2.6. Schematic view of combustion. Process α is adiabatic combustion. The energy released goes into heating and expanding the products. After subsequent processes β, the products reach volume V_f at ambient conditions of temperature and pressure.

energy unit RT_0. We write

$$\overline{C_p} = aR, \tag{2.36}$$

$$-\Delta H = hRT_0. \tag{2.37}$$

The dimensionless enthalpy change is $h = 324$ for $T_0 = 298\,\mathrm{K}$ ($RT_0 = 0.592$ kcal/mole). Working backward from the known flame temperature, we find the dimensionless average specific heat to be $a \cong 4.75$ (for comparison, it would be 3.5 for a diatomic gas near room temperature). In terms of these dimensionless quantities, Eq. 2.35 takes the form

$$h = na\left(\frac{T_c}{T_0} - 1\right);$$

solving for the flame temperature, we get

$$T_c = T_0\left(1 + \frac{h}{na}\right). \tag{2.38}$$

In this example,

$$h = 324, \quad n = 10.5, \quad a \cong 4.75, \quad \frac{h}{na} = 6.50.$$

These numbers lead to $T_c = (298)(7.50) = 2235\ K$, which confirms the choice of a. (It can be remarked, incidentally, that the flame temperatures for other hydrocarbon fuels do not differ greatly from this value because h and n scale in approximate proportion to the molecular weight of the fuel while a is approximately the same for all fuels.)

Now we look at available work. In adiabatic combustion, the entropy of the atmosphere does not change. We can therefore write (from Eq. 2.19)

$$\Delta B = -T_0 \Delta S_{\mathrm{sa}} = -T_0 \Delta S, \tag{2.39}$$

where ΔS_{sa} is the entropy change of the atmosphere plus the system and ΔS is the entropy change of the system. This can be written

$$\Delta S = \Delta S_1 \text{ (from change of composition)}$$
$$+ \Delta S_2 \text{ (from isobaric heating of products)}. \tag{2.40}$$

The first of these two contributions to entropy change can be further subdivided:

$$\Delta S_1 = \Delta S_{1a} \text{ (from summing individual component categories)}$$
$$+ \Delta S_{1b} \text{ (from mixing of components)}. \tag{2.41}$$

The second contribution, in an ideal-gas approximation, can be written

$$\Delta S_2 = n\overline{C_p} \ln(T_c/T_0). \tag{2.42}$$

With the help of thermodynamic tables, we find contribution 1a to be

$$T_0 \Delta S_{1a} = -0.61 R T_0 = -0.36 \text{ kcal/mole}. \tag{2.43}$$

(This is for water in gaseous form.) The mixing contribution 1b can be written

$$T_0 \Delta S_{1b} = -RT_0[1 \ln(1/10.5) + 2 \ln(2/10.5) + 7.5 \ln(7.5/10.5) - 2 \ln(2/9.5) - 7.5 \ln(7.5/9.5)] = +3.3RT_0 = 1.95 \text{ kcal/mole}. \quad (2.44)$$

Within the brackets, the first three terms are mixing terms for the products CO_2, H_2O, and N_2, respectively (see Eq. 2.33); the remaining terms, preceded by negative signs, are mixing terms for the O_2 and N_2 initially in the air. The fuel is assumed to be unmixed with air initially. The mixing entropy increases ($\Delta S_{1b} > 0$) because the products are "more mixed" than the reactants. (If the product water were condensed out in liquid form, the mixing entropy would decrease.) We include these remarks only for qualitative insight. In fact, the dominant contribution to entropy change comes from isobaric heating in the flame. With the help of Eq. 2.43, we find, for contribution 2,

$$T_0 \Delta S_2 = +101RT_0 \cong 60 \text{ kcal/mole}. \quad (2.45)$$

The net loss of available work (Eq. 2.40) is

$$\Delta B = -T_0 \Delta S = -104RT_0. \quad (2.46)$$

The available work left in the flame is

$$B_c = B - T_0 \Delta S. \quad (2.47)$$

Consistent with the present level of approximation, we can set $B \cong -\Delta H = hRT_0$ and $\Delta S \cong \Delta S_2$. Then, with the help of Eqs. 2.42 and 2.38, Eq. 2.47 takes the simple approximate form

$$B_c \cong B\left[1 - \frac{na}{h} \ln\left(1 + \frac{h}{na}\right)\right]. \quad (2.48)$$

For the specific numbers of this example ($h/na = 6.50$),

$$B_c \cong 0.69B; \quad (2.49)$$

this means that 31 percent of the available work has been lost in the combustion process. Applying the simple approximations of this example to other hydrocarbon fuels, we find values of $|\Delta B|$ ranging from 29 to 31 percent of the heat of combustion.[31] As a grand general approximation for the hydrocarbon fuels,

$$h \cong 19.5M.W., \quad a \cong 4.8, \quad n \cong 0.6M.W.,$$

where $M.W.$ is the molecular weight of the fuel. This gives $h/na \cong 6.8$ and

$$\frac{|\Delta B|}{B} = \frac{1}{6.8} \ln(7.8) = 0.30. \quad (2.50)$$

[approximation for all hydrocarbon fuels]

---

[31]More exact calculations, taking account of the temperature dependence of specific heats, are, of course, possible. As an example, Keenan *et al*. (1973) treat a hypothetical fuel CH_2 with a heat of combustion of 20,000 Btu/lb and find $|\Delta B|$ to be 29 percent of the heat of combustion or 27 percent of the initial available work.

A more traditional way of looking at the loss of available work in combustion. The Carnot efficiency associated with a temperature $T_c \cong 2250\,\mathrm{K}$ is $\eta_c = 0.87$. Accordingly, it may at first seem surprising that the available work in the flame is only 0.70 times the available work in the fuel. One way to clarify the reason for the difference is to note that the flame is a finite, not an infinite, heat reservoir. Imagine the change from state c to state f in Figure 2.6 to occur by having the hot reservoir (the "flame") drive a perfect Carnot engine as the temperature of the reservoir falls from T_c to T_0 during isobaric expansion. In the approximation of constant specific heat the work delivered is

$$W = n\overline{C_p} \int_{T_0}^{T_c} \left(1 - \frac{T_0}{T}\right) dT, \tag{2.51}$$

which gives

$$W = n\overline{C_p}(T_c - T_0) - n\overline{C_p}\, T_0 \ln(T_c/T_0). \tag{2.52}$$

The first term on the right is (at the level of approximation in this treatment) the same as B, the fuel's initial available work (see Eq. 2.35), and the second term is approximately ΔB, the loss of available work in combustion (see Eq. 2.42; also Eq. 2.48), about 30 percent of B in this example.

A remark on catalytic combustion. The low-temperature combustion of fuel with the help of catalysts has some advantages if one wants to use the fuel only as a source of low-temperature heat. The process can avoid most flue losses and duct losses, and it can provide efficient zone control for space heating. Nevertheless, it must not be imagined that decreasing the combustion temperature decreases the loss of available work inherent in the combustion process. On the contrary, from a second-law viewpoint, catalytic combustion is less efficient than ordinary high-temperature combustion. Consider a catalytic reactor that is perfectly efficient in a first-law sense ($\eta = 1$), providing all of its heat at $T_1 = 110\,^\circ\mathrm{F}$ ($43\,^\circ\mathrm{C}$) when the outdoor ambient temperature is $T_0 = 32\,^\circ\mathrm{F}$ ($0\,^\circ\mathrm{C}$). Its second-law efficiency is $\epsilon = [1 - (T_0/T_1)] = 0.14$. Eighty-six percent of the fuel's available work has been "wasted." An ordinary furnace is, of course, even more wasteful of available work because it takes no advantage of its high flame temperature and has additional losses.

Available work of various fuels

Various energies associated with methane combustion are shown in Table 2.5, for burning in oxygen and in air, and for product water in vapor and in liquid form.[32] Rows (1) and (2) divide the heat of combustion into its internal-energy part $E - E_f$, and its non-useful-work part, $P_0(V - V_f)$. The former is 192 kcal/mole ("low heating value") or 212 kcal/mole ("high heating value"). The latter

---

[32] These energies are essentially exact. The large terms are taken from standard tables (see, for example, Kauzman, 1967). The small terms associated with mixing and diffusion are calculated in an ideal-gas approximation. For the diffusion contribution, see Eq. 2.29. For an example of a mixing contribution (which, in its nature, is the same as a diffusion contribution), see Eq. 2.44.

Table 2.5. Energies associated with methane combustion. [All energies in kcal per mole of CH_4.]

Energy terms	Burning in oxygen: H_2O vapor product	Burning in oxygen: H_2O liquid product	Burning in air: H_2O vapor product	Burning in air: H_2O liquid product
(1) $E - E_f$	191.8	211.6	191.8	211.6
(2) $P_0(V - V_f)$	0	+1.2	0	+1.2
(3) $-T_0(S - S_f)$ (a)	−0.4	−17.3	−0.4	−17.3
(b)	+1.1	0	+2.0	−1.1
(4) Heat of combustion $-\Delta H = (1) + (2)$	191.8	212.8	191.8	212.8
(5) Available work without diffusion, $B = (1) + (2) + (3)$	192.5	195.5	193.4	194.4
(6) Percentage change from $\|\Delta H\|$ to B	+0.4%	−8.1%	+0.8%	−8.6%
(7) Additional diffusion contribution B_d	9.1 (5%)	4.8 (2.5%)	6.5 (3%)	4.0 (2%)
(8) Total available work, $B_{total} = (5) + (7)$	201.6	200.3	199.9	198.4

Notes on Table:

(3) (a) This is the entropy term excluding the mixing contribution to entropy.
(3) (b) This is the mixing contribution to entropy (the fuel is initially unmixed).
(7) This is calculated for diffusion to the following mole fractions in air: $x(N_2) = 0.8$, $x(H_2O) = 0.01$; $x(CO_2) = 3 \times 10^{-4}$. No diffusion for liquid H_2O.

is very small (about 0.5 percent). Row (3) gives the entropy contribution to available work, $-T_0(S - S_f)$. Row (3)(a) is the entropy contribution associated with change of composition from reactants to products; Row (3)(b) is the entropy contribution associated with mixing of the components (the fuel itself is assumed to be unmixed initially). Row (4) sums the first two rows to give the heat of combustion and Row (5) sums the first three rows to give the available work apart from diffusion. Finally, Row (7) gives an estimate of the additional diffusion contribution to available work; it amounts to 2 to 5 percent of the "basic" available work.

Table 2.5 shows that condensation of water vapor adds much more to the heat of combustion than it adds to the available work. There is a simple physical reason for this: The energy released in condensation, although substantial in magnitude (9.7 kcal per mole of H_2O), is low in quality. It adds relatively little to the fuel's ability to do work. Note, for example, in the last column of Table 2.5 that the larger energy release [Row (1)] is largely compensated by a negative entropy term [Row (3)(a)]. The high and low values of available work are nearly the same, and both are close to the low heat of combustion.

Table 2.6 presents similar data for other fuels, all for combustion in air with

Table 2.6. Energies associated with combustion of various fuels. [Calculations are for combustion in air with gaseous H_2O product ("low heating value"). Energies in the upper part of the table are in kcal per mole of fuel. Available work is given in other units in the lower part of the table.]

Energy terms	Hydrogen H_2	Carbon C (to CO_2)	Carbon monoxide	Methane CH_4	Ethane C_2H_6	Propane C_3H_8	Ethylene C_2H_4	Liquid octane C_8H_{18}
(1) $E - E_f$	57.5	94.1	67.3	191.8	341.5	489.1	316.2	1216
(2) $P_0(V - V_f)$	+0.3	0	+0.3	0	−0.3	−0.6	0	−3
(3) $-T_0(S - S_f)$ (a)	−3.2	+0.2	−6.2	−0.4	+3.3	+7.2	−2.1	+35
(b)	+0.4	0	+0.4	+2.0	+3.3	+4.5	+2.5	+11
(4) Heat of combustion, $-\Delta H = (1) + (2)$	57.8	94.1	67.6	191.8	341.2	488.5	316.2	1213
(5) Available work without diffusion, $B = (1) + (2) + (3)$	55.0	94.3	61.8	193.4	347.8	500.2	316.6	1259
(6) Percentage change from $\|\Delta H\|$ to B	−4.8%	+0.2%	−8.6%	+0.8%	+1.9%	+2.4%	+0.1%	+3.8%
(7) Additional diffusion contribution B_d	1.9 (3.5%)	3.9 (4%)	4.0 (6%)	6.5 (3%)	11.3 (3%)	16.2 (3%)	9.8 (3%)	41 (3%)
(5)′ Available work in other units [a] kcal/gm	27.3	7.86	2.21	12.06	11.58	11.35	11.29	11.03
MJ/kg	114	32.9	9.23	50.5	48.4	47.5	47.3	46.1
kBTU/lb	49.1	14.1	3.97	21.7	20.8	20.4	20.3	19.9
kBtu/lb-mole	99.0	170	111	348	626	900	570	2266
eV/amu	1.18	0.341	0.096	0.523	0.502	0.492	0.490	0.478

[a] Available work in other units does not include the diffusion contribution B_d.

the water product (if any) in gaseous form. Overlooking the diffusion terms, one sees that the available works for hydrocarbon fuels are within a few percent of the (low) heats of combustion. For carbon, B and $|\Delta H|$ are nearly identical. For the benefit of readers accustomed to other units, the last five rows of the table restate the available work B in other units. In Table 2.7, heats of combustion and available work for several fuels are stated and compared for the alternative assumption of condensed water product. As for methane (Table 2.5), one sees that the available work is generally less than the heat of combustion when high heating value is used. The difference is especially marked for hydrogen, whose available work—with or without water condensation—is some 18 percent less than its higher heat of combustion.

F. Second-law efficiency: A national perspective

To assign a second-law efficiency to any type of energy use on a national scale is possible in principle, although, of course, difficult in practice. Berg (1974a) has estimated a few such efficiencies. In Table 2.8 we set forth very rough estimates of ϵ for major energy uses in the United States. For any one of the areas, an informed expert could undoubtedly calculate a more accurate number. Nevertheless, we consider the overall perspective provided by such a table to be valuable. Despite the uncertainty in some of the efficiencies, the table emphasizes the fact that there is room for great improvement in most areas. Transportation and low-temperature use of heat are notably inefficient.

Remarks on individual entries follow. Reference to another part of the report means that a calculation of average second-law efficiency can be found there.

Space heating, water, heating, and air conditioning. See Chapter 3, Section D, especially Table 3.5, p. 82.

Cooking. We have not attempted to calculate an overall average second-law efficiency for cooking.

Refrigeration. For the room temperature near the refrigerator coil we take $T_0 = 300$ K (81 °F). As a rough average of the temperatures inside freezers and

Table 2.7. Available work and heat of combustion compared for water product in liquid form (high heating value). [Energies are in kcal per mole of fuel. The columns correspond to Rows (4), (5), and (6) in Table 2.6. For methane, see Table 2.5.]

Fuel	Heat of combustion $-\Delta H$	Available work B [a]	Percentage change from $\|\Delta H\|$ to B
Hydrogen, H_2	68.2	55.9	−18%
Ethane, C_2H_6	373	349	−6.4%
Propane, C_3H_8	531	502	−5.5%
Ethylene, C_2H_4	337	317	−5.9%
Liquid octane, C_8H_{18}	1307	1261	−3.5%

[a] This is the "basic" available work, not including a diffusion contribution.

Table 2.8. National perspective on efficiency of energy use.

Use	Percent of U. S. fuel consumption (1968) [a]	Estimated overall second-law efficiency ϵ
Space heating	18	0.06
Water heating	4	0.03
Cooking	1.3	...
Air Conditioning	2.5	0.05
Refrigeration	2	0.04
Industrial uses		
process steam	17	~0.25
direct heat	11	~0.3
electric drive	8	0.3
electrolytic processes	1.2	...
Transportation		
automobile	13	0.1
truck	5	0.1
bus	0.2	...
train	1	...
airplane	2	...
military and other	4	...
Feedstock	5	...
Other	5	...
	100	

[a] Sources: SRI (1972), Mitre Corporation (1973).

refrigerators we take $T_3 = 265\,\mathrm{K}$ (18 °F). From square 7 in Table 2.3, the second-law efficiency is $\epsilon = \eta_t[(T_0/T_3) - 1]$. We evaluate this using $\eta_t \cong \eta_e/3$ and η_e (= COP) $\cong 1$ (typical of existing refrigerator-freezer combinations). These numbers give $\epsilon = 0.044$. The fact that this is very nearly the same as the value of ϵ calculated for air conditioning in Chapter 3 is not surprising. The economics and technology of refrigerator and air-conditioner cycles are very similar; therefore similar second-law efficiencies (*not* COP's) are to be expected.

Industrial process steam. From square 5 in Table 2.3, the efficiency is $\epsilon \cong \eta_t[1 - (T_0/T_2)]$, where η_t is the boiler efficiency (the fraction of the heat of combustion transferred to the steam). Our approximate calculation uses $\eta_t = 0.8$, ambient temperature $T_0 = 286\,\mathrm{K}$ (55 °F), and steam temperature $T_2 = 422\,\mathrm{K}$ (300 °F). These numbers give $\epsilon = 0.26$. This is only a rough estimate because we lack data on the relative quantities of steam produced at different temperatures.

Industrial direct heat. The efficiency formula is the same as that given above for process steam. Using data provided by Berg (1974b) for typical operating temperatures ($T_2 \sim 500$–1000 °C) and typical first-law efficiencies ($\eta_t \sim 0.4$), we calculate second-law efficiencies ϵ of 0.25 to 0.31.

Industrial electric drive. If the electric efficiency η_e is around 0.9, the thermal efficiency is $\eta_t \cong \eta_e/3 \cong 0.3$, and the second-law efficiency (square 2 in Table 2.3) is $\epsilon \cong \eta_t$.

Industrial electrolytic processes. We have not calculated an average efficiency for electrolytic processes.

Transportation. For the automobile, see Chapter 4, Section B, especially Table 4.4. Nearly two thirds of the automobile mileage and more than two thirds of the automobile fuel consumption occur in urban driving, so the average efficiency is $\epsilon \cong 0.1$. In over-the-road driving, diesel trucks are more efficient than gasoline-powered automobiles. However, two thirds of all truck mileage is accumulated in urban driving, much of this with Otto-cycle, not diesel, power. The overall average efficiency of trucks is therefore probably less than 0.15. We estimate it simply as 0.1. We have not calculated second-law efficiencies for other forms of transportation.

Table 2.8 is "conservative" in that it is based on narrowly defined tasks: current demand for heat or work is assumed to reflect actual need. The efficiency of an automobile, for example, is calculated not with respect to the broadly defined task of moving people comfortably and speedily from one place to another, but with respect to the narrowly defined task of providing exactly as much energy at the drive wheels of the vehicle as is currently provided in the average American car. And no account is taken of unoccupied seats. The efficiency of using direct heat in industry (to pick another example) is defined without consideration of alternative ways to achieve industrial goals. The value $\epsilon \approx 0.3$ in this row of the table does *not* mean that the ultimate possible reduction in fuel use for direct-heat processes is by a factor 3. In certain processes, at least, the ultimate saving could be more—if, for example, specific radiation were used to heat the working substances, or if alternative lower-temperature processes were developed.

There are, of course, technical as well as economic obstacles, some of them formidable, that prevent the achievement of second-law efficiencies close to 1. In economic terms, there may be no *reason* to press for efficiencies beyond a certain level. Nevertheless, as physicists, we must think in terms of what nature permits.

3. ENERGY INDOORS

A. Introduction: Some systems considerations

This chapter of our report describes the furnace, the water heater, and several other energy systems in the house. "House" is short-hand for any dwelling unit, including apartments, row houses, and detached houses; there are 60 million dwelling units in the United States. Much of the discussion is generalizable to apartment buildings and to the stores and public buildings which make up what is commonly called the commercial sector. Care must be taken in such generalizations, of course: a dominating factor in the heating and cooling of commercial buildings, the large and predictably varying internal electric load, is quite distinct from the corresponding relatively smaller and partially random internal electric load in a house. The SRI compilation (see Appendix I) indicates that 18 percent of the United States energy consumption in 1968 went to space heating, 4 percent to water heating, and 2.5 percent to air conditioning (residential and commercial sectors combined).

The most significant environmental parameter for energy consumption in a house is average daily outside temperature. Accordingly, the degree day[33] is a useful measure of the need for heating a house, to which fuel consumption is closely proportional.[34] A median value for U. S. homes is near 5000 Fahrenheit degree days per year, but the range is from 14,000 in Fairbanks, Alaska, to below 2000 in some southern states. Fuel consumption for heating, by far the largest energy use in most residences, is close to 20,000 Btu per Fahrenheit degree day in modest (1500-ft^2 floor area) houses heated by gas or oil in average climates.[35] One good

---

[33]The number of degree days in a single day is found by subtracting the average of the high and the low temperature for that day from a reference temperature, usually 65°F in the U.S. The number of degree days in an interval of several days is then found by summation, including only *positive* values in the sum. In Great Britain, the base for degree days has been 60°F.

[34]The degree day is often not a useful measure for commercial buildings, where lighting and machinery may contribute larger amounts of heat than the heating system.

[35]A Celsius degree day is 1.8 times larger than a Fahrenheit degree day; 20,000 Btu per Fahrenheit degree day is equivalent to 9100 kcal per Celsius degree day.

measure of performance of the system of house plus furnace is therefore Btu per Fahrenheit degree day.[36] This value is larger in warmer climates, reflecting less attention to insulation and (perhaps) a greater casualness about costs when total heating costs are lower.

After average daily temperature, the wind speed is the most important environmental parameter determining the fuel consumption in winter in today's houses. Wind effects on energy consumption in a house are only now being studied, and it is becoming clear that wind dominates the dynamics of the exchange of air between inside and outside, known as air infiltration. One-third or more of the total heat lost from the interior of the house is associated with this convective exchange with outside air. (These dynamics have been a neglected area of building technology, in part because large commercial buildings tend to have mechanically driven air-exchange systems and tend to have tighter construction. Accordingly, the mechanical system dominates the wind effects. Most of the computer codes to predict energy consumption in commercial buildings do not incorporate wind.)

The solar energy incident on an unshaded house on a clear day in winter, averaged over day and night, is comparable to the energy required to heat a typical house on the coldest day in a 5000 Fahrenheit degree day climate. But winter sun is largely decoupled from the interior environment of a house today. The sun is a more critical environmental parameter in summer time, because the typical temperature difference between outdoors and indoors is rarely more than 20 °F (11 °C). This is one reason why no "cooling degree day" measure based on temperature alone is as reliable a predictor of energy consumption as the "heating degree day." Also, preferences and behavior with respect to air conditioning are more variable and more random from house to house. We have restricted this report to end uses of energy and thus do not systematically consider the sun as an energy source. This field is beginning to be vigorously investigated by physicists and engineers.

Other environmental parameters of a site represent opportunities for an interactive system. Ground water undergoes little fluctuation in temperature over a season, being part of a system whose time constant is measured in years, and hence reflecting an average yearly temperature; in many situations ground water has attractive possibilities as a component of heating and cooling systems. Other interactive solutions to the problem of the house as an energy system involve passive components, such as high heat capacity systems to take advantage of a comfortable daily *average* temperature in summer.

Another type of opportunity involves locating the crucial stage of energy conversion at the site of end use. One suggestion is that fuel-based heating systems might be best controlled, and operated with reduced flue and duct losses, if fuel could be burned catalytically at low temperature in room radiators (see Chapter 2, Section E; also APS, 1974b). The catalytic combustion in itself is not an efficient process in the second-law sense; but the advantage of location (also enjoyed by electric resistive heating) might mean a system with overall improved efficiency compared to

---

[36]A "degree day per gallon" meter would be useful. To deliver the most degree days per gallon, a house should be *compact* and/or *energy efficient*. The analogy to automobile miles per gallon is obvious.

present furnace-based systems. There are more mundane and perhaps more important examples of this principle—location of conversion at point of use—such as spot water heaters, for example at the dishwasher.

Both air conditioners and furnaces are usually designed to run intermittently at full capacity, with the capacity being chosen to handle the severe weather of "design conditions," which are met or exceeded, typically, only about 50 hours per year. The advantages and disadvantages of running more nearly continually at reduced output in more typical winter or summer weather do not appear to be well understood from a theoretical standpoint.

A particularly important aspect of energy use, especially electricity use, is the time dependence of demand. The minimum demands occur during spring and fall. In some localities with cold winters and much installed electric heating equipment, the maximum peak load occurs during the winter. In many places (New York City is one of the more extreme examples) the peak occurs on hot summer afternoons. The summer peak load for Consolidated Edison Company of New York was 8700 MW in 1973 (Citizens for Clean Air, 1973), of which 39 percent was for air conditioning equipment. This load just about matches their generating capacity. Any changes that could be made in the methods of air conditioning to reduce electricity demand during peak hours would be most important. Such changes might include increased heat capacity of the house, changeover to other fuels, more efficient cooling cycles, energy storage, improved architectural design, use of shading and reflecting materials, cooling with ground water, etc.

The following sections of this report look more quantitatively at the opportunities for improved energy utilization. Section B attempts to give a sense of the magnitudes of the various energy gains and losses in the winter heating of a typical house made of typical materials and located in an average climate. Several important components and subsystems relevant to energy use in buildings are examined in Section C. In Section D, water heating, space heating, and air conditioning are re-examined, this time from the standpoint of the second law of thermodynamics. A brief description of the opportunities and obstacles related to producing low-temperature heat and electricity by combined systems is presented in Section E. Finally, principal recommendations and conclusions are summarized in Section F.

B. Winter heating requirements

We will investigate the energy required to convert a region of space into a comfortable place. What constitutes comfort is a matter of personal choice, but there are not very wide disagreements. As an illustrative introduction, we consider the house's prehistoric precursor.

A cave dwelling

If we lived in a cave in a tropical climate, the temperature would be stable and we might have to expend no energy at all on space conditioning. If, however, we lived in a cave in a temperate climate, heating would be required all year round, since the ground temperature remains at about 55 °F (13 °C). To estimate the energy required for this underground lifestyle, we assume that the cave size is 50 $\times$30$\times$10 ft (15.2$\times$9.1$\times$3.0 m) and calculate the heat energy that must be supplied to hold the air temperature at 70 °F (21 °C) under the assumption that the earth's temperature stays constant. The surface area of the walls, roof, and floor is 4600 ft^2. The rate at which heat is transported into the earth ($\dot{Q}$) is approximated in engineering calculations (ASHRAE, 1972) by the expression[37]

$$\dot{Q} = UA(T_0 - T_1).$$

In this example, we may set $U = 0.1$ Btu/ft^2 hr °F, or 0.57 W/m^2 °C (approximately the heat conductance of ten feet of rock); A is the surface area in contact with the ground, T_0 is the interior temperature, and T_1 is the ground temperature, measured, to be sure, several feet away from the structure. The rate of heat loss into the ground surrounding the cave is $0.1 \times 4600 \times 15 = 6900$ Btu/hr or about 2 kW.

Living above ground: Walls

If we live above ground we are faced with temperature variations: daily and yearly cycles. We are now coupled to the environment in several ways: heat loss to the ground; heat loss to the atmosphere; heat input from solar radiation. We have a choice as to whether to take advantage of the natural thermal environment or to try to insulate ourselves from it.

In the temperate parts of the United States, heating is required during the winter. To estimate the amount of heat required to keep the interior of a structure warm we consider a "standard house" of the same volume and floor area as the cave, but two floors high; its dimensions are $25 \times 30 \times 20$ ft ($7.6 \times 9.1 \times 6.0$ m). Our "standard winter temperatures" are 40 °F (4 °C) outside and 70 °F (21 °C) inside.[38]

To calculate the effective heat transfer properties of a surface containing several layers, one treats the layers as a set of resistances in series. The resistivity of wood is about 1 (Btu/hr ft^2 °F)$^{-1}$ per inch of thickness [0.069 (W/m^2 °C)$^{-1}$ per cm of thickness]. The thermal resistivity of fiberglass insulation is about four times larger.[39] A careful calculation would include the non-uniformities in cross section brought about, in particular, by studs. To the thermal resistances of solid materials and trapped air we add surface resistances which describe the heat transfer from the wall to the air; these depend on the air velocity. Typical values for the surface resistances (for unit area) are 0.8 and 0.2 (Btu/hr ft^2 °F)$^{-1}$ [0.14 and 0.035 (W/m^2 °C)$^{-1}$], inside and outside respectively. Our standard house has walls with 2 in. of insulation and a roof with 4 in. of insulation. Unit areas of wall and roof have thermal resistances of 10 and 18 (Btu/hr ft^2 °F)$^{-1}$ [1.8 and 3.2 (W/m^2 °C)$^{-1}$], respectively. The heat load through walls and roof is 7850 Btu/hr (2.3 kW). This

---

[37]The "U-value" is the standard parameter representing the heat transfer through a shell including its boundary layer. It is the reciprocal of the integral of the resistivity along the path of heat flow.

[38]Our "standard house" and "standard winter temperatures" have many features in common with the houses and weather in the planned community of Twin Rivers, New Jersey, under study by a group at the Center for Environmental Studies, Princeton University (Fox, 1973a, 1973b). These houses conform to minimum acceptable construction practices, as codified, for example, in the pre-1974 Minimum Property Standards of the Federal Housing Administration.

[39]Insulation is discussed further in Section C, Part 5.

heat load, of course, would be less if some "outside" walls were shared, as in row houses and apartments. Conductive losses to the ground add an additional 1100 Btu/hr (0.3 kW) if a ground temperature of 55 °F (13 °C) and a sandwich of materials equivalent to what is in the walls are assumed. A reasonable goal for improved insulation, probably requiring some new materials and design practices, would provide total resistances of 30 $(\text{Btu/hr ft}^2\,°\text{F})^{-1}$ [5.3 $(\text{W/m}^2\,°\text{C})^{-1}$] for all opaque surfaces, reducing this class of heat losses by about 60 percent. We imagine that this goal, and comparably difficult goals for other aspects of design and materials, are met in a "target house," whose energy balance we develop in parallel with that of the standard house.

Windows

When we replace 200 ft^2 (18.6 m^2) of opaque surface by windows, we increase the heat transfer out of the house by conduction, convection, and radiation, and we also add cracks which increase air infiltration. On the other hand, we permit the warming effect of sunlight, which, we shall see, can sometimes cancel the losses.[40] A typical U-value for single-pane windows is 1.1 $\text{Btu/hr ft}^2\,°\text{F}$ (6.2 $\text{W/m}^2\,°\text{C}$), and it is dominated by the behavior of the two surface resistances associated with the boundary layers of air inside and outside. (The temperature drop across the glass is usually only about 1 °F, or 0.5 °C.) Our standard house now loses 6600 Btu/hr (1.9 kW) by conduction through the windows, but the wall losses are reduced by 600 Btu/hr (0.15 kW) to 7200 Btu/hr (2.1 kW). Clear double glass windows today have quoted U-values in the range of 0.5 to 0.7 $\text{Btu/hr ft}^2\,°\text{F}$ (2.8 to 4.0 $\text{W/m}^2\,°\text{C}$). A reasonable future goal would be windows with U-values of 0.3 $\text{Btu/hr ft}^2\,°\text{F}$ (1.7 $\text{W/m}^2\,°\text{C}$) (APS, 1974a).

Fresh air

Only rarely does a house require deliberate ventilation; in general, adequate air enters through pores and cracks. The needed rate of air exchange depends upon the number of occupants and their activity. It is a topic in need of coherent investigation and standardization. Our standard house experiences one air change per hour.[41] To heat the incoming air to the temperature of the interior of the house (which contains 15,000 ft^3, 1100 lb, or 500 kg of air) requires that energy be supplied at a rate of 8000 Btu/hr (2.4 kW).

If the fresh air we let into the house has a relative humidity of 60 percent outside (at 40 °F), and if no moisture is added, it has a relative humidity of 20 percent inside (at 70 °F). If we want to raise the relative humidity indoors to 40 percent, we must supply an additional 3800 Btu/hr (1.1 kW) to evaporate water. This heat of vaporization is usually not recovered as the water vapor recondenses outside. The total energy supplied to heat and humidify the incoming air is almost as large as

---

[40] Also, natural daytime lighting might replace some electric lighting.

[41] The measured air infiltration rates in Twin Rivers houses, when all windows and doors are closed, vary from 0.2 to more than 2 exchanges per hour, depending most sensitively on wind speed and wind exposure.

(75 percent of) the energy supplied to compensate for the conductive losses through the shell.

The infiltration rates, of course, increase with opening of doors and windows, and the *amount* by which these rates increase is sensitive to building design. Vestibules, for example, will reduce energy losses from open doors. In most houses, an adequate supply of fresh air in winter is provided without opening any windows. Nevertheless, many people do open windows even in cold weather. The perceived benefits of this practice need to be better understood. Perhaps alternative less costly approaches could be devised for controlled admission of fresh air.

A reasonable goal for a target house would be an air infiltration rate of 0.2 exchanges per hour. This is one parameter which it will be critical to remeasure periodically, since the consequences of faulty construction and differential settling appear over time, in the form of cracks around window frames and elsewhere. (In commercial buildings, where most of the air exchange typically is brought about by forced ventilation, the increased use of heat exchangers could recover much of the lost heat.) As we discuss in Section C, part 6, there is a need for more extensive measurements of infiltration. Regional field tests could raise the level of awareness of these problems and reduce the losses.

Energy balancing

The components of heat loss in our standard house and in our target house are gathered in Table 3.1. In our standard house, as in most recently built housing, the heat losses through opaque surfaces, through windows, and as a result of air infiltration, are roughly (within a factor of 2) comparable to one another. The calculations the reader has just been led through, known as heat load calculations,

Table 3.1. Heating budget for a standard house and a target house on a mild winter day.[a]

Source of energy expenditure	Standard house Btu/hr	Standard house kW	Standard house % of total	Target house Btu/hr	Target house kW	Target house % of total
Heat transfer through walls and roof	7200	2.1	27	2800	0.8	38
Heat transfer through floor to ground	1100	0.3	4	400	0.1	5
Heat transfer through windows	6600	1.9	25	1800	0.5	24
Warming of incoming air	8000	2.3	30	1600	0.5	22
Humidification of incoming air	3800	1.1	14	800	0.2	11
Total	26,700	7.8	100	7400	2.2	100

[a] A "mild winter day" is defined by an outdoor temperature of 40 °F (4 °C) and an outdoor humidity of 40 percent. This temperature is approximately the six-month winter average in climates characterized by 5000 Fahrenheit degree days. Annual average rates of energy consumption are thus estimated by dividing the table entries by 2.

are routinely performed by those responsible for sizing furnaces and utility supply systems. They will generally be interested in the heat load on a day of extreme climate, and hence will use a temperature difference of, typically, 70 °F (39 °C) between inside and out. (Extreme temperature differences are about double the average temperature differences.) The energy to humidify and the energy lost to the ground are effectively constant. The other energies in the heat load calculation are simply proportional to the temperature difference. Using these assumptions, we can scale the results in Table 3.1 to predict loss rates on the extreme day ($\Delta T = 70\,°\mathrm{F} = 39\,°\mathrm{C}$) to be 55,800 Btu/hr (16.3 kW) for our standard house, and 15,600 Btu/hr (4.6 kW) for our target house.

It appears that our standard house requires a heating system capable of delivering at least 55,800 Btu/hr (16.3 kW) into the house. Actually, the nominal capacity of the system would have to be greater than this, to provide for losses in the distribution system, degraded performance arising from improper or infrequent maintenance, etc. On a typical mild day, when the indoor-outdoor temperature difference is 30 °F (17 °C), the heating system will operate at irregular intervals, perhaps 25 percent of the time overall. The fraction is low not only because the furnace is designed conservatively and the weather is far from extreme, but also because there are other sources of energy contributing to the heating. The appliances and the people themselves are internal heat sources, and the sun coming in the windows is an external heat source. The sum of these heat inputs and the furnace heat input must closely balance the losses we have just examined.

Solar energy input

Ordinary houses (not provided with special collectors) are directly heated by the sun, mainly by radiation influx through the windows. The heat influx through the roof, attic, and walls is typically small, because of their low thermal conductance and rapid convective cooling. The energy influx through windows is substantial, because the heat energy is released *inside* the house. A window on a house is a superb solar heat collector.

The solar constant, I_0, is the incident radiant energy flux above the earth's atmosphere. Its value is about 1.36 kW/m^2, or 430 Btu/hr ft^2. On a sunny day the radiant energy flux at the surface of the earth peaks at about 1 kW/m^2. Some solar fluxes at ground level are listed in Table 3.2.

The 200-ft^2 (18.6-m^2) of windows in our standard house are evenly distributed over the four exposures (50 ft^2 per exposure). On an average day in December, assuming that clouds and reflections at the window surfaces reduce the incident solar energy by a factor of 2,[42] the solar input, easily estimated from Table 3.2, is 0.77 kW, or 2600 Btu/hr. This compensates roughly five percent of the heat loss on a very cold day. All the north-facing windows on the standard house have been shifted to the south-facing wall of the target house. Again reducing the solar energy by a factor of 2, we calculate the heat gain through the windows of the target house to be 1.22 kW, or 4200 Btu/hr. This compensates roughly a quarter of the heat loss from the target house on a very cold day.

---

[42]See Petherbridge (1967).

Table 3.2. The solar flux at 40° latitude on horizontal surfaces and on vertical surfaces with four orientations: north, east, south and west.[a] (The tabulated values assume a clear day.)

	Average flux (24-hour average) (kW/m^2)[b]					Flux at noon (kW/m^2)[b]				
Date	Horizontal surface	Vertical surface *N*	*E*	*S*	*W*	Horizontal surface	Vertical surface *N*	*E*	*S*	*W*
June 21	0.295	0.064	0.158	0.082	0.158	0.842	0.120	0.129	0.300	0.129
Dec. 21	0.074	0.013	0.057	0.205[c]	0.057	0.356	0.054	0.057	0.748	0.057

[a] Source: ASHRAE (1972), p. 390.
[b] 1 kW/m^2 =317 $Btu/hr\,ft^2$.
[c] For the sake of comparison, a single pane of glass (U-value of 1.1 $Btu/hr\,ft^2\,°F$) transmits heat at a rate of 0.243 kW/m^2 when the temperature difference is 70 °F; thus a south window almost breaks even on a cold, clear winter day.

As noted in Table 3.2, a single-pane, south-facing window, averaged over a 24-hr day, transmits almost as much solar energy into the house as it conducts heat out of the house when the temperature is 0 °F outside (and when the skies are not cloudy all day). Figure 3.1 shows the solar flux and net heat flux through a south-facing window, hour by hour, for an assumed daily temperature cycle and an assumed factor of 0.5 for clouds and window reflections. A more detailed study would need to treat reflection more carefully and take account of drapes, other interior window treatment, and exterior shading by overhangs. It would also need to investigate the best ways to store some of the solar heat and to distribute this heat uniformly around the house (perhaps by improved controls for the blower in a hot-air system).

Internal inputs other than the heating system

When people are in a house, they are sources of about 100 W each, a factor which needs to be taken into account in designing auditoria and which account for open windows in houses having large parties in mid-winter. Our standard house has an average of three people inside (0.3 kW). Well endowed with appliances but not using electricity for space heating, it uses 1500 kW hr of electricity in a winter month, a rate of 2.1 kW, nearly half of it for hot-water heating. Most of this energy (perhaps 80 percent) ends up as heat somewhere in the house, but often in kitchen, laundry, and basement areas where it is not appreciated. The 20 percent (0.4 kW) which leaves the house as hot air vented from dryers and hot water down drains is a loss at the five percent level on an average winter day. This loss is avoided in our target house, where heat recovery systems permit the integration of appliances with the heating system, and the even distribution of the heat generated by the appliances. But the target house would also use perhaps one-third as much electricity (0.7 kW) to provide the same services (hot water, dish and clothes washing, lighting), as a result of improved design.

If internal inputs are provided by fossil fuels instead of electricity, as in gas ranges, dryers, and hot water heaters, somewhat more energy is deposited in the

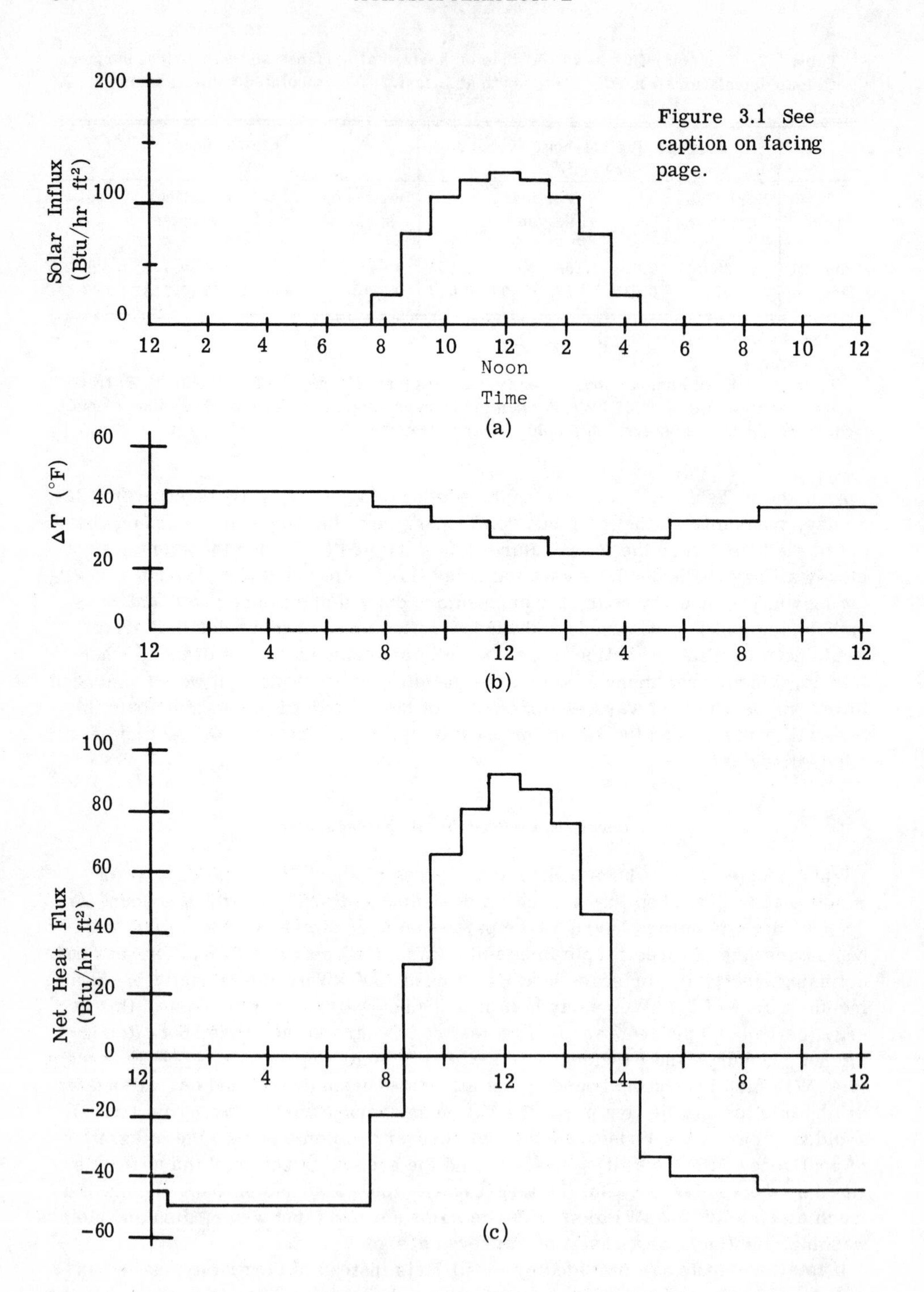

Figure 3.1 See caption on facing page.

Figure 3.1. (a) The solar influx through windows on a south wall. We have assumed an hourly variation characteristic of 40° latitude on December 31. (The data are taken from ASHRAE, 1972.) The rate of solar energy influx as tabulated has been multiplied by a factor of 0.5, which is intended to account for average cloud cover and the transmission properties of single-pane glass (see Petherbridge, 1967). (b) A specific model for indoor-outdoor temperature difference. The average temperature difference through the day is 37.5 °F (21 °C). The shape of the daily variation is similar to that used by the NBSLD computer program for simulating dwelling heat loads. (c) Net heat flux into and out of a single pane window as a function of time of day. We have assumed a *U*-value of 1.13 Btu/hr ft^2 °F (6.41 W/m^2 °C) for the window. This curve is the difference between curve a and 1.13 times curve b. The net daily flux through the window under the conditions stated here is 230 Btu/ft^2 (0.73 kW hr/m^2) outward. Of course if blinds are drawn in the evening the outward flux is reduced.

house for the accomplishment of the same task[43]; in the case of the hot water heater, for example, the gas pilot burns continuously, and the exchange of heat between hot air and the water in the tank is less efficient than the exchange of electric resistive energy, where coils are inside the tank. The systems characteristics of the energy inputs associated with appliances merit closer study; these characteristics bear heavily on the suitability of complexes of residential housing as sites for on-site electric power generation and the utilization of byproduct heat. Hot water heaters are discussed in greater detail in Section C, and district energy systems in Section E.

The furnace

For definiteness, our standard house has a gas furnace, with a "bonnet capacity" of 100,000 Btu/hr (29 kW$_t$, or 100 ft^3 of natural gas consumption per hr) and a forced-air heat delivery system. The rated ("first-law") efficiency of 80 percent for the transfer of chemical energy in the fuel to heat energy in the forced air exaggerates its performance; it is known in the trade that the actual efficiency is more like 75 percent. The heat distribution further degrades the efficiency: nearly half of the heat added to the air stream is dissipated sideways through the uninsulated ducts, as the air temperature drops from 160 °F at the furnace to between 110 and 130 °F at the room registers and as warm air finds its way through the duct linkages. Only a portion of the energy leaking out of the ducts actually helps to heat the house; another portion is delivered to unused areas, and to spaces between the wall studs, and from there to the outside. Overall, we assume that about 60 percent of the energy in the gas, or 60,000 Btu/hr (18 kW), actually

---

[43]Two large sets of identical houses in Twin Rivers, New Jersey, differ principally in the replacement of electric hot water heater, range, and dryer by their gas-fired counterparts. The energy content of the associated gas consumption is 1.5 times the energy content of the associated electric consumption. Some of the gas consumption does not enter the energy balance, because the heat is exhausted through flues. (The fossil fuel energy required at the power plant is then twice the fossil fuel energy required at the house for the same service, assuming a conversion efficiency of one third at the power plant, and neglecting any differentiation among fuels—for example, neglecting the fact that coal or nuclear fuels can today be used at power plants but not in homes.)

Table 3.3 Energy balance in a standard house and in a target house [a] [power in Btu/hr (with kW in parentheses)].

	Standard house		Target house	
	Mild, clear day (ΔT = 30 °F = 17 °C)	Cold, partly cloudy day (ΔT = 70 °F = 39 °C)	Mild, clear day (ΔT = 30 °F = 17 °C)	Cold, partly cloudy day (ΔT = 70 °F = 39 °C)
Energy losses [b]				
walls and roof	7,200 (2.1)	16,800 (4.9)	2,800 (0.8)	6,500 (1.9)
floor	1,100 (0.3)	1,100 (0.3)	400 (0.1)	400 (0.1)
windows	6,600 (1.9)	15,400 (4.5)	1,800 (0.5)	4,200 (1.2)
incoming air (heating)	8,000 (2.3)	18,700 (5.5)	1,600 (0.5)	3,700 (1.1)
incoming air (humidification)	3,800 (1.1)	3,800 (1.1)	800 (0.2)	800 (0.2)
Total	26,700 (7.8)	55,800 (16.3)	7,400 (2.2)	15,600 (4.6)
Energy gains				
solar through windows	5,200 (1.5)	2,600 (0.8)	8,300 (2.4)	4,200 (1.2)
3 people	1,000 (0.3)	1,000 (0.3)	1,000 (0.3)	1,000 (0.3)
electric load	5,700 (1.7)	5,700 (1.7)	2,400 (0.7)	2,400 (0.7)
gas furnace load to balance [c]	14,800 (4.3)	46,500 (13.6)		
solar collector load to balance			−4,300 (−1.3) [d]	8,000 (2.3)
Total	26,700 (7.8)	55,800 (16.3)	7,400 (2.2)	15,600 (4.6)

[a] The assumptions underlying each of the numbers in this table are found within Section B.
[b] The losses in the first and third columns are also found in Table 3.1.
[c] The actual fuel input is larger; the value here does not include flue losses and duct distribution losses.
[d] Presumably, the windows would be shaded to reduce the solar input! The heat collected by the roof-top collector would be exported to some suitable user of low-temperature heat.

heats the house. Insulation of ducts is an energy conservation strategy worthy of further consideration.

The furnace runs intermittently, and an energy balance, including the other inputs we have just discussed, permits us to estimate the number of minutes the furnace is functioning on a mild, clear day (temperature difference of 30 °F, full solar load) and a very cold, partly cloudy day (temperature difference of 70 °F, half solar load). The energy balance for the standard house is shown in Table 3.3. The furnace, delivering 60,000 Btu/hr (18 kW) to the house, functions 15 min per hr on the mild sunny day and 46 min per hr on the very cold, cloudy day.

Houses heated with oil furnaces and/or houses with water distribution systems have similar energy budgets. Electricity can also be the heating source within the house, in which case there are three principal alternatives: (1) a central resistive coil and a heat exchanger, with air or water distribution system; (2) resistive heating throughout the house, with energy transport through wires; (3) an electric heat pump, driving a fluid through a thermodynamic cycle which (when evaporating) cools either water or air outside the house and (when condensing) heats an interior region, and an associated air or water distribution system. The first of the three alternatives combines the power plant inefficiencies and the ducts distribution inefficiencies—yet it is frequently used. The second alternative offers considerable convenience, simplified zone control, low capital expense, and reduced distribution losses within the house, but lower limits in cost and upper limits in overall efficiency are set by the properties of the power plant, which (today, at least) is a large central generating station. If the full 60,000 Btu/hr (18 kW) is to be supplied electrically, a 1000 MW (electric) plant can supply the energy for 52,000 homes, assuming 10-percent line losses between the power plant and the homes. (In reality, it can supply a larger number of homes, since there will be some phasing in the time of consumption of different houses.) The third alternative, the heat pump, is increasingly recommended for its energy-conservation properties. Heat pumps can be driven by locally burned fossil fuels instead of electricity, may be solar-assisted, may use ground water as the low-temperature source, and have still other intriguing variations. We discuss heat pumps in more detail in Section C, Part 2.

Our target house has a solar collector on the roof and receives whatever additional heat is required via a district system which is also generating electricity. When the solar collector produces more heat than is needed in the house, the district system conceivably has a use for it, or it is stored. (District systems are discussed in Section E.) There is no furnace in the house at all; just hot water pipes. In Table 3.3, we also show the energy balance for the target house. Assuming that 1 ft^2 (0.093 m^2) of a future solar collector can provide an average of 40 Btu/hr (12 W) of heat at an appropriate temperature (over a 24-hour day and taking weather into account), the target house with 200 ft^2 (19 m^2) of collector surface on the roof will stay warm on the cold, partly cloudy day, even if there is no contribution from an internal electric load. Most roofs have, or could have, a pitched, south-facing side bigger than that.

If we assume that the 40-°F outside temperature of the mild clear day in Table 3.3 is the average of the high and low temperatures, and if we adopt 65 °F as the reference temperature, the mild clear day represents 25 (Fahrenheit) degree days. If 60 percent of the fuel energy enters the energy balance of the house, the furnace in the standard house uses $14,800 \times 24/0.60 = 592,000$ Btu on the mild day, or 24,000 Btu per degree day. Under parallel assumptions, the furnace uses 29,000 Btu per degree day on the cold cloudy day (a day representative of 65 degree days). The proportionality between furnace consumption and degree days is close but not precise. Two other indices of performance of the furnace are: (1) Btu per degree day per ft^2 of floor area (16 on the mild day, 19 on the cold day); (2) degree days per gallon of fuel oil equivalent (5.9 on the mild day, 4.9 on the cold day, for a fuel-oil energy of 140,000 Btu/gallon).

The standard house, in a 5000-degree-day climate, consumes 135 million Btu of fuel in a year (at an assumed rate of 27,000 Btu per degree day). If the 60 million dwellings in the United States consumed fuel at this rate on the average, residential space heating would account for 8.1×10^{15} Btu per year. The 1968 figure (SRI, 1972) was 6.7×10^{15} Btu. This comparison should give us some confidence that the numerical estimates in this section are reasonable ones.

C. Six subsystems of the house energy system

Virtually every subsystem of the house energy system presents interesting physics. In this section, we briefly discuss six of these subsystems: air conditioning, heat pumps, hot water heaters, lighting, temperature control, and aerodynamic control. Further development in each of these areas can lead to significant energy savings.

1. *Air conditioning*

If a house like the standard house in a standard (northeast U. S.) climate, considered in Section B, is equipped with a central air conditioner, it will typically have a nominal capacity to extract heat at a rate of 24,000 Btu/hr, or 7.0 kW_t or 2 "tons" (the rate required to freeze 2 tons of water per day). If it is used consistently to reduce indoor temperatures to around 75 °F, it will run about 300 hr per month for the three hottest months (a duty factor of 0.4 for the season, 0.1 for the year). The most widely sold residential models, having an EER (energy efficiency ratio)[44] of about 6, will consume electricity at a rate of about 4.0 kW_e when running, or about 3600 kW hr over the summer. As a result, in summer, the monthly electricity consumption will be approximately double what it is in winter, if the house has electric hot water heating but non-electric space heating. The growth in the summer air conditioning load is the principal reason for the shift in the peak electricity consumption from winter to summer, which has occurred in most electric utilities in recent years.[45]

---

[44]The EER, an unfortunate unit, is the rate at which heat is removed in Btu/hr divided by the electric power consumption in W. It is 3.414 times the coefficient of performance (COP), or first-law efficiency.

[45]A considerable part of the air conditioning load is attributable to dehumidification. In a typical instance, in which air at 90°F and 70-percent humidity is cooled to 55°F and 100-percent humidity before injection into the house, the enthalpy difference of 22.2 Btu/lb of air (12.3 kcal/kg) can be thought of as a sum of 13.0 Btu/lb (7.2 kcal/kg) for dehumidification at 90°F and 9.2 Btu/lb (5.1 kcal/kg) for subsequent cooling at constant water vapor content. The rate of air infiltration is also an extremely important variable. For a house containing 1000 lb of air and for the conditions just stated, one air exchange per hour demands 22,000 Btu/hr, nearly the full capacity of a two-ton air conditioner.

Commercial buildings use far more elaborate procedures to air condition internal spaces. To simplify the temperature controls, conditioned air is sometimes sent through ducts at temperatures a bit too cold for most purposes, and "terminal reheat" (usually electric resistive heating) is used to obtain the desired temperature. Frequently, in mild weather, interior portions of a structure will be air conditioned to remove the internal energy load (from elevators, Xerox machines, etc.) while perimeter portions are heated. Neither of these practices is thermodynamically attractive. Both are a reflection of a time of relatively low energy costs and minimal societal emphasis on energy conservation. The present period deserves (and is beginning to get) some clever new technology.

One area where new technology may have significant impact is in the improvement of heat exchangers; currently, there is some promising activity exploiting the properties of fluidized beds. With improved heat exchangers, it would be possible to reduce the effective temperature difference across which the air conditioner must operate. That temperature difference is dominated by the temperature drop across the heat exchangers at both the hot and cold reservoirs; it is much larger than the inside-outside temperature difference. In the absence of new technical concepts, to reduce the temperature drop across the heat exchanger while keeping the heat transfer rate constant requires increasing the area of heat exchange surface, and this means more costly components.

Air conditioning based on desiccation. One particularly intriguing idea is the adsorption air conditioner. It is currently being discussed as a way of exploiting solar energy for cooling, but the thermodynamics of adsorption cooling allow the use of any moderate temperature (200–300 °F, 90–150 °C) energy source to drive the device. Consider, as an example of a class of such devices, the Munters Environmental Control (MEC) system, invented in Sweden and currently being developed by the Institute of Gas Technology (IGT).

The MEC cycle, illustrated on the psychrometric (moist-air) chart in Figure 3.2, runs entirely at atmospheric pressure. Warm, moist room air (a) passes over a molecular sieve, where moisture is adsorbed, and yields up its heat of condensation, roughly 1800 Btu/lb (1000 cal/gm) for a molecular sieve, to the air. The warmer air (b), which emerges with less than 3 parts per thousand, by weight, of water, is then cooled by heat exchange to about room temperature (c). We now have a supply of "dry, desert air" which can be further cooled without any energy input, by a humidifier, so that it emerges at about 55 °F and 80-percent relative humidity (d).

The essential parts of the MEC system, displayed in Figure 3.3, are two rotating wheels which transport the water-laden desiccant and heat from the cooling system to a second air stream for regeneration. In almost a reverse cycle, outside air (e) is cooled evaporatively to condition (f) and then passed through the heat wheel to remove the sensible heat absorbed in the cooling system. This 176-°F air (g), however, is not warm enough to dry the molecular sieve, and a portion of it must be boosted in temperature by a flame to condition (h), to drive the water from the drying wheel. Although the present configuration uses natural gas burners to boost the air temperature to 290 °F (143 °C) to regenerate the molecular sieve, IGT estimates that $\frac{4}{5}$ of the needed energy could be supplied by present-day solar collection technology. Using solar heat, the "2.7-ton cooler" shown removes 32,500 Btu/hr (9.5 kW_t), with only 9000 Btu/hr (2.6 kW_t) supplied by non-solar

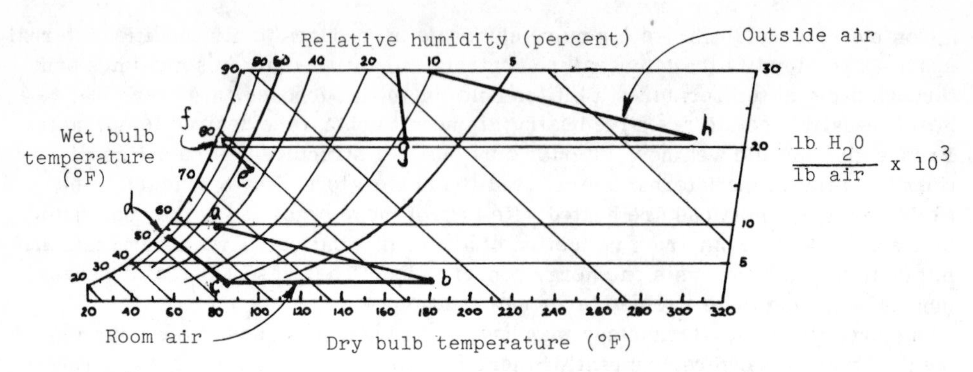

Figure 3.2. The MEC cooling cycle (a to d) and wheel regeneration process (e to h). [Source: Rush (1973).]

sources.

2. Heat pumps

As stated in Table 2.1, the first-law efficiency of an ideal heat pump (the thermodynamic maximum of heat delivered per unit of work input) is

$$\eta_e = (1 - T_0/T_2)^{-1},$$

where T_0 is ambient (outdoor) temperature and T_2 is the temperature of the warm region to which the heat is pumped (both on an absolute scale). Given the problem

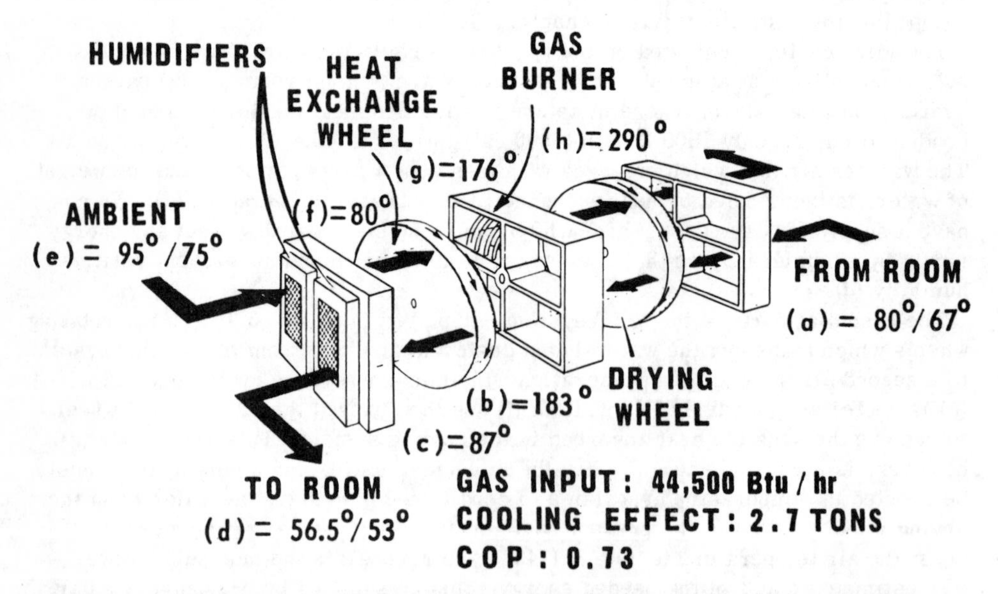

Figure 3.3. MEC unit in recirculation mode. The cycle is described in the accompanying text. The drying wheel turns at about 0.2 rpm, the heat exchange wheel at about 7 rpm. [Source: Rush (1973).]

of pumping heat from an outdoor temperature of 45 °F (7.2 °C)[46] to an indoor temperature of 68 °F (20 °C), one calculates a maximum (or Carnot) COP $\eta_e = 23$. But residential heat exchangers typically work at a 20-°F temperature difference for efficient exchange, and heating engineers specify hot air from the ducts at 90–110 °F (32–43 °C) to minimize the sensation of chilling by the air current. For example, if the working fluid of the pump operates between 25 °F (–4 °C) and 120 °F (49 °C), the Carnot efficiency is $\eta_e(\max) = 6.1$. A decrease in the required temperature difference at the heat exchangers from 20 °F to 10 °F increases the maximum COP for this case to 7.6. Once again, as with air conditioners, the heat-exchange problem is of central importance in improving efficiencies.[47]

Off-the-shelf electric heat pumps currently operate at $\eta_e \approx 2$ to 3 at 45 °F ambient. When the ambient temperature drops further, heat pumps become increasingly inefficient in a first-law sense, until, when $\eta_e = 1$, they are designed to operate as resistive heaters. Typical heat pumps now in use operate like electric-resistance-heated air furnaces when the outdoor temperature goes below about 30 °F. Because of the duct losses discussed in Section B, this puts an even larger demand on the electric system than is associated with a comparable baseboard electric heating system. This extra demand occurs at the winter peak. Further research directed towards the development of cascading or compound-compression cycles could eliminate the need for switching to resistance heat at all outdoor temperature conditions.

How to drive a heat pump. Heat pumps may be electrically driven or engine driven. A diesel engine typically has an efficiency for the production of work from chemical energy about equal to that of a power plant, approximately $\frac{1}{3}$. Thus, ordinarily, both kinds of heat pumps will have about the same overall efficiency. When the outside temperature is such that the COP of the heat pump is 2, the system efficiency of both the engine-driven and the electrically driven heat pump will be about $\frac{2}{3}$, comparable to that of a well maintained gas or oil furnace, and about double that of electric resistive heating. It is possible, however, to fit the diesel engine with heat exchangers to recover heat from its cooling water and its exhaust. If the fraction of the heat recovered and used for heating the house is $\frac{1}{2}$, and the coefficient of performance of the heat pump alone is η_p, the coefficient of performance of the diesel-engine-driven heat pump system, η_s (defined as the heat added to the house divided by the energy in the diesel fuel), is

$$\eta_s = \frac{1 + \eta_p}{3}. \tag{3.1}$$

For $\eta_p = 2$, this gives a system COP of 1.0 (about 50 percent better than an oil-fired furnace). In general, the exhaust heat is most useful when the heat pump alone is functioning least well (low COP, or large indoor-outdoor temperature differences).

The major problem with engine-driven heat pumps is the size of the engine. A typical truck diesel may have an output of about 250 brake horsepower (bhp), or 186 kW. With 50 percent exhaust recovery, the system ($\eta_p = 2$, $\eta_s = 1$) would deliver

---

[46]This is the industry standard outdoor temperature for heat-pump specifications.

[47]In some climates, the outdoor heat exchanger has an extra inefficiency: energy must be expended to prevent ice from building up on it.

1 Btu of heat for every Btu of fuel input—about 1.9 million Btu/hr (558 kW), or enough to heat about 24 of our standard houses (at 80, 000 Btu/hr per house). Smaller diesels are less efficient and more expensive per bhp; a 10-bhp (7.5-kW) diesel for an individual home is an unfavorable proposition. Engine-driven heat pump systems appear, at the moment, to be suitable mainly for heating large buildings or groups of houses.

Borrowing heat and cold from ground water. Consider a heat pump using ground water rather than air as a heat source. Ground water frequently remains within 1 °F of a constant temperature, typically 55 °F (13 °C). Using ground water (or surface water, as from a lake) as the source for a heat pump is one way to avoid the serious on-peak inefficiencies associated with very cold days. It has other advantages as well.

A heat pump having a COP of 3 for the outdoor design temperature of 45 °F will have a COP of about 4 for a ground-water source temperature of 55 °F.[48] If the pump is diesel driven, as above, with fifty-percent heat recovery, the system coefficient of performance will be $\frac{5}{3}$. The energy flows for the system are shown in Figure 3.4(a). For comparison, Figure 3.4(b) shows the flows in a 75-percent efficient gas furnace. The heat pump and ground water system uses only 45 percent as much fuel for the same amount of heating.[49]

A system of this kind has associated capital costs, and ought to have a high duty cycle. It can be supplemented by low-capital-cost auxiliary heating on the coldest days. Referring once again to our standard house, we might size the heat pump to supply 25, 000 Btu/hr to the house ducts, implying the extraction of 15, 000 Btu/hr from ground water (and the combustion of fuel at the same rate, by an accident of our numerical assumptions). If the ground water is cooled 10 °F (5.5 °C), it must flow at a rate of 1500 lb/hr, or 3.0 gal/min, or 11 kg/min, approximately the normal flow from a full open tap.

Assuming the heat pump operates for 2500 hr during the winter, 60, 000 ft^3 (1700 m^3) of water will be cooled. Taking a typical aquifer to be 100 ft thick and one-fourth water by volume (three-fourths sand and gravel), the water volume is that of a cylinder of radius 28 ft (8.4 m), which is not excessively large. (The cylinder, of course, isn't "drained," but is replenished by more ground water.)

The energy for pumping is not likely to be substantial. Once an adequate well is drilled in a permeable aquifer, the water level in the well does not generally drop significantly during pumping. Consequently little work is needed to raise

---

[48] Our translations of first-law efficiencies (COP's) to different temperatures assume a temperature-independent second-law efficiency.

[49] Workers at the Centre d'Études Nucléaires in Grenoble, France, have built and studied a series of heat pumps using ground water. The smaller units, driven electrically, are sized for single rooms. They use domestic water, deliver 1.25 kW_t (4250 Btu/hr), and operate with a COP of 3. They are said to run 60,000 hr without maintenance and to compete economically with oil heat if water costs less than 8 cents/ton (Dumont, 1974). Larger, building-sized units have been operated with both electric and diesel-engine drive (de Cachard, 1974). These systems meet base-load demand; supplementary oil heat meets peak demand.

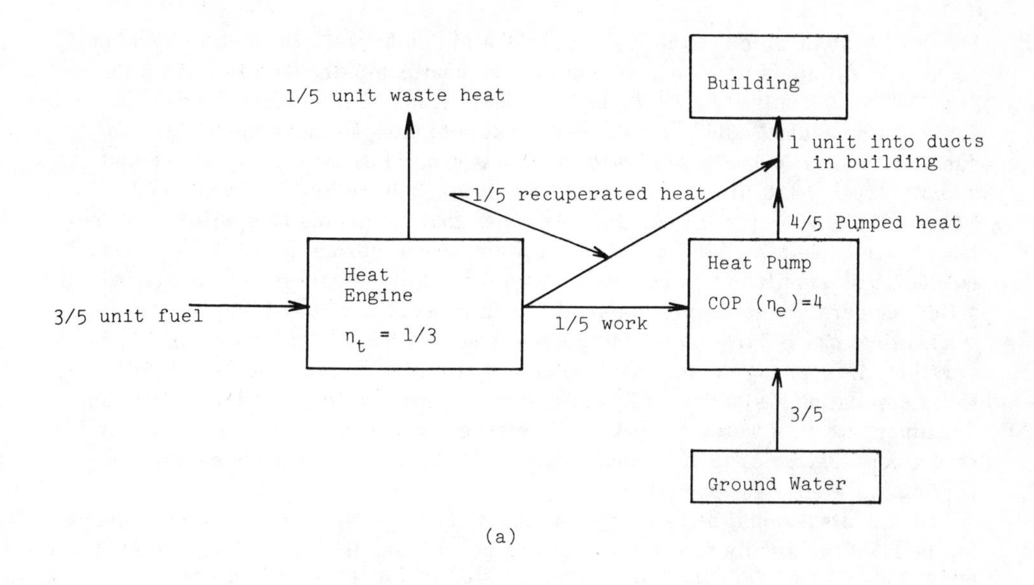

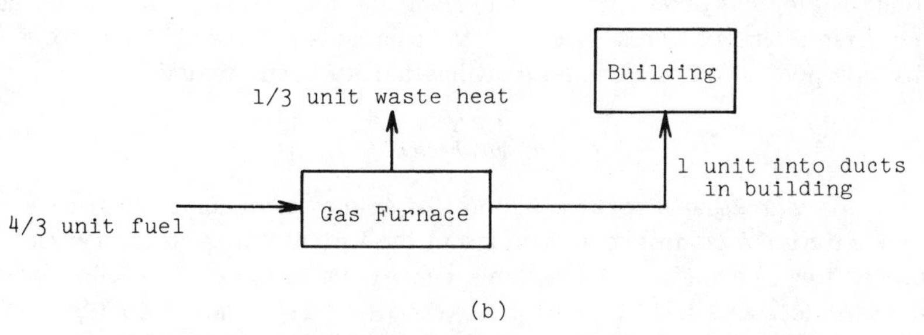

Figure 3.4. (a) Building heated by a diesel-driven heat pump with heat recuperation using ground water as the low-temperature reservoir. (b) Building heated by a gas furnace.

the water to the surface. When the potential energy surface is above ground level an aquifer may even supply work.

The same ground water, one would expect, could be used for summer cooling. Indeed, there are many old (mainly abandoned) cooling systems using ground water for direct cooling. The technique may deserve renewed consideration.[50]

A more sophisticated system than direct cooling would use the ground water as a heat sink for a reversed heat pump, i.e., an air conditioner. We envision a

---

[50]Even surface water exhibits examples of untapped cooling potential. Shasta and Oroville lakes in California are fed by mountain water, and their temperature varies between 42°F and 48°F. The city of Oroville is two miles from Oroville Dam and swelters with an average July high of 97°F; Redding, 9 miles from Shasta Dam, has an average July high of 99°F. Why do both these cities use only conventional air conditioning?

two-well system which could heat and cool a single house. The wells should be roughly 100 ft apart or more. In winter, the heat pump drains water from the "winter" well, cools it by 10 °F, and reinjects it into the "summer" well. If some of this added "cold" is still there next summer, so much the better. In summer the heat pump is reversed, withdraws water from the slightly cooled summer well, heats it by 10 °F, cools the house, and reinjects the warmed water into the winter well. Again, one hopes that not all the heat will be conducted or convected away before the next winter. Such thermal storage, however, is much less significant an advantage than the relative nearness of the ground water temperature to room temperature. In most of the United States, the winter heating load is larger than the cooling load, so the system is only quasireversible, i.e., ground water is slowly piped from the "winter" to "summer" well, simulating a slow flow of ground water. This continual pumping (into the "summer" well) of water chilled 10 °F below ground temperature may actually cool a house-sized area of ground water by 10 °F, thus making the summer cooling more efficient than ever.

Such a system might be practical immediately for a single house if we replace the thermodynamically attractive heat engine with an electric motor. We seriously suggest that the complete system be studied for a planned unit development or a city. Such an effort should include a careful examination of possible geological and ecological problems. The present methods of heating and cooling may already have such problems: Beneath the atmospheric "heat island" of a modern city is a pool of warmed water that has hardly been studied.[51]

3. *Hot water heaters*

Hot water heaters are the largest consumer of energy after space heating in both the residential and commercial sectors in the United States (SRI, 1972). They accounted for 4 percent of the nation's energy consumption in 1968 as compared, for example, with 2.5 percent for air conditioners. They accounted for 15 percent, or 58×10^6 kW_t, of residential sector consumption and 7 percent, or 22×10^6 kW_t, of the commercial sector consumption. The electric energy consumption was about half the direct gas and oil consumption, although there are estimated to be three times as many gas units as electric units (among a total of 55×10^6 units). The average rate of consumption of energy for hot water was 1.4 kW_t per household.

A hot water heater is an energy storage device. Consider a mass M of hot

---

[51]An interesting variant on the two-well scheme has been suggested to us (Pinder, 1974), which exploits the temperature excursions of surface water stored under ground for six months. Consider a river or lake in hydraulic connection with a suitable aquifer, and a *linear* "ground water" collection system (for a town), parallel to the bank and say 300 ft inland, from which water is withdrawn steadily (and returned, cooled in winter, warmed in summer, to the river downstream). Under conditions of steady pumping, the collection system, if it is the right distance from the river, will be withdrawing water that has left the river six months before and has (under many circumstances) been well insulated in the interim, i.e., it will be warm in winter and cool in summer. The central heat pump then has an easier job in all seasons.

water at temperature T_2. The heat added to the water to warm it from ground-water temperature[52] is

$$Q' = CM(T_2 - T_{ground}), \tag{3.2}$$

where C is the specific heat; but in the house it must be viewed as storing a smaller amount of heat,

$$Q = CM(T_2 - T_{room}). \tag{3.3}$$

A 100-gal (378-kg) hot water tank at 150 °F (65 °C), relative to 70 °F (21 °C) room temperature, stores 67, 000 Btu (20 kW hr) of heat, about as much as the space heating system will supply to the house in one hour of continuous operation, and about as much as the thermal content of one day's use of electricity in the house for all other purposes.

The characteristic time for thermal energy storage depends on the average heat transfer rate through the shell surrounding the water. Consider the 100-gal (380-l) tank to be a cylinder 6 ft (1.8 m) high. It will then have a radius of 0.84 ft (0.26 m). Assuming 1 in. (2.5 cm) of insulation, we may estimate $U = 0.2$ Btu/hr ft^2 °F (1.1 W/m^2 °C). The tank at 150 °F (65 °C) will then initially lose energy at a rate of 580 Btu/hr (0.17 kW), so that the characteristic thermal storage time is 67, 000 Btu/580 Btu/hr = 120 hr.

This storage time is more than long enough for there to be little difference between heating hot water at night and during the day. Taking advantage of this, some electric utilities provide a reduced off-peak rate for hot water heaters, making financially attractive the use of a two-element hot water heater, in which the bottom element has a timer circuit and runs only on off-peak current. The top element, with its own proximity thermostat, then heats the top water for immediate consumption. This design utilizes the convective flows so that the peak heating element is working against a smaller ΔT. Element ratings have been growing as manufacturers try to decrease the recovery time; this leads to greater peak demand in an individual house (with consequences for combined systems to be considered in Section E), but makes considerably less difference in a large (incoherent) assemblage of houses.

The continual loss of 580 Btu/hr (0.17 kW) through the walls of each of more than fifty million hot water heaters represents 10 percent of the energy consumption associated with home hot water heating. It is a decided inefficiency in the national energy system, for little of this heat is added to useful places in the house. This loss can be substantially reduced by either of two strategies: (1) Improve the insulation of the water heater. Of all the "retrofits" to a house, this is one of the easiest to perform. Researchers at Princeton[53] have found a reduction by a factor of 2 in standing losses from a gas water heater by simply wrapping the body of the water heater (not even the ends) with R-7 insulation[54] and taping the wrapping so that it fits snugly. Currently, domestic electric wa-

---

[52]The conventional (first-law) efficiency is defined as the ratio of this heat "content" Q' to the energy expended in the house to provide the heat.

[53]D. Harrje, private communication (1974).

[54]R-7 has a thermal resistance of 7 (Btu/hr ft^2 °F)$^{-1}$, or 1.2 (W/m^2 °C)$^{-1}$.

ter heaters are manufactured with considerably more insulation than gas water heaters [about 2 in. (5 cm) for electric, 1 in. (2.5 cm) for gas]. (2) Reduce the temperature of the water stored in the water heater. This is another simple retrofit, but it may possibly be inconvenient if the thermostat setting is lowered too far. From an entropy standpoint, it makes little sense to mix hot and cold water, and there is no reason to store water at a temperature above that at which it will be used. If a water tank is small, however, the user may become more frequently aware of exceeding the system's storage capacity.

There are two systems for hot water heating which involve no hot water storage at all. One system, the "side-arm" unit, runs parasitically off the furnace and is highly inefficient, as one has to turn on the space heating system to obtain hot water. Another system, popular in Europe, employs a local coil and an unvented gas furnace at the point of use. Although we do not know the burner efficiency, we may calculate the energy to heat water by 65 °F (36 °C), from 55 °F (13 °C) to 120 °F (49 °C), at the typical rate of flow of a wide open faucet, 3 gal/min (11 l/min): 98, 000 Btu/hr (29 kW), comparable to the heat supplied by an entire house furnace. Evidently lesser flow rates must be accepted.

Some dishwashing machines have a built-in heater for boosting the internal hot-water temperature. As the dishwasher is usually the only appliance in a house demanding water hotter than about 120 °F (49 °C), the presence of such a boosting coil may be an inducement to turn down the thermostat setting at the water heater.[55]

There are smaller losses of energy from the system supplying hot water, associated with the cooling off of the water standing in the pipes. In 200 ft (61 m) of 0.5-in. (1.3-cm) diameter pipe, there stands 2.0 gal (8 l) of water cooling off, in this case quite rapidly. For uninsulated copper pipe, the time constant for heat loss (time for the temperature difference between water in the pipe and air in the room to fall by a factor e) is about 20 min (ASHRAE, 1972, p. 372, Table 12). This loss is dominated by the surface heat-transfer coefficient. Insulating the pipe with 2 in. of styrofoam will increase the time constant to about 100 min. The strategy of insulating the pipes makes sense, from an energy-efficiency standpoint, therefore, only if intervals of intermediate duration between hot water uses are expected to be common. More direct attacks on this standing loss are mounted by reducing the diameter of the pipe and, again, lowering the temperature of the hot water.

There is a small heat loss associated with the conduction of heat along the copper pipe connected to the water heater. Assuming the pipe reaches ambient temperature in 10 ft, we estimate the rate of heat loss to be roughly 70 Btu/hr (20 W); a small loss but an unnecessary one. It could be much reduced by inserting an insulating section between the tank and the line.

Electric hot water heaters have a typical first-law efficiency (net heat added to water divided by electric energy consumed) of about 75 percent. Gas hot water heaters have lower first-law efficiencies, typically about 50 percent. The

[55]One wonders whether there can be some way, if appliances are to be "rated" for their energy efficiency, to give bonus points for dishwashing machines that have such boosting coils.

additional losses for gas come from several sources: (1) The combustion of the gas flame is not a perfect conversion of chemical energy into heat, and the heated air, in turn, which passes up through a core region in the center of the tank, does not cool all the way down to ambient temperature before leaving the top of the water tank. Contrast this process with the direct heating by electric resistive elements immersed in the water. (2) The gas pilot light burns continuously, using chemical energy at a rate of about 600 Btu/hr (175 W). This energy loss can be reduced by reducing the length of the pilot light flame or by using some form of electric substitute for a pilot light. (3) The hollow core of the water tank, and the open flue above it, form a chimney which entrains room air, serving both to add to the air infiltration rate and to cool the interior cylindrical boundary of the water heater. The gas pilot may, in fact, not be heating the hot water at all in some geometries. A damper on the flue would reduce the infiltration losses, though it will permit greater entry into the house of the exhaust products from the pilot light combustion.

We return to the hot water heater, to discuss its efficiency from the standpoint of the second law of thermodynamics, in Section D.

4. *Lighting*

Lighting consumed about 20 percent of U. S. electric energy, or about 5 percent of all U. S. energy, in 1973. Of the energy going into lighting, about 20 percent is used in residences, 40 percent in schools, offices, stores and other public buildings, 10 percent out of doors, and 10 percent elsewhere.[56] Split another way, two-thirds of the lighting energy is used by incandescent lamps and one-third by fluorescent and vapor lamps. This means that fluorescent and vapor lamps, which provide roughly 4 times as much useful light per unit energy as incandescent lamps, supply about two-thirds of the useful light.

Units: lumens and lumens/watt. The basic unit for measuring the response of the human eye to light is the lumen (lm), a unit that blends physiology, psychology, and physics.[57] A 100-W incandescent bulb, as an example, yields a "luminous flux" (effective light per unit time) of 1800 lm. The ability of the eye to perceive and resolve objects is dependent on the effective light per unit time per unit area, or "illuminance", measured in lm/ft^2 or lm/m^2. 1 lm/ft^2 is also called a foot candle.[58]

---

[56]These are estimates by the General Electric Co. We wish to thank R. T. Dorsey and W. S. Fisher for much information and help. The SRI tables, used widely elsewhere in this report, estimate lighting at 1.5 to 3 percent of fuel energy, but SRI has not made as careful a survey of lighting as General Electric.

[57]In general, if $P(\lambda)$ is the spectral radiant power of a light source (in W per unit wavelength interval), the luminous flux Φ (in lm) is given by $\Phi = 680 \int_0^\infty P(\lambda) v(\lambda) d\lambda$, where the weighting function $v(\lambda)$, called the "relative luminosity factor" (or "spectral luminous efficacy"), is a standardized measure of eye response. A graph of $v(\lambda)$ appears in Figure 3.5 (curve B), and tabulated values can be found in Boast (1953). It is normalized to $v_{max} = 1$. Note that the lumen, although called a unit of luminous "flux," is *not* a "per-unit-area" unit. The physicist would find it more natural to refer to illuminance as a light flux.

[58]Some common units that we shall *not* need in this section: 1 candela (standard international candle) emits 1 lm/steradian; a flux of 1 lm/m^2 is called 1 lux, so 1 lm/ft^2 (= 1 foot candle) = 10.76 lux.

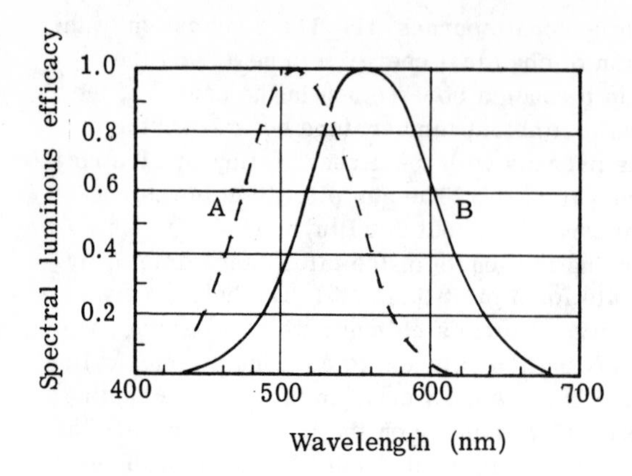

Figure 3.5. Spectral luminous efficacy. Curve A is the "scotopic" response of rods (in dim light); curve B is the "photopic" response of cones in normal vision. [Source: Stevens (1969), Figure 1.4.] Curve B is the same as the "relative luminosity factor" used in the definition of the lumen.

Figure 3.5, curve B, shows the relative visual response of the cones of the human eye to unit incident energy as a function of wavelength. A monochromatic light source at $\lambda = 555$ nm has the maximum possible efficacy,[59] 680 lm/W for the internationally defined standard observer. Sunlight has an efficacy of 92 lm/W; this factor relates solar illumination and solar heating.[59a] Phosphors can be optimized to emit "acceptable" white light with an average efficacy of 400 lm/W, as in the case of the yellowish-green light of "cool-white" fluorescent bulbs.

Specifications for the illumination of a given task are usually given in $\mathrm{lm/ft^2}$ (or $\mathrm{lm/m^2}$), but for a quantitative understanding of a curve of visual acuity (Figure 3.6) we must distinguish between illuminance ($\mathrm{lm/ft^2}$) and the brightness of the light *reflected* from an object, which is called luminance and is commonly measured (as in Figure 3.6) in foot lamberts (ft-L). Conveniently a ft-L is defined such that a white card (perfect uniform diffuser) illuminated by 1 $\mathrm{lm/ft^2}$ has a luminance of 1 ft-L.

In Figure 3.6, the ordinate, called "visual acuity," is just $1/\theta$, where θ, measured in minutes of arc, is the angle subtended by the gap in a broken printed circle. Thus an acuity of 2 (at 10 ft-L) means that one can see any gap larger than 0.0017 in. (0.004 cm) at a viewing distance of 12 in. (30 cm). Note the logarithmic nature of the acuity curve; one can see a gap of 0.003 in. (acuity = 1.2) with only 0.5 ft-L instead of 10 ft-L, and one cannot see a gap of 0.001 in. (acuity = 3.5) at all. As to the horizontal scale, we have already seen that a luminance of 1 ft-L is roughly the brightness of a white card exposed to an illuminance of 1 $\mathrm{lm/ft^2}$. A horizontal white surface outdoors by day will generally lie between 500 and 5000 ft-L. The surface of a common fluorescent lamp is about 2500 ft-L.

---

[59] Lighting engineers prefer to reserve the word "efficiency" for the dimensionless quantity (energy out/energy in), and use "efficacy" for human response/energy in.

[59a] This result is obtained by performing the integration in footnote 57, using the spectrum of a 5800°K black body for $P(\lambda)$. *Visible* sunlight has an efficacy of about 240 lm/W, since 39 percent of sunlight is "visible" (between 400 and 700 nm).

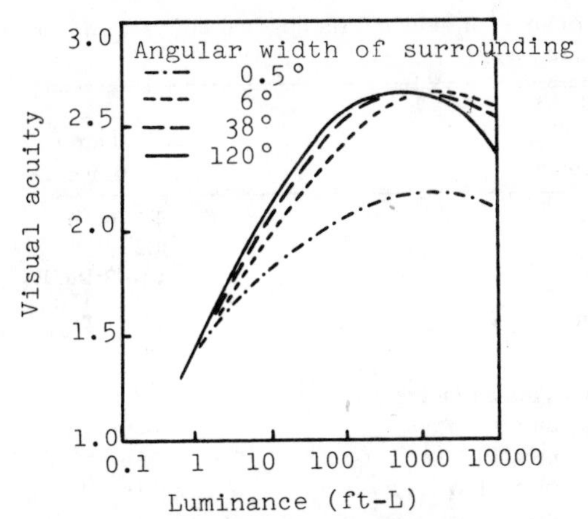

Figure 3.6. Visual acuity as a function of luminance. For white paper, luminance in ft-L is about the same as illuminance in lm/ft^2. The various curves are for different sizes of surrounding, illuminated the same as the central field. The labeled angular widths include the 0.5-deg angular width of the central field. For the lowest curve, the surrounding is black. [Source: Stevens (1969), Figure 1.5.]

Lamps. Fluorescent lamps are often cited as being 4 to 5 times more "efficient" than incandescent lamps. The factor of difference depends on the "wattage"; for both kinds of lamps, efficacy increases with power. Table 3.4 gives efficacies for several common sources of light.

We note that gas flames (still used for lawn lights and other decorations) are extremely inefficient. We note also that putting several small lamps in a fixture instead of one large one produces less light per unit energy. Another source of inefficiency is the solid-state "dimmer," an inexpensive circuit using semiconductor rectifiers, combined with a light switch. A factor-of-2 reduction in power supplied to an incandescent bulb produces a factor-of-5 reduction in the lumen output [see Stevens (1979), Figure 3.3]. The Bureau of Standards is currently preparing labels for major appliances. Perhaps it should also recommend an informative label for "dimmers."

Fluorescent lights are presently 10–15 percent more efficient if powered by high-frequency rather than 60-Hz current (NATO, 1973). Basic research in kHz sources and lamps might reveal opportunities for overall improvement in lighting efficiency.

Other losses in lighting. A bare 40-W fluorescent lamp, suspended 2 ft above a desk top, and equipped with a mirror so that the upwards half of its light is directed back down, provides 120 lm/ft^2 at the desk top. This is almost twice current lighting standards. A modern office with one desk contains five or ten such lamps. Clearly, most of the light is not reaching work surfaces.

Lighting in most commercial buildings is not, as one might expect, concentrated at the site of the task (such as the desk), but is nearly uniform. Moreover, more than half of the lumens are not available for tasks performed on horizontal surfaces, no matter where they are placed; the light is absorbed in fixtures or at the walls.

TABLE 3.4. Efficacies of selected flames and lamps. (All entries except for flames and mantles are in lm/W_e.)

Light source	Efficacy: Lamp alone [a]	Efficacy: Lamp plus ballast
Sunlight	92 lm/W_t	
Open gas flame [b]	0.2 lm/W_t	
Gas mantle [b]	1 to 2 lm/W_t	
Incandescent lamps [c], 40 W	12 lm/W_e	
100 W	18	
Fluorescent lamps,[c] plus ballast[d]		
20 W, 24-in. *T*12, plus 13 W	(65)	39
40 W, 48-in. *T*12, plus 13.5 W	(79)	59
75 W, 96-in. *T*12, plus 11 W	(84)	73
Metal halide,[e] 400 W, plus 26 W	(80)	75
Sodium[e]		
400 W high pressure, plus 39 W	(120)	109
180 W low pressure, plus 30 W	(180)	154
Mercury,[e] 400 W, plus 26 W	(57)	53.5

[a] Numbers in parentheses are for the bare lamp, without ballast losses.
[b] Source: Handbook of Chemistry and Physics (1961), p. 2849.
[c] Source: Handbook of Chemistry and Physics (1971–72), p. *E*-185.
[d] These are typical ballasts; however, 5-W ballasts are available for an extra 10 to 20 percent in price.
[e] Source: NATO (1973).

The fluorescent lamps are usually placed in the ceiling, shielded from direct view by louvres or plastic to prevent glare. As we mentioned earlier, the surface brightness (luminance) of a standard 40-W cool-white fluorescent lamp is about 2500 ft-L, which is no brighter than the surface would look if we turned off the bulb and carried it outside on an average day. But most people find it too bright to look at directly when inside. The plastic prismatic lenses that are frequently used to shield the direct light of the lamps actually absorb about $\frac{1}{4}$ of the light, even when the plastic is new, clear, and clean.

The familiar "egg-crate" louvre also wastes light, and can easily be improved. In some such louvres, for example, the baffles are not white surfaces, but mirrors, shaped to reflect all the incident light downwards.

Other lighting losses result from deterioration of fluorescent lamps (they drop slowly in light output to a plateau of 80 percent of their original output), and from accumulations of dirt on fixtures.

Lighting standards and area lighting. In the United States, standards have risen by a factor of 2 to 4 in the last 20 years (Stein, 1972). For example, in 1952 the New York City Board of Education Manual of School Planning called for 20 lm/ft^2 at the desk; this was raised to 30 in 1957, and to 60 in 1971. Modern office buildings have until recently been designed for 80–100 lm/ft^2, and for uniform illumination even in corridors and on stairways. Figure 3.6 suggests that at this lighting level, gains in visual acuity from increases in illumination are minimal. [For typical tasks at a desk, the measures of incident light

(lm/ft^2) and reflected light (ft-L) are roughly equivalent.]

Energy conservation standards for building lighting are being written in terms of installed power per square foot of building floor area. Modern office buildings typically install 2.5 to 5 W_e/ft^2, whereas the new "1974 General Services Administration Public Service Building Energy Conservation Guidelines" recommends 1 to 2 W_e/ft^2 (Dubin, 1974, p. 6-4 and Figure 9-5), and the energy-conserving Federal Office Building in Manchester, New Hampshire, actually uses 0.7 W_e/ft^2 (Dubin, 1974, Figure 4.1).

If we juxtapose lighting practices given in W/ft^2 of floor area and lighting standards measured in lm/ft^2 of task surface, and add the assumption that interior lighting is uniform, we can estimate the efficiency with which light is conveyed to *any* horizontal surface, as opposed to being absorbed on fixtures and walls. A building equipped with 2.5 W_e/ft^2 in the form of 40-W fluorescent bulbs will emit 197 lm per ft^2 of floor area (see Table 3.4), at least double the 80 to 100 lm/ft^2 figure quoted above. Thus about half of the lumens don't find their way down, either directly or via reflections.

Patterns of electricity consumption and lighting use in the United States[60] make clear that average illumination in homes, where lighting is governed by personal choice, is well below the average level in commerce and industry, where lighting is governed principally by established standards. On the average, each person encounters 10 times as much artificial light (in lm hr) outside the home as inside the home (even though most of us work during the day and spend our nights at home!). At home we use moderate levels of general lighting, and employ desk lights and reading lights for special tasks. Clearly a re-examination of industrial and commercial lighting standards and practices is in order.[61]

Controls. Timed light switches are generally used in halls and stairways in Europe, and are found in some library stacks in the United States. We suggest that there should be a large commercial market for light switches with hand-set timers, so that one could dial light for one hour in a class-room, or for up to eight hours in an office. Dubin (1974) and others suggest photo-cell cutoff of lights near windows in office buildings.

We have observed that many of our friends are surprised by two facts:

1. The lights in a typical office (0.5 to 0.8 kW_e), if left on 16 hours overnight, consume half as much energy as an individual uses altogether (i.e., in home and car) during those same 16 hours.

2. It "pays" to turn off the modern fluorescent light for even a few minutes, certainly for an hour.

Accordingly we suggest a label near light switches in stores, schools, offices, etc., filled in by hand for the individual switch, saying:

TURN OFF THE LIGHTS, even for 5 minutes. The lights on this switch consume 0.8 kilowatts. If left on 16 hours overnight, the power plant will consume the equivalent of 1 extra gallons of oil.

---

[60]Source: The General Electric Co.; see footnote 56.

[61]See Appel and MacKenzie (1974).

Apart from basic research in the conversion of electric energy to light and applied research in human response to lighting patterns and intensity, there are obvious near-term steps for energy conservation connected with lighting. The most dramatic way to save power in residential lighting would be to switch from incandescent to fluorescent lamps; the biggest commercial saving would be to use bright light only where it is needed rather than to illuminate uniformly a whole office or a whole floor, to switch more completely to fluorescents, and to use timed controls.

5. *Temperature control*

Thermal stability—back to the cave. Some houses have been deliberately built with a long time constant for responding to temperature variations, in order to take advantage of a comfortable average temperature. (72 °F is the 24-hr average daily temperature for Los Angeles in July and 76 °F is the average in New Jersey in July.) Familiar examples are adobe houses in the Southwest. The use of masonry building materials provides a large thermal mass to reduce fluctuations.[62] Recent implementation of a similar idea occurs in the "Hay houses" (Hay and Yelott, 1969) in which pools of water are supported on the roof. One such house was built in Phoenix, Arizona, with roof ponds containing 5900 lb (2700 kg) of water covering 170 ft^2 (16 m^2) of roof surface and in turn covered by retractable polyurethane insulation. The system is completely passive except for the retractable insulation. The ponds collected solar energy during winter days and stabilized the building temperature. During the summer the pond was covered and insulated during the day but was allowed to cool by evaporation and radiation at night. The interior remained at temperatures between 70 °F and 80 °F all year round with *no* auxiliary sources of heat or cooling. This provides an interesting example of a nearly passive system for temperature regulation (one, of course, that is especially well adapted to the Phoenix climate).

Much work is under way to develop heat storage media for use with solar energy (Lorsch, 1973; General Electric, 1974; TRW, 1974; Westinghouse, 1974). Two heat storage media which seem promising at present are water, in which sensible heat can be stored at the rate of 62 Btu/ft^3 °F, and paraffins. Two examples of paraffins are (1) Sunoco P116, with a heat of fusion of 50 cal/gm, melting point of 116 °F (47 °C), and storage density of 4400 Btu/lb (10 MJ/kg); and (2) Enjay C_{15}–C_{16}, with a melting point of 40–50 °F (4–10 °C) and storage density of 3200 Btu/lb (7.4 MJ/kg). An advantage of storage in the form of latent heat rather than sensible heat is the constancy of the temperature, affording nearly constant efficiency for heat exchangers. Problems with storage as latent heat are associated with the poor heat transfer properties of the solid phase.

Time control and zone control of heating and cooling. Most office buildings are unoccupied during 14 hr of the day, and it has become common practice to reduce the heating or cooling requirement during off hours. In houses, the time constant for thermal relaxation ranges from 4 hr to 60 hr depending on construction, siting, and weather conditions. Substantial savings in fuel would be possible if the thermo-

---

[62] For quantitative studies, see Peavy (1973).

stat setting were reduced during daytime hours (if the house is vacant) and during the night. This type of thermostatic control works well because the air within a house heats extremely rapidly: Referring again to our standard house, a heating system supplying heat at a rate of 60,000 Btu/hr will raise the temperature of 15,000 ft^3 of air 3.4 °F (1.9 °C) per minute. The air gradually equilibrates with a building structure about 100 times more massive. The furnace could be restarted about one hour prior to waking or returning home. Figure 3.7 shows a possible temperature cycle for a house vacant during working hours.

An estimate of the fuel savings will be made for a simple case. We assume that the entire heat supply to the house is from the furnace and that the rate of heat flow to the outside is proportional to the temperature difference ΔT between inside and outside. (We completely neglect thermal energy storage.) τ is the thermal relaxation time constant for the house. If the outside temperature is constant, the temperature difference is given by $\Delta T = \Delta T_0 e^{-t/\tau}$, where ΔT_0 is the initial equilibrium temperature difference. The instantaneous rate of heat flow outward is $K \Delta T_0 e^{-t/\tau}$, where K is the thermal conductance of the house in Btu/hr °F. The total loss of heat during a time t_1 during which the furnace is off is

$$Q = K \Delta T_0 \tau (1 - e^{-t_1/\tau}). \tag{3.4}$$

Assume that the furnace is off for 8 hr and also that $\Delta T_0 = 30$ °F, $K = 600$ Btu/hr °F, and $\tau = 12$ hr. The heat lost over the 8-hr interval is 105, 000 Btu, and ΔT drops by half (to 15.4 °F). When the furnace is turned on, approximately the same amount of heat is added, while the internal temperature climbs back to its original value. If the furnace remained on dı 'ng the 8 hr, the furnace would have kept a nearly constant temperature within tne house, compensating a loss of $600 \times 30 \times 8 = 144{,}000$ Btu. The strategy of turning the furnace off, therefore, saves 38,000 Btu in that period, or 26 percent of the fuel. If the furnace is programmed to turn on before the occupants return home, they need not experience discomfort. At night, one might prefer a minimum of 60 °F; this will be reached after 5 hr. The savings are still large.

A closely related type of control system (zone control) involves heating selected portions of the house at particular times. The potential savings here, which depend on the house layout and the habits of the occupants, have been estimated to be often as large as 25 percent (Kurtz, 1974a). The additional instrumentation re-

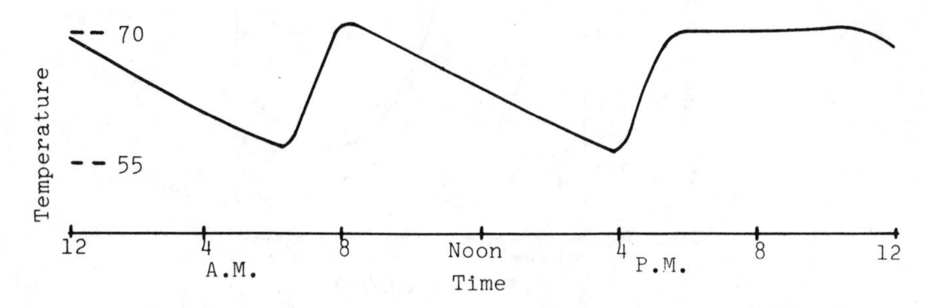

Figure 3.7. Possible daily temperature cycle in a house with a time-controlled thermostat.

quired for zone control includes remotely actuated air dampers for a forced air system and a programmable logic systems to actuate a heating sequence. Although such controls systems are available for very large commercial installations, there is a need for more sophisticated but inexpensive control and logic devices for home use. The scale of application is so large that integrated circuit controls should be economic. The reliability and performance of the mechanical aspects of the system might be more of a problem than the electronics.

Insulation. The heat transfer rate through uninsulated walls is appreciable. Placing standard (R-11) insulation in a 3-in. air space reduces the heat conduction of an outside wall by about a factor of 3; the nominal U-value drops from 0.23 to 0.07 Btu/hr ft^2 °F in a wall made of siding, sheathing, and wallboard. The actual heat conductance through any surface depends in part on the surface heat transfer rates, which in turn depend on wind velocity. The smaller the other resistances in series, the more important the surface heat transfer coefficients; the surface coefficients are therefore most important for windows and somewhat less important for walls, especially insulated walls.

The problem of producing a material of high thermal resistance is complicated by convective heat transfer modes within the insulation. Among the best solutions to this problem which are practical near room temperature are fibers and foams: mineral wool, fiber glass, and styrene and urethane foams. At low densities the thermal conductivity increases with *decreasing* density, due to heat transfer through the gas; at high densities the conductivity increases with *increasing* density, because of conduction through the material itself. Materials are prepared at densities near the minimum of conductivity. Figure 3.8 shows the dependence of conductivity on density for fiber glass.

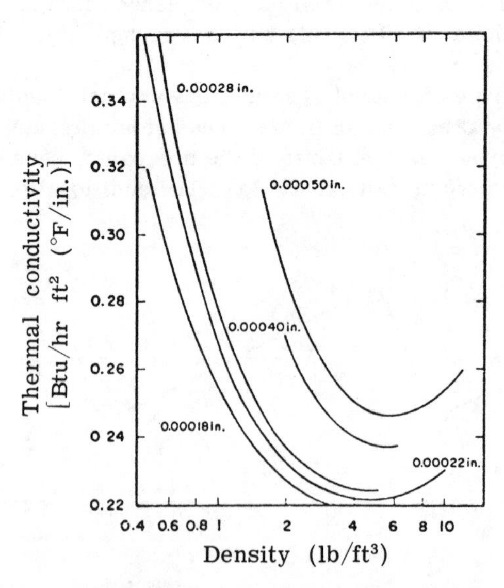

Figure 3.8. Thermal conductivity of fiber glass as a function of fiber size and density at 75 °F mean temperature. [Source: ASHRAE (1972), p. 293.]

It should be possible to design optimum cell sizes or size distributions for a particular insulating material. The cells of the urethane foams are often filled with a low conductivity, heavy gas, such as a "freon," during the foaming process. Unfortunately, the heavy gas exchanges with air after some months or years, and the thermal properties are degraded. Another extremely important mechanism for long-term degradation of performance is absorption of water vapor. Vapor barriers generally prevent this: materials research may suggest other approaches. Techniques for establishing vapor barriers to use with foamed insulation are required. Instrumentation for measuring the thermal conductivity of installed insulation would be useful for diagnosing problems in older buildings.

The thermal conductivity of insulation increases with temperature; for example, the conductivity of fiber glass at 200°F is about 50 percent greater than at 70°F. Optimal materials for insulation vary with the temperature range. It is likely that more imagination has gone into developing insulation at very high and very low temperatures than into developing insulation for room-temperature applications. Materials research on insulation useful at all temperatures is significantly constrained by fire safety considerations; some organic foams will exacerbate a fire by contributing their heat of combustion.

Since convection is an important component of heat transfer, it should be possible to design insulation, using either cells or multiple layers, which has different heat conductivities in the two directions, i.e., a thermal diode. An insulation material or structure whose conductivity could be easily varied, perhaps by controlling gas flow, might also be valuable for space conditioning applications and for solar energy collection, where daytime and night-time requirements are different.

6. Aerodynamic control

Several studies (Moyers, 1971; Hittman, 1973; Dubin, 1974; Kurtz, 1974b) have concluded that infiltration losses are one of the largest mechanisms degrading space conditioning in the home. Infiltration is thought to be responsible for approximately one-third of the heating and air conditioning load, accounting for several percent of the national energy consumption. Clearly improvement in the air flow patterns around and through the house has the potential for major energy savings.

In commercial buildings ventilation is designed according to empirical rules established many years ago and badly in need of revision. New studies of suitable ventilation rates are needed. Great energy savings would be possible if methods were developed to determine air quality "on-line," including moisture content, impurities, odors, etc., and procedures were used to "rehabilitate" the air quality without drawing in large quantities of air at the wrong temperature from outside.

Designed ventilation is rarely used in residences. Rather, one depends on inadvertent air flow through doors, cracks, etc. The resulting infiltration rates in houses are always far in excess of what is needed for ventilation, often exceeding one air exchange per hour. The infiltration rates are strongly dependent on wind velocity and direction.

In addition to determining heat loss rates due to infiltration, wind and air currents affect the heat loss through convection. Convection losses from a material are particularly significant when the material is a good thermal conductor or when the thermal energy is deposited at its surface, as in solar radiation heating.

Careful studies of the air flow patterns into, within, and around houses are needed. Although much work has been carried out on wind pressure loading on structures, only very recently have studies of heat loss mechanisms begun.

Investigators have recently begun to use a sulfur hexafluoride (SF_6) detector to measure infiltration rates in houses. A small amount of SF_6 (amounting to about 20 parts per billion) is released into the house, and the concentration in the air is monitored as a function of time with a leak detector. Good precision is possible, and the measurements have been carried out to study the effect of wind velocity and other variables. The next step would be to locate and determine the flows rates. A building-sized leak detector would be valuable; the SF_6 instrument might be modified for such use.

An air-velocity measuring device on an extremely small scale (a few mm), perhaps utilizing the hot wire method, would be useful both for air-flow work and surface-film coefficient studies. Given convenient low-velocity devices, one could study the aerodynamics of flow within a house and modify window frames, radiators, ducts and other surfaces to optimize heat exchange rates and decrease losses.

The aerodynamics of the exterior of a house can be studied from the point of view of reducing infiltration and convective heat losses at roofs, windows, etc. It may even be possible to retrofit properly shaped air-flow surfaces to houses to reduce the effect of wind on the heat load. Studies with wind tunnels and in the field are badly needed in this new area of research.

Models of the microclimate surrounding a building would be useful. A computer-driven simulation that included the effects of terrain, other buildings, and prevailing winds would be an aid to the aerodynamic design of energy-efficient shells with minimal leakage and controlled ventilation. Such a model would provide, also, an opportunity to study the aerodynamic design of walls, windows, and roofs. The location and aerodynamic shape of heat-exchangers and ventilators could be studied to maximize the effect of prevailing air flow patterns. Such designs would certainly be dependent upon the regional climate.

There is also a need at the room-size level and smaller to be able to include the effects of air flow. The location and design of windows, doors and thermostats would benefit from an understanding of these effects. The design problem that requires careful modelling is to describe a ventilation system that provides a uniform temperature distribution with variable volume flow rates. Registers and ventilators need to be designed to eliminate "dumping" of air at low flow rates and the creation of hot spots. A result of such aerodynamic design might be the possibility of reducing the required inlet air temperature in rooms from, say, 110°F to 80°F. This would reduce duct losses, permit smaller furnaces and improve the coefficient of performance of heat pumps used for space conditioning by about 50 percent.

D. Efficiencies from the standpoint of the second law of thermodynamics

In this section we calculate second-law efficiencies for hot water heating, space heating and air conditioning. This is one way to provide insight into the potential for future energy savings in these areas. A second-law efficiency, as discussed in Chapter 2, is the ratio of the minimum energy required to perform a specified task to the energy actually used to perform that task. Hence, the second-law efficiency depends, in part, on precisely how a given task is defined. Some arbitrariness and some idealization are necessary.

An "ambient" temperature must be specified. We choose to use typical outdoor winter air, summer air, and ground water temperatures, rather than to add (for air conditioning) or subtract (for space heating) the temperature drop across a heat exchanger.

Hot water heating

We define the task of heating hot water to be bringing the water, as a finite reservoir, up to a temperature T_2 at a central heater. It would be possible to make a more forward looking task definition, which in principle allows greater efficiency improvement, by specifying the task as delivery of water at points of use as desired (i.e., at lower temperatures than T_2). However, this would be a complicated formalism, and for present purposes it seems more practical simply to specify a temperature T_2 which is not unnecessarily high. We choose $T_2 = 120\,^{\circ}\mathrm{F}$ (49 °C).

The minimum available work required to add a differential amount of heat dQ to a finite reservoir at temperature T, withdrawing heat from an infinite ambient reservoir at the lower temperature T_0, is $dB = dQ(T - T_0)/T$. The added heat will raise the temperature of the finite reservoir by $CM\,dT$, where C and M are its specific heat and mass, respectively. By elementary integration, therefore, assuming C is independent of temperature, the minimum available work required to raise the temperature of a finite reservoir from T_0 to T_2 is

$$B_{\min} = CM[T_2 - T_0 - T_0 \ln(T_2/T_0)]. \tag{3.5}$$

We will use this expression several times in this and the following section.[63]

The first-law efficiency η for a hot water heater is equal to its heat content at T_2, relative to ambient, divided by E, the actual chemical or electrical energy expended in the house to provide the hot water:

$$\eta = CM(T_2 - T_0)/E. \tag{3.6}$$

The total available work expended is approximately 3 times larger than E for electric energy, to take into account the inefficiency in the power plant, and approximately equal to E for oil or gas heaters. For the second-law efficiency, $\epsilon = B_{\min}/B_{\mathrm{actual}}$, of a finite water heater, we may therefore write

$$\epsilon = f\eta\left[1 - \frac{T_0}{T_2 - T_0}\ln(T_2/T_0)\right], \tag{3.7}$$

where $f \cong \frac{1}{3}$ for electricity and $f \cong 1$ for gas or oil.

We have estimated the first-law efficiencies η of electric and fossil-fuel hot water heaters in Section C, Part 3. They are 75 percent and 50 percent respectively. Table 3.5 gives the associated, *far smaller*, second-law efficiencies ϵ. From Table 3.5, we can compute the national average second-law efficiency: Since one-fourth of the water heaters are thought to be electric, the average efficiency is 0.025. This forty-fold difference between the actual and the ideal shows that there

---

[63] A closely related result is derived in Chapter 2 (see Eq. 2.26). For small values of $\lambda = (T_2 - T_0)/T_0$, $B_{\min}$ is quadratic in λ: $B_{\min} \cong \frac{1}{2} CMT_0\lambda^2$.

TABLE 3.5. The efficiencies of household energy systems.

Household task	Ambient temperature T_0 [°F (°C)]	Task temperature T_2 [°F (°C)]	First-law efficiency η	Second-law efficiency ϵ
Hot-water heating				
Electric	55 (13)	120 (49)	0.75 [a]	0.015
Gas	55 (13)	120 (49)	0.50	0.029
Space heating				
at the room	40 (4)	70 (21)	0.60	0.028
at the register	40 (4)	110 (43)	0.60	0.074
at the furnace plenum	40 (4)	160 (71)	0.75	0.145
Air conditioning	90 (32)	55 (13)	2.0 (COP) [a]	0.045

[a] The first-law efficiency is a factor of 3 smaller if one refers back to the fuel at the electric power plant. (To compute the second-law efficiency, one *must* refer back to the power-plant fuel.)

is enormous potential for increasing the thrift with which we use fossil fuels for heating water.

Space heating

The idealization of the task of space heating is more complex and more arbitrary; it requires the specification of boundaries to the system. It requires also that the characteristics of the protective shell and the demands and use patterns of the occupants be specified. The house as a system can be split into two subsystems: (1) the shell, and (2) the heating system. We idealize only the heating system and do not further consider the shell.

There are three reasonable idealizations of the heating system, differing in the task the system performs. The heating system may be conceived to transfer heat from an ambient heat reservoir (at the temperature of outside air) either (1) to a reservoir at the temperature of the interior of the house (about 70 °F, or 21 °C), or (2) to warm air at a temperature typical of air flowing out from room registers in forced air systems (about 110 °F, or 43 °C), or (3) to air at a temperature typical of air as it flows through the heat exchangers in the plenum of a furnace (about 160 °F, or 71 °C). The second-law efficiency of existing systems looks successively more favorable as we shift from the first, to the second, to the third perspective—as we postulate higher intermediate temperatures successively further removed from the final task temperature, that of the room.[64]

---

[64] Referring an efficiency calculation to the plenum temperature is equivalent to stating that air at 160°F is *necessary* to heat a room to 70°F, that duct losses and high (110°F) register temperatures are *unavoidable* features of space heating. This clearly leads to an underestimate of the ultimate potential for improvement.

We specify one further element of the idealized task, in the first case only. One-third, approximately, of the task of space heating is associated with the warming and humidifying of outside air that passes into and then out of the house. We idealize this portion of the task by considering the infiltrated air to be a finite reservoir. The second-law efficiency of an oil or gas furnace is then

$$\epsilon \cong \eta \left\{ \frac{2}{3} \left[\frac{T_2 - T_0}{T_2} \right] + \frac{1}{3} \left[1 - \left(\frac{T_0}{T_2 - T_0} \right) \ln \left(\frac{T_2}{T_0} \right) \right] \right\}. \qquad (3.8)$$

[Compare Eq. (3.7); here $f \cong 1$.] For idealizations 2 and 3, the task is delivering heat to an infinite reservoir.

In Section B, we estimated the first-law efficiency of gas furnaces to be 0.75 at the plenum and 0.60 at the room. Corresponding second-law efficiencies are given in Table 3.5.

Air conditioning

The idealization of the process of air conditioning adopted here is the continuous extraction of heat from air at 55 °F and the transfer of that heat to an ambient summer reservoir. We do not choose a higher interior temperature than 55 °F because we assume that a portion of the task of air conditioning is dehumidification, particularly straightforwardly accomplished by cooling outside air (usually to below its dew point) to a temperature where even when saturated it can hold little water. Other dehumidification methods exist, which in principle could permit working across a smaller temperature difference.

In Section C, part 1, we estimated typical air conditioners to have a COP near 2.[65] In Table 3.5, we state the second-law efficiency corresponding to $\eta(\mathrm{COP}) = 2$ and for the task of air conditioning as just defined.

Remarks on efficiency

All the second-law efficiencies calculated here are extremely small; that is, present devices are far from ideal. The basic reason for this is that water heating, space heating, and air conditioning are all associated with small percentage changes in absolute temperature. Thermodynamics tells us that relatively small amounts of chemical energy, electricity, or other forms of high-quality energy are sufficient, in principle, to transfer large amounts of heat through small temperature differences.

Departures from the idealized performance occur throughout a practical device. It is within the reach of present technology to come considerably closer to the idealized systems, and it is the task of physicists and engineers to discover new ways of coming closer still.

E. Combined electric and heating systems

In this section we examine the possibility of fuel savings arising from a hypothetical restructuring of the energy industry and of energy use such that a major

[65]We specified an EER of 6, or a COP of 1.8.

TABLE 3.6. Fuel consumption for electricity and low-temperature heat in 1968[a] (percent of national fuel consumption).

	Residential	Commercial	Industrial	Total
(a) Space heat, hot water, process steam	14	8	17	39
(b) Air conditioning	1	2	b	3+
(c) Other low-T heat	2	1	b	3+
(d) Electricity (excluding air conditioning)[c]	6	3.5	9	19
Ratio $\frac{(d)}{(a)+(b)+(c)}$	0.35	0.32	< 0.53	< 0.42

[a]Source: SRI(1972).
[b]Not known.
[c]Other less significant low-temperature uses of electricity are not subtracted.

portion of the supply of low-temperature heat and electricity is produced by *combined systems*, systems in which both the electricity and the low-temperature heat generated in a fuel-consuming device are put to use. The problems of temporal variation and spatial dispersal of demand are discussed, and sample, tentative, fuel savings are estimated. These potential savings are large. Indeed, in favorable circumstances, one unit of fuel in a combined system can produce about as much low-temperature heat and electricity as two units of fuel used separately. In more typical circumstances, the savings are not quite as favorable.

The yearly consumption to produce electricity is roughly one-half the fuel consumption to produce low temperature heat.[66] Remarkably, as Table 3.6 shows, approximately the same ratio of 0.5 applies separately to the residential, the commercial, and the industrial sector.[67] Yet in only rare instances are combined systems in evidence. The name "total energy system" has been used for a residential and/or commercial centralized system. In general, total energy systems are *not* connected to external power lines; rather, all the electricity is consumed on site. The name "district energy system" has been used for a combined system serving

[66]Low-temperature heat is defined to be heat delivered at about 500°F or less. For the applications included in Table 3.5, the required heat is not far from ambient.
[67]Perhaps a more relevant ratio to calculate is the energy to generate electricity exclusive of its direct use for low-temperature applications divided by the energy to generate both low-temperature heat and electricity for low-temperature applications. This is because electricity used for air conditioning and resistive heating *could* be replaced by low-temperature heat sources.

TABLE 3.7. Sample performance of energy conversion devices supplying usable heat and electricity.

Device	Useful output per unit of fuel (available work) input: Electricity	Usable heat	Heat Temperature of heat, °F (°C)	Second-law efficiency ϵ [a]	Units of fuel required to produce the same electricity and heat separately [b]
Diesel [c]	0.30	0.33	250(120)	0.44	1.70
		0.07	700(370)		
Gas turbine [d]	0.20	0.70	400(200)	0.49	1.40
Steam turbine [d]	0.13	0.78	400(200)	0.46	1.31
Fuel cell [e]	0.25	0.65	300(150)	0.47	1.72

[a] The second-law efficiency is the sum of the electricity fraction and $(\Delta T/T)$ times each of the usable heat fractions, where $T = T_0 + \Delta T$ is the temperature of the heat and T_0, the ambient temperature, is taken to be 40 °F, or 278 K. Background theory appears in Chapter 2.

[b] We assume first-law efficiencies of $\eta_e = 0.33$ for electricity, $\eta_t = 0.67$ for small-scale heat-delivery (diesel and fuel cell), and $\eta_t = 0.85$ for large-scale heat delivery (gas and steam turbines).

[c] Only a small fraction of the energy in the 700- °F exhaust stream can be usefully extracted, unless the exhaust is cleaned up.

[d] Source: Thermo Electron (1974).

[e] Fuel cells can operate at higher electric efficiency than 0.25 but in combined heat and electricity application that may not be optimal. The temperature of fuel cell coolant may be in the range 200 to 400 °F.

a more varied community of residential, commercial and industrial users and a district energy system is more often interconnected with utility grids.[68]

In Table 3.7, we present sample performance data for energy conversion devices which could be used as generating units in combined systems. The useful low-temperature heat output is seen to be somewhat larger than the electric output. The last column of Table 3.7 gives the ratio of the fuel required to produce heat and electricity separately, relative to the fuel required to produce them together. Thus, for example, the diesel combined system uses fuel 1.7 times as efficiently as separate systems.

Variations in time

Based on space-time average demands, the consumption of low-temperature heat and electricity is fairly well matched to the performance characteristics of the available energy conversion devices. However, it is important to consider space-

[68] There is an extensive literature on combined systems (Diamant, 1970; Gamze, 1972; Federal Council on Science and Technology, 1972; Decision Sciences, 1973; Leighton, 1973; Lusby, 1974). There are about 500 total energy installations in the United States (Federal Council on Science and Technology, 1972).

time variations. Consider, first, time-of-year variations in a temperate climate. In the United States, a typical ratio for the maximum use of electricity to the use averaged over a year is 1.6 (Federal Power Commission, 1971). Much of this important extra, or "peak," use is associated with electric resistive space heating or air conditioning, which could, following ideas developed in this report and elsewhere, be economically provided in part in other ways.

Compared with electricity, heating demands vary much more strongly month to month, as illustrated in Figure 3.9. In our standard house, which is representative of much housing in the northeast, we saw that a furnace will operate only 27 percent of the time on a mild day (see Section B). More refined calculations could

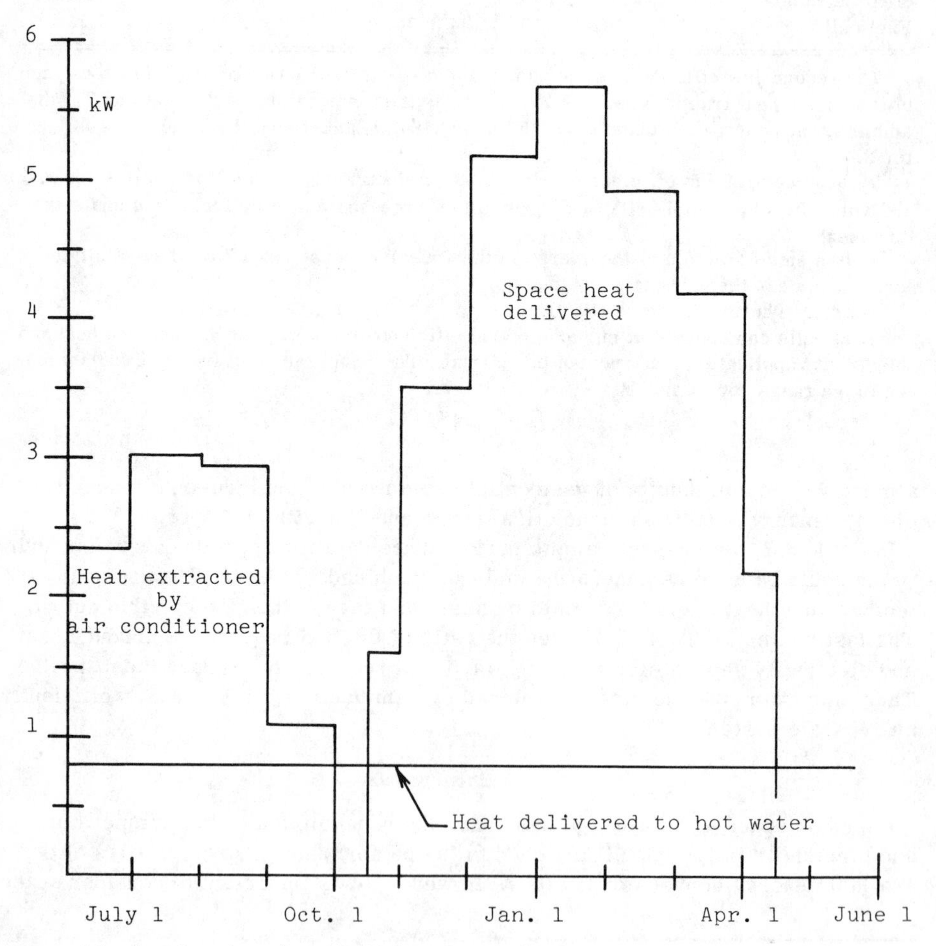

Figure 3.9. Sample rates of residential energy involved in space heating, air conditioning, and heating water for Middle Atlantic region, with central air conditioning (Fox and Socolow, 1974). The residential demand for electricity in 1973 (national average per residence) was 1.2 kW_e (Bureau of Mines, 1974). The relative areas under the curves are from SRI (1972).

TABLE 3.8. Typical distribution of *winter* days as a function of temperature in the Middle Atlantic region.

Average temperature °F	Number of days	Degree days based on the following reference temperatures 65 °F [a]	40 °F [b]	30 °F [b]
5–10	2	110	60	40
10–15	3	160	85	55
15–20	6	280	130	70
20–25	10	420	170	70
25–30	16	600	200	40
30–35	28	910	210	0
35–40	35	960	85	0
40–45	34	770	0	0
45–50	32	560	0	0
50–55	15	190	0	0
Total	181	4960	940	275

[a] This is the conventional reference temperature.
[b] The point of these columns is to emphasize how little boosting by an auxiliary source of heating will be required when a base-load system can take care of all heating needs above a modest temperature. A base-load system sized for 40- °F reference temperature can handle approximately all but 940/4960 = 19 percent of the heating over the winter.

be performed for a given location using a nearby weather tape, which will include both temperature and cloud-cover histories. A feel for such data can be obtained from Table 3.8, which shows the number of winter days that fall within various temperature ranges for a typical (5000 Fahrenheit-degree-day) climate. In summer, an air conditioner is unlikely to run more than 40 percent of the time (see Section C, Part 1). Any proposal to use equipment which produces heat and electricity in a fixed ratio for homes or commercial buildings must contend with these seasonal disparities in demand.

A second time dependence occurs over the day. For example, sample Minneapolis hourly winter and summer residential demand data are shown in Figures 3.10 and 3.11. (Fluctuations within any hour are not included). There is some tendency for the electric and the heating demand to follow each other. The demand for electricity shown in the figure varies less than a factor of 2 during the winter day, and that required for heat less than a factor of 1.5. The largest peak is in the early evening after supper, and there is a smaller peak in the late morning. This is a typical pattern and corresponds roughly to electric utility peaks (Federal Power Commission, 1970). Thus, the ratio of heating to electric needs is fairly steady in residences during the day in winter. In the summer, air conditioning creates a large demand which is far from constant over the day. The variability could be dealt with by a balanced choice of equipment: electrically-driven or engine driven compressor, absorption air conditioning, etc.

The data in Figures 3.10 and 3.11 do not show fluctuations within an hour. Yet these exist. For example, household water heaters are currently marketed with 2.5 kW heating elements, double the peak hourly rate in Figure 3.10. Practical

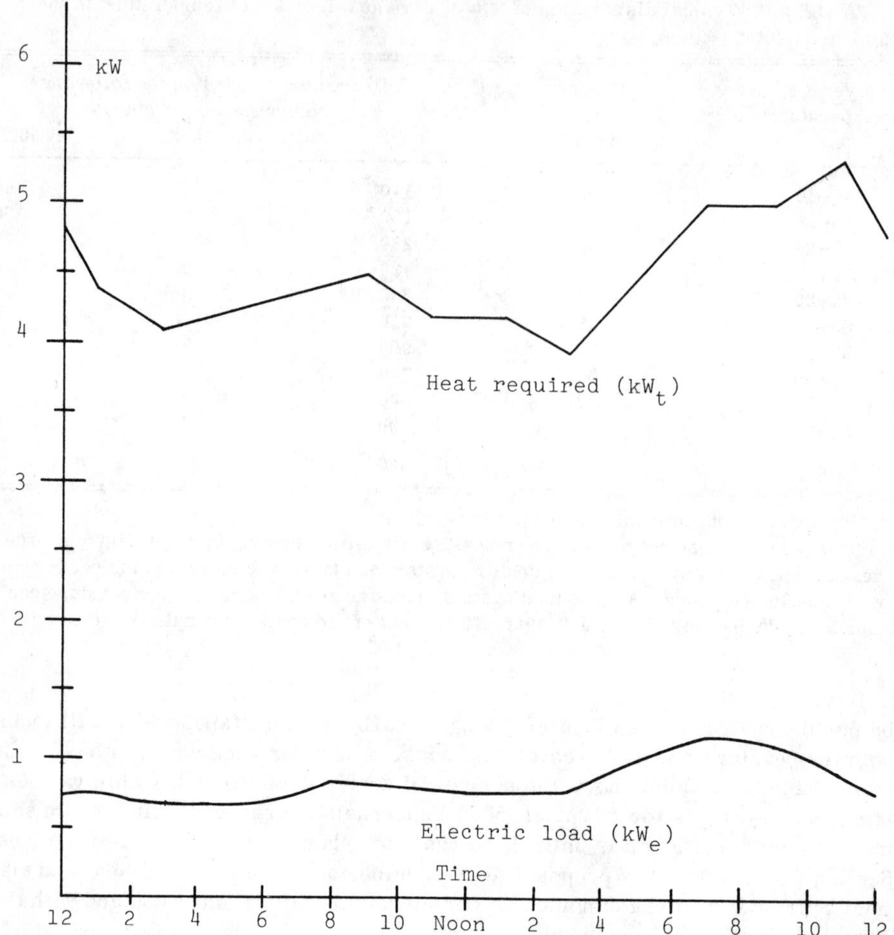

Figure 3.10. Winter diurnal loads for a sample Minneapolis residence. (Adapted from Decision Sciences, 1973).

single-house combined systems may turn out to be incompatible with some very high wattage "conveniences." (Such systems would, in any event, not heat water electrically; a self-cleaning oven might be a better example of a device with an occasional high power demand.)

One has a choice among several strategies in designing a combined system. The cogeneration unit can be sized (1) to provide on-site electricity needs, with a supplementary furnace, (2) to provide on-site base-heating needs with sale off-site of excess electricity, or (3) to provide all heating needs. Which strategy will make best use of capital investment is a difficult problem requiring detailed study. One usually will want to employ a combined device whose electric generating capacity is used with a high duty factor; this is why the ability to place excess electricity onto a power grid can become a critical factor.

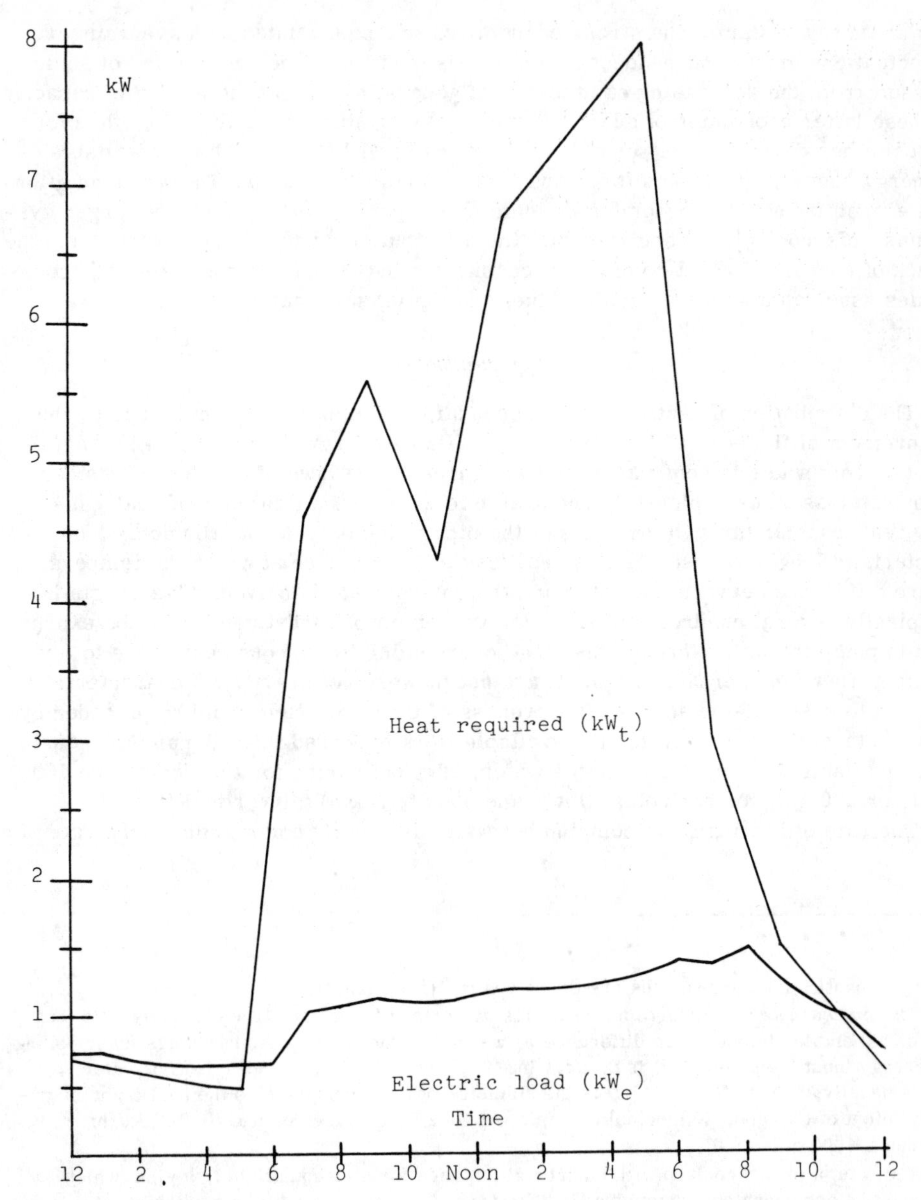

Figure 3.11. Summer diurnal loads for a sample Minneapolis residence with absorption air conditioning. (Adapted from Decision Sciences, 1973.)

A further question concerns the applicability of the combined heat and electricity system to the single dwelling unit or small commercial establishment. One problem is large, short-term fluctuations in demand. Fluctuations above twice the local daily average in heating fuel or electric consumption, while rare in typical residences, do occur, and they present serious design problems for a single-house total energy system. The strong economy of scale associated with averaging of fluctuations among independent customers is obvious. Other economies of scale result from the decreasing capital cost of generating equipment per kW of capacity. These latter economies of scale are much less significant for fuel cells than for most other devices; it is possible that fuel cells will become the most suitable energy conversion devices for house-sized combined systems. There are additional economies and diseconomies of scale in comparing single house and larger systems: diseconomies associated with the distribution system (pipes, wires) and the lack of consensus about comfort, necessitating individual control systems; economies associated with the supply of fuel and the maintenance of equipment.

Heat distribution

The distribution of heat to distant secondary users can be accomplished by the movement of fluids.[69] The transport of available energy in the form of steam or hot or cold water is constrained in two senses by the laws of physics. Consider hot water as an example. (1) The available work in the fluid is continually lost by heat transfer through the walls of the pipe to the outside environment; a characteristic length to describe this heat loss is the distance at which the temperature difference between the water and the environment is halved. This length is typically several hundred miles.[70] (2) Available work must continually be expended to pump the water through the pipe, overcoming its viscous resistance to (turbulent) flow (and perhaps raising it against the force of gravity); a characteristic length to describe the energy effectiveness of transportation would be provided by the distance (height) at which the available work expended through pumping equals the available work contained in the water. Typical values for this length are 100 mi, or 200 km, for horizontal flow[71] and 6000 ft (2 km) for vertical flow.

The rate of heating from pumping hot water is usually comparable to the rate of

---

[69]More exotic, chemical means are discussed by Häfele (1974).

[70]*Note on characteristic length for heat loss by pumped hot water:* The characteristic length for halving the temperature difference between the water and its surroundings (overlooking energy input from pumping) is $L = C\rho Dv \ln 2/4U$, where C is the specific heat of water, ρ is its density, v is its flow rate, D is the diameter of the pipe, and U is the heat transfer coefficient of the pipe. Numerical example: if $D = 2$ ft, $v = 8$ ft/sec, and $U = 0.2$ Btu/hr ft^2 °F, then $L \cong 600$ miles.

[71]The rate of pressure drop with horizontal distance for a viscous fluid in a pipe (which must be compensated by pumping) is $dp/dx = (f/2D)\rho v^2$, where D is pipe diameter, ρ is water density, v is flow velocity, and f is a dimensionless parameter approximately equal to 0.02. The available work content of the water per unit volume, b, is

$$b = C\rho\Delta T\left[1 - \frac{T_0}{\Delta T}\ln\left(1 + \frac{\Delta T}{T_0}\right)\right],$$

where the water is a finite reservoir of heat at absolute temperature ΔT above ambient (T_0); C is its specific heat. The ratio $b/(dp/dx)$ is the distance at which the pumping work equals the available work content. For $v = 8$ ft/sec, $D = 2$ ft, $T_0 = 70$°F, and $\Delta T = 100$°F, the distance is 120 miles.

cooling through the walls of the pipe.[72] It is not obvious, accordingly, whether the hot water moving through the pipe will get hotter or colder. We conclude from energy considerations alone that a distribution network extending at least 10 miles or about 20 km would be reasonable.

Industrial combined systems

Industrial demands are often rather time independent from day to day and over the 24-hr day, and this has facilitated the introduction of combined heating and electric systems in industry. More combined systems would be deployed if off-site sale of the produced electricity were easier to arrange. We have estimated the fuel economy factor for gas and steam turbines to be 1.40 and 1.31, respectively (see Table 3.7, last column). Thus (0.40/1.40), or 29 percent, and (0.31/1.31), or 23 percent, respectively, of the fuel currently used to provide the industrial process steam and the electricity separately could be saved. The national potential for economically justifiable on-site combined electricity and process steam generation by industry has been examined by Thermo Electron (1974) and is under study by the Dow Chemical Co. (Decker, 1974) for the National Science Foundation. The potential is clearly large.

Less optimism is warranted for the potential use by industry of waste heat from large central power plants. A modern power plant designed for maximum electric efficiency discharges just over one-half the input fuel energy through the condenser cooling water. This cooling water acquires a typical temperature of $T = 100\,^\circ\mathrm{F}$ (40 °C), and contains only 1.5 percent of the input fuel available work.[73] Thus, failure to use the warm cooling water from power plants is not in itself an especially wasteful process; power plants are very efficient in this respect. The interest in using this warm water is, instead, associated with the gross inefficiency of the usual independent systems designed to produce this kind of warmth.

Proposals to modify central power stations to use the heat discharged must be viewed cautiously. The inflexibility of a large power plant constrains its maintenance schedule, its part-load performance, and its ability to function at variable efficiency, as would be associated with variations in rates of withdrawal of steam or hot water or variations in the quality of the steam withdrawn. Constraints associated with the transportation of steam and hot water have already been discussed. The long planning time and its associated uncertainty complicate the process of finding a suitable user of the low quality heat. In one recent instance, a large nuclear plant in Midland, Michigan, has been located next to an

---

[72]Apparently, the Alaska oil pipeline, which must carry heated oil to reduce the viscosity, nonetheless does not have to provide heat along its route because the degradation of the energy of pumping keeps the oil warm (Marsden, 1974).

[73]The available work in this water (a finite reservoir) is

$$B = Q\left(1 - \frac{T_0}{\Delta T}\ln\frac{T}{T_0}\right) \cong \frac{1}{2}\left(\frac{\Delta T}{T_0}\right)Q;$$

for $\Delta T = 30°\mathrm{F}$ and $T_0 = 70°\mathrm{F} = 530°\mathrm{R}$, $B = 0.027Q \cong 0.015B_0$, where B_0 is the input available work.

industrial user, the Dow Chemical Co. Uses of the warm water from power plants for agriculture, mariculture, the melting of harbor ice, and other purposes, have received considerable study (Yee, 1972).

Sample calculations for combined systems

We conclude this section by illustrating some of the technical issues which are raised when a combined system is examined quantitatively. We discuss briefly two slightly unconventional systems which could serve a group of residences and/or commercial establishments. System 1 uses a large diesel-driven generator of sufficient capacity that the heat recuperated can serve "base" space heating demands. Peak heating demands are met by an auxiliary combustion process. The system capacity is sized to the heating needs. There would typically be extra electricity which is fed into utility lines and used by off-site customers. This extra electricity would be most plentiful at times of general peak demands for electricity (i.e., winter heating and summer air conditioning peaks). System 2, which could be available in the near future, uses a large bank of fuel cells for electricity and heat. It too requires an auxiliary furnace to provide further heating needs. However, the fuel cell capacity is sized to cover peak on-site electric demand. The diesel and fuel-cell characteristics are specified in Table 3.7; the climate is that of Table 3.8. Consider now what happens at a given house.

For both systems, we assume that the house requires space heat at a rate of 0.15 kW_t per degree day. From Table 3.8, the average number of degree days per day during the 181-day winter is 4960/181 = 27.4 degree days/day, so the average space heating demand is $27.4 \times 0.15 = 4.1$ kW_t. The peak space heating demand is 9.0 kW_t (5 °F outside). Each system is assumed to provide a certain amount of usable heat at a central location (40 percent of the fuel energy for the diesel, 65 percent of the fuel energy for the fuel cells, according to Table 3.7). Of this usable heat, 1.0 kW_t per residence is used continuously for water heating and 90 percent of the rest is delivered for space heating (thus not quite all of the usable heat is used). For the auxiliary furnaces, we assume that 67 percent of the fuel energy is delivered as space heat. Finally, we assume that the peak demand for electricity (per residence) is 2.0 kW_e.

The diesel in system 1 is arbitrarily sized to supply enough heat for water and for space heating down to 30-°F outside temperature. This choice of base-load system covers 94 percent of the winter heating requirement (all but 275 out of 4960 degree days, according to Table 3.8). At 30-°F outside temperature, the space heating demand is (35 degree days)$\times$(0.15 kW_t/degree day) = 5.25 kW_t. To *deliver* this much space heating, the diesel must *produce* 5.8 kW_t. To this, we add 1.0 kW_t for hot water to get the usable heat provided by the diesel at full power: 6.8 kW_t. Scaling from Table 3.7, we then find that the diesel produces electricity at a maximum rate of 5.1 kW_e and consumes fuel at a maximum rate of 17.0 kW_t. These numbers, and others to be discussed below, are summarized in Table 3.9 (keep in mind that all of the powers are per residence).

On the *average*, system 1 meets 94 percent of the 4.1-kW_t demand for space heating, or 3.85 kW_t. This requires the production of 4.3 kW_t before losses in delivery, to which again the hot-water demand of 1.0 kW_t is added, for a total of 5.3 kW_t, the average production rate of usable heat. Again scaling from Table 3.7,

Table 3.9. System characteristics of two combined systems over a six-month winter.

	System 1		System 2	
	kW		kW	
House characteristics				
space heating needs per degree day	0.15		0.15	
average space heating	4.1		4.1	
average hot water heating	1.0		1.0	
electric peak load	2.0		2.0	
Base-load system [a]		Diesel [a]		Fuel cell [a]
minimum temperature without auxiliary		30 °F [b]		40 °F
fraction of space heating		0.94 [c]		0.81 [c]
heat capacity	6.8		5.2	
electric capacity	5.1		2.0 [b]	
maximum fuel consumption	17.0		8.0	
duty factor		0.78		0.90
average production rate				
space heat	4.3 [d]	40% [e] (space heat and hot water)	3.7 [d]	65% [e] (space heat and hot water)
hot water	1.0		1.0	
electricity	4.0	30% [e]	1.8	25% [e]
average fuel consumption rate	13.2		7.2	
Auxiliary system				
average fuel consumption rate	0.4		1.2	
Total fuel consumption rate	13.6		8.4	
Fuel consumption rate for separate generation	19.6		13.0	
Fraction of fuel saved		0.31		0.35
Second-law efficiency of system		0.34		0.28

[a] System characteristics from Table 3.7.
[b] Design parameter for the system.
[c] From Table 3.8.
[d] To get *delivered* space heat, multiply the figure for *produced* space heat by 0.9.
[e] Percent of fuel energy (from Table 3.7).

we find that on the average, the diesel produces 4.0 kW_e of electricity and consumes 13.2 kW_t of fuel. Its duty factor (ratio of average power to maximum power) is 0.78. (We are assuming here that the diesel characteristics are the same at part load and full load.) Since the electric peak load is only 2.0 kW_e, electricity must be put on the utility grid almost all winter.

For system 1, the auxiliary furnace burns fuel at an average rate of 0.4 kW_t, bringing the total fuel consumption rate for the system to 13.6 kW_t. If the 5.1 kW_t of space and hot water heating had been supplied by a separate system, with an assumed first-law efficiency $\eta = 0.67$, while the 4.0 kW_e of electricity had been supplied by a power plant at an assumed efficiency of 0.33, fuel would have been consumed at a rate of 19.6 kW_t. Thus, system 1 uses 31 percent less fuel than separate systems. The second-law efficiency of system 1 can be calculated: If we assume that 7 percent of the hot-water thermal energy,[74] 12 percent of the space heating thermal energy,[75] and 100 percent of the electric energy qualify as available energy, system 1 uses fuel at a rate of 13.6 kW_t to produce available work at a rate of 4.6 kW_t, an overall second-law efficiency of 0.34.[76]

System 2, based on fuel cells, is sized to provide the peak electric load of 2.0 kW_e. The other fuel cell characteristics are given in Table 3.7. We leave it to the interested reader to work through the analogous calculations for this system; results are tabulated in Table 3.9. Not surprisingly, the auxiliary furnace is used more often in system 2. The two systems are about equally successful in saving fuel, relative to producing the heating and electricity in separate devices.

The summer operation of systems 1 and 2 could also be quite favorable. The proportion of absorption to compression air conditioning is an important design parameter. If the systems can operate at comparable power levels at winter and summer peaks, this makes effective use of capital. Operation in spring and fall will be somewhat slack with hot water the only dependable demand for heat (see Figure 3.9, which shows roughly two months in spring and two in fall having low heat needs). The fuel-cell system has greater flexibility for this situation.

The above discussion of combined energy systems is mainly intended to indicate some possibilities which should be analyzed in addition to the usual "total-energy" modes. There is a range of interesting possibilities in system components and parameters. Any serious evaluation of alternatives requires more detailed specification of needs and system optimization.

F. Recommendations

Our recommendations for further technological improvement embrace both design of systems and research which could lead to new design.

A new direction for design problems having an energy component is to develop passive systems capable of implementing energy conservation. The development of thermal diodes and of new building materials with larger specific heats[77] for thermal stability might offer new options (see Section C, part 5). The improved utilization of ambient air flow and of ground water involve both design and research problems. Sun-shielding through surface treatment is an important problem, with potential for the contribution of physics. A feel for physics led one house designer, who knew his house would have snow-covered fields to the south, to plan a more steeply pitched roof for his solar collectors than the usual calculation based on direct sunlight would have suggested, to take advantage of reflected light.

As we have emphasized throughout this chapter, dramatic improvements in the

---

[74]The hot water tank is treated as a finite reservoir at 120°F (580°R) sitting in an environment at 40°F (500°R); $[1-(500/80)\ln(580/500)] \cong 0.07$.

[75]The space heat is treated as an infinite reservoir at 110°F sitting in an environment at 40°F; $70/570 \cong 0.12$.

[76]The system has not been chosen to maximize the second-law efficiency. Such an optimization could be done.

[77]The ASHRAE guide's tables of properties of building materials omits specific heats for over a third of the materials listed.

performance of active systems are possible in principle. One should expect the development of more efficient heat pumps, better heat exchangers, and direct fuel-to-electricity conversion (fuel cells). The efficiency of the heating system can also be greatly improved, on perhaps a faster time scale, by "plugging leaks," by using more sophisticated control systems and diagnostic instrumentation, and by taking advantage of ambient conditions whenever possible. Some architects (Olgyay, 1963) have recognized for a long time that individual houses can be designed to take maximum advantage of the natural conditions of summer, winter, day, and night. We must now produce designs which minimize total energy use (or loss of available work), averaged over the entire year, for a complex of dwellings.

Empirical data are needed on virtually all components of the energy consumption system. The degree-day is our best predictor of energy consumption for heating (Fox, 1973a); consequently, a Btu per degree-day heating index, measured over a large population, would identify performance characteristics. Development of a similar parameter for air-conditioner use would be valuable. This may require investigation of multiple dimensions, e.g., air temperature, humidity, and radiation temperature. Investigations of comfort indices based on such parameters and of trade-offs between the various parameters would be of interest for both winter and summer space conditioning. For example, radiant heating may provide a favorable trade-off under certain conditions.

The use of infrared technology, from survey meters to sophisticated systems, locates sources of heat loss. An easy-to-use local (of order 1-cm^2) heat-flux meter would also be valuable for this purpose.

Studies of the aerodynamics of houses, using scale models in wind tunnels and using houses in the field, should be undertaken. Associated studies of heat transfer at a more "micro" level are also needed, in particular, studies of the rate of heat transfer from the inside air to walls and windows, described by surface film coefficients. Detailed properties of the surfaces and of internal air movements determine these coefficients. It is possible that a detailed experimental and theoretical examination of air flows and heat exchange across typical air-wall configurations would give fruitful insights into heat losses. A local air velocity instrument would be useful here.

Heat-driven air conditioning is in fairly extensive use, most of it based on absorption cycles. Successful research to improve the efficiency of such devices, in particular to permit the efficient use of lower-temperature heat, would widen their applications, and permit better systems integration. Among newer ideas is the adsorption air conditioner, such as the Munters Environmental Control System (MEC) discussed in Section C, Part 1. Since the key to the adsorption cycle is the efficient drying of air, the development of effective desiccants and adsorption techniques with low regeneration temperatures appears to be a fruitful area of research. Other air conditioning systems which, like MEC, use room air as the working fluid, and thus eliminate the heat exchanger between the device and room air, need to be explored.

Mathematical models of heat flow processes are a valuable tool for understanding the dynamics of energy consumption. The National Bureau of Standard's computer simulation of the heating of a building, NBSLD, was run at our Summer Study. It needs development and calibration to be more useful for investigating the heating of a residential structure. It would be helpful if the program could be modified to accept a wider range of formats of weather data. Another modification, the inclusion of air infiltration as carefully as solar heat is presently treated, awaits improved modeling of the dependence of air infiltration on external weather conditions.

Hot-water heating merits a dynamic model. The 50- to 75-percent first-law efficiency quoted above does not include the effects of pipe losses and varying demand schedules. Such a model would be useful to evaluate the potential of point-of-consumption heaters.

We make two recommendations with respect to combined heating and electric systems:

(1) Systems studies should be carried out to determine optimum characteristics for combined electric and heating systems and to determine which improvements in device characteristics would most influence the efficiency and viability of such systems. Technical parameters and considerations which require study include: fraction of on-site demand for electricity to be met by the combined on-site units, size of units, role of heat pumps and heat-driven air conditioners, possibilities for energy storage, distance of transport of heat, interconnection with central power stations and control systems at the interconnections, and reliability and maintenance. These considerations must be related to spatial and temporal distribution of demand and economic and institutional constraints.

(2) The fuel cell as the fuel-consuming unit of a combined system should be studied. In such an application, maximum efficiency for electricity generation is unnecessary. This permits different optimization for reliability and cost than when fuel cells are designed for central power stations. (A similar remark applies to the specifications for diesel engines intended for combined systems. Tractor engines are not optimal.) The fuel cell characteristics under heavy load, with decreased electricity to heat ratio, may turn out to be favorable for combined-system operation.

4. THE AUTOMOBILE

Fuel consumption for transportation amounted to almost 25 percent of the total U. S. energy consumption in 1970 (Mutch, 1973). More than half of this energy, 66 billion gallons of gasoline with an energy content of 9 quadrillion Btu (9.4×10^{18} J), was directly consumed by automobiles. This accounted for 13.3 percent of the 1970 U. S. energy budget of 67.4 quadrillion Btu (71×10^{18} J), and 30 percent of the U. S. petroleum consumption (Hirst, 1974). Adding the direct fuel cost for trucks and buses—3.3 quadrillion Btu (3.5×10^{18} J)—and the 1970 military vehicle consumption of about 0.3 quadrillion Btu (0.3×10^{18} J) (Mutch, 1973), the rubber-tired, internal combustion engine-powered highway vehicle has a direct fuel consumption which accounts for about 18.5 percent of the total U. S. energy consumption and 40 percent of its petroleum. The opportunities for energy savings in this sector by technical means alone have been estimated at 25 to 30 percent (TEP, 1972), a figure which we regard as conceptually conservative. When this is combined with possible new transportation strategies to reduce U. S. dependence on the automobile as the dominant mode of personal transportation (EPA, 1973), an ultimate reduction in fuel consumption by road vehicles down to less than half of projections of present use appears to be quite possible (Lees and Lo, 1973).

A. Power train analysis

The internal combustion (IC) engine

Almost all of the automobiles, trucks, motorcycles, and buses in use today are powered by either Otto-cycle or diesel-cycle internal-combustion engines. As shown in Table 4.1, merely shifting from the present engine mix to improved diesel engines and ultimately to advanced external-combustion engines could in itself reduce fuel consumption by over a third (TEP, 1972; NAS, 1973).

We begin our examination of where the engine losses occur when fuel energy is converted into transportation with the Otto-cycle engine, the single largest energy user in the transportation sector.

The theoretical Otto cycle is an air cycle, in which the working fluid is treated as an ideal gas and combustion is replaced with a heat transfer to the engine from an infinite reservoir at the flame temperature. Under these conditions, the maximum

TABLE 4.1. Outlook for improved engine performance.[a] [E = excellent, G = Good, F = Fair, P = Poor.]

Type of engine	Fuel economy[b] (mpg)	Weight[c] (lb/bhp)	Emissions	Strategic materials	Novel fuel compatibility	Projected cost	Earliest availability[d]
1969 Otto	12[e]	3	P	E	F	E	Pre-1969
1976 Otto	10[f]	4	F	P	P	G	1976
Stratified charge	13	4	F	G	F	G	1980
Light weight diesel	18	4–5	F	G	F	G	1982
Advanced gas turbine	16[g]	3–4	G	F	G	G	1984+
Advanced Rankine	18+[h]	4–6	E	G	E	G	1984+
Stirling	18+	5–7	E	F	E	F	1984+

[a] Source: TEP (1972).
[b] Based upon a 4300-lb automobile driven over the Federal Emissions Urban Driving Cycle (see Federal Register, Vol. 35–219, November, 1970, Vol. II).
[c] Specific weight of the prime mover excluding transmission.
[d] Date when 10 percent of annual production could be achieved.
[e] Based upon EPA test data on unmodified 1969/1970 vehicles in this weight class.
[f] Based on a limited amount of unpublished EPA data from a 1971 Oldsmobile with 1976 emission control systems (8 percent exhaust gas recycling plus dual-bed catalyst).
[g] This is somewhat more optimistic than the estimates of NAS (1973).
[h] This is considerably more optimistic than the estimates of NAS (1973).

engine efficiency[78] is

$$\eta_{\text{air}} = 1 - \left(\frac{1}{r}\right)^{\gamma - 1}, \tag{4.1}$$

where r is the compression ratio and γ is the adiabatic compressibility C_p/C_v (1.4 for air at room temperature). The ideal air-cycle diesel has a more complex formula giving slightly lower efficiencies at all compression ratios.

Further decreases in efficiency are brought about by the increase of C_p and C_v with temperature and by the decrease in γ (the former lowers the maximum temperatures, and the latter decreases the work obtained in the power stroke). There are also chemical equilibrium losses (even in chemical equilibrium, combustion may be incomplete at ignition owing to the high temperature and pressure; therefore, some of the fuel energy is released late in the power stroke). Figure 4.1(a) shows the thermal efficiency η_t of an ideal fuel-air cycle Otto engine as a function of the ratio of fuel to air at various compression ratios, with "100-percent theoretical fuel" being a ratio of 0.067 kg of fuel per kg of air, the chemically correct (stoichiometric) ratio for complete burning. As the mixture is made lean (<100 percent) the temperature rise at ignition will be less [as shown in Figure 4.1(b)] and losses due to the factors discussed above will be smaller. As the mixture is

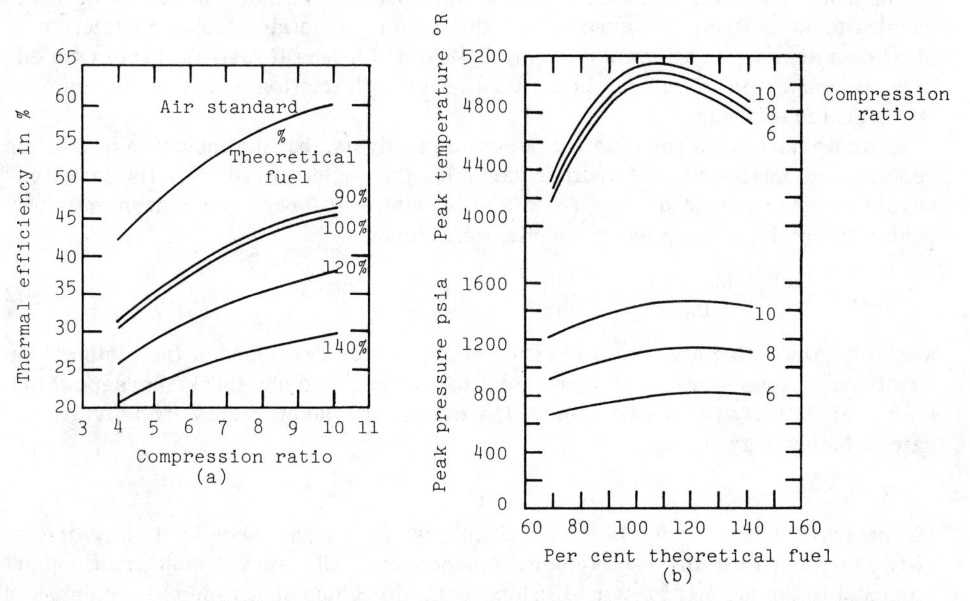

Figure 4.1. Effect of fuel-air ratio on thermal efficiency, peak cylinder pressure, and peak combustion temperature for an ideal fuel-air cycle Otto engine. [Source: Rogowski (1953).]

[78]We follow the convention in this field in defining efficiencies relative to the "low heating value" of the fuel (water not condensed). The fuel available work B is about 4 percent higher than $|\Delta H|_{\text{low}}$, and $\epsilon \cong 0.96\eta_t$ throughout.

made rich (>100 percent), there will be an additional decrease in efficiency owing to incomplete combustion of the fuel.

In theory, a positive-displacement piston engine draws in a fixed volume of air per piston stroke. With increasing fuel-air ratio, the quantity of fuel burned per stroke increases. Therefore, although the efficiency of converting fuel energy to mechanical work drops as the fuel-air ratio increases, more fuel energy is released per stroke. The net mechanical work per stroke peaks at a mixture slightly richer than stoichiometric. Engines have always tended to run most smoothly at slightly rich fuel-air ratios, owing to better fuel-air mixing and better flame propagation, but current engines are set to run somewhat lean at cruise to compensate for reduced efficiency at lower compression ratios and to minimize CO and hydrocarbon emissions. For purposes of analysis, we shall assume a stoichiometric mixture which, according to Figure 4.1(a), reduces the thermal efficiency to about 0.75 of air-standard even for an *ideal* fuel-air cycle Otto engine.

For a *real* fuel-air cycle engine, there are additional losses due to heating of the cooled cylinder walls and head (which quenches combustion somewhat and absorbs heat from the expanding gases), delayed burning, and exhaust blowdown owing to the necessity of opening the exhaust valve before all the energy has been extracted from the hot gases. These losses reduce the efficiency by roughly 20 percent (Taylor and Taylor, 1961).

The efficiency of an Otto engine with a compression ratio of 8 is quoted in most physics texts as 0.56. But even for an ideal, stoichiometric, fuel-air cycle, the maximum achievable efficiency is only 0.42, and for a real fuel-air cycle Otto engine, the maximum efficiency is $\eta_i = 0.33$, even with friction totally neglected. (η_i is "indicated efficiency.")

So far we have performed only a per-cycle analysis, but the inclusion of friction requires the introduction of another variable, the engine speed, usually quoted in revolutions per minute (rpm). The effect of frictional forces can conveniently be dealt with by introducing the mechanical efficiency

$$\eta_{\text{mech}} \equiv \frac{\text{bhp}}{\text{ihp}} = \frac{\text{ihp} - \text{fhp}}{\text{ihp}} = 1 - \frac{\text{fhp}}{\text{ihp}}, \tag{4.2}$$

where ihp (indicated horsepower) is the engine power developed at the piston,[79] fhp (frictional horsepower) is the power lost to friction, and bhp (brake horsepower) is the power delivered to the flywheel. The overall engine efficiency (bhp divided by rate of fuel energy input) is

$$\eta = \eta_i \times \eta_{\text{mech}}. \tag{4.3}$$

Mechanical friction, which is the dominant loss at low and medium rpm, is the sum of two terms having different velocity dependences. Classic Coulomb friction, proportional to the normal force and independent of rubbing speed, is characterized by a constant coefficient of friction which depends upon the nature of the rubbing surfaces. When the sliding surfaces are separated by a film of lubricant, however,

---

[79]It is equivalent to a mean pressure on the piston: ihp ~ pressure × displacement × N, where N = total number of power strokes/sec for the engine. It follows that ihp ~ rpm ~ intake air mass at constant pressure. η_i is defined as ihp/(fuel energy × fuel flow rate).

the frictional force is produced by viscous shear, the sliding of layers of oil molecules over each other. In this case, the frictional force is proportional to the relative velocity of the two surfaces and inversely proportional to the film thickness. There is no term proportional to the normal force. Even in a well-lubricated engine, however, there is some Coulomb friction at the piston, especially near the top where temperatures are high. The work done to pump air into the engine and exhaust out is also generally lumped with mechanical friction, since it is comparable to other contributions and increases with increasing engine speed. In general, then, frictional forces in an engine increase roughly in proportion to engine speed, except at low rpm, where they tend toward a constant value.

The brake horsepower (bhp) is always determined by the load on the engine, while the indicated horsepower (ihp) at full load is determined by the intake air mass per unit time (at constant fuel-air ratio). Using Eq. 4.2, the net result of the shape of the friction curve is to cause η_{mech} to decrease at low rpm, where the ratio of friction to engine output increases because of the relatively constant low-speed friction. At high engine speeds, "frictional" forces associated with high velocity airflow through restrictions and inertial effects where rapid reciprocating flow exists rise more rapidly with rpm than power does, and η_{mech} decreases again as rpm increases. The mechanical efficiency thus has a definite maximum at some intermediate value of engine rpm. Finally, at very high engine speeds, the amount of air which can flow through the intake plumbing approaches a constant value. Then ihp, which is directly proportional to the air mass, becomes nearly constant while fhp continues to increase with increasing rpm. The high rpm limit on engine power is generally set by this factor.

At the most favorable rpm, these losses combine to decrease the engine efficiency by a factor $\eta_{mech} = 0.8$ (for a typical engine). The efficiency of converting fuel energy to useful mechanical work has now dropped to

$$\eta \cong 0.33 \times 0.8 \cong 0.26 \tag{4.4}$$

for an 8:1 compression-ratio Otto-cycle engine under optimal conditions, and with no accessories mounted. This corresponds to a specific fuel consumption (sfc) of 0.54 lb per brake horsepower-hour (0.33 kg/kW hr). Figure 4.2 illustrates these points. For an ideal, frictionless Otto engine, both torque and specific fuel consumption would be constant; horsepower would be directly proportional to engine rpm.

Such curves, however, only represent the engine efficiency at full load. To operate the engine at a desired speed under partial load, the power must be reduced at that speed. In a conventional engine, this is done by throttling the intake. This decreases the amount of air taken in on each stroke below the maximum, decreasing the ihp. The friction in the engine, however, depends only on rpm, and will remain constant, while the "frictional" loss due to pumping air through the restricted inlet increases. The specific fuel consumption will rise at part load because the ratio of ihp to fhp drops, and the mechanical efficiency decreases (Eq. 4.2). The effects of these several factors on engine performance and efficiency can be combined to give a generalized "map" of engine performance as shown in Figure 4.3. The engine torque, which is a measure of force and therefore of the load on the engine (or the accelerating ability under lesser loads), is measured in terms of brake mean equivalent pressure (bmep), the equivalent force per unit area on the pistons during the power stroke. Efficiency is expressed as brake specific fuel consumption (bsfc), in lb of fuel per bhp-hr, while engine power is measured in terms of bhp per unit piston area. Engine speed is expressed in terms of piston speed to give a generalized picture applicable to most engines of similar design.

Evidently, the product of indicated and mechanical efficiency is highest in the region of relatively low engine speed and relatively high load (bmep). Moving from the region of maximum efficiency along a line of constant engine speed, the mechanical efficiency decreases rapidly as the load is reduced, since friction remains nearly constant while fuel energy supplied decreases; moving towards higher loads, the ratio of fuel to air increases, which decreases the indicated efficiency. Moving from the region of maximum efficiency along a line of constant bmep, efficiency decreases at higher engine speeds because of increased engine friction; the decrease towards lower engine speeds is due largely to increased relative heat loss as the time per stroke increases.

Figure 4.3 also contains an estimate of typical level-road power demand in high gear. The allocation of fuel energy and net efficiency for steady cruising is shown schematically in Figure 4.4. The choice of gearing which allows high acceleration, very high speed, and good high-gear hill-climbing ability in most current domestic passenger automobiles also prohibits their operation at anywhere near the region of best fuel economy at constant speed on a level road.

The brake horsepower as described thus far is, as is conventional domestic prac-

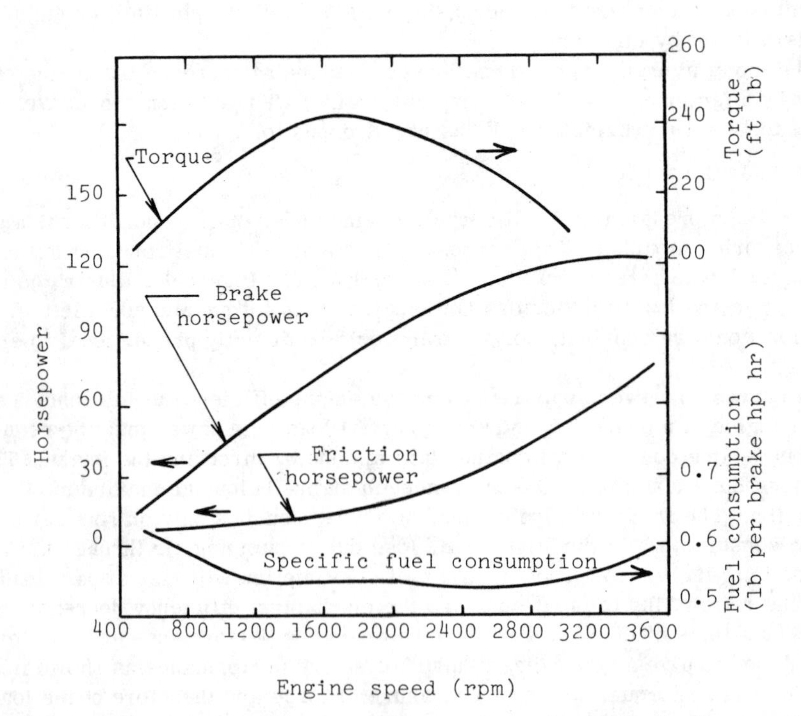

Figure 4.2. Performance of a typical Otto-cycle engine under full load. [Source: Gill, Smith and Ziurys (1954).]

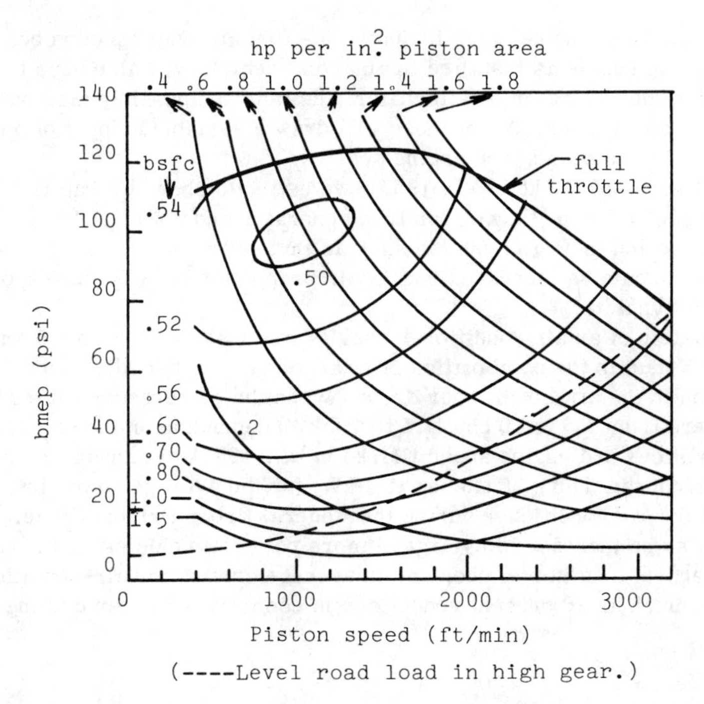

Figure 4.3. Performance map of a typical domestic passenger car engine. [Source: Taylor and Taylor (1961).]

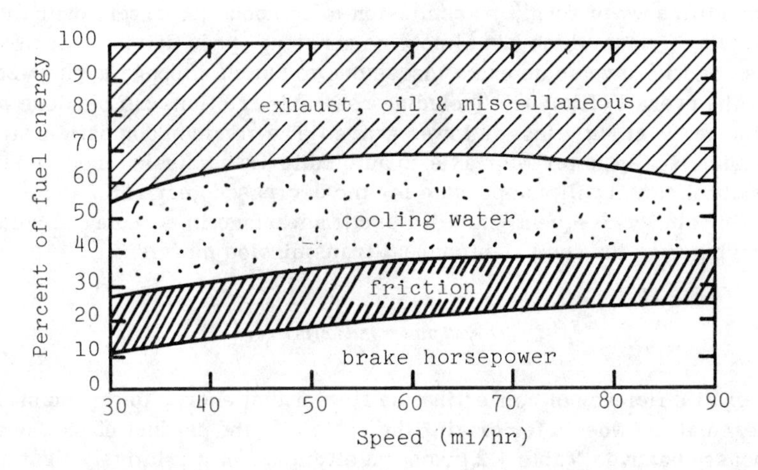

Figure 4.4. Energy balance and efficiency of a typical domestic automobile engine at cruising speed. [Source: Burke *et al*. (1957).]

tice, that of the bare engine only, including the fuel and water pumps but no other accessories. An engine as installed in an actual vehicle will also have to supply power to drive the generator or alternator, the fan, and possibly also power steering and an air conditioner. All of these will draw a variable amount of power depending on design, demand, and engine speed.

Using the figures of Burke *et al.* (1957), we estimate about 1.5 bhp (1.1 kW) for the fan and about 1.5 bhp (1.1 kW) for both generator and power steering for a relatively small and low-powered vehicle such as that described by Kummer (1974). These figures should be increased roughly in proportion to brake horsepower for a more powerful vehicle.

The installation of an air conditioner results in a loss of between 0.5 and 1.5 bhp (0.4 to 1.1 kW) due to the combination of a larger fan, higher idle speed, and additional friction in the air-conditioner drive. When the compressor cycles on, it will draw anywhere from 1.5 to 10 bhp (1.1 to 7.5 kW) depending on the unit size and the ambient conditions and engine speed (Burke *et al.*, 1957). Following Kummer (1974), we adopt the figure of 1.5 bhp (1.1 kW) for the average power loss due to operation of the air conditioner during the Federal Urban Driving Cycle. This figure should also be increased for larger, more powerful vehicles. EPA (1972) estimates a 9 percent loss in fuel economy in average driving for air-conditioned vehicles, and as much as 20 percent reductions in economy for urban driving on hot days.

Transmission losses

Most domestic cars are currently fitted with a fluid-coupled automatic transmission. The constant-speed efficiency of such a transmission (including the rear-axle friction losses) is about 90 percent in top gear, less in lower gears. During acceleration, however, the efficiency drops well below this figure because of slippage and power loss to the fluid during torque multiplication. Kummer (1974) estimates the efficiency of such a transmission to be about 75 percent over the Federal Urban Driving Cycle, which has almost no constant-speed driving. Even a manual transmission will have some extra losses due to clutch slippage during acceleration, but the overall efficiency (including rear axle friction) can be close to 95 percent in top gear, slightly less in lower gears. The common domestic practice of fitting a numerically lower rear axle to automatic-transmission models (since the torque multiplication will compensate for the decreased gear ratio and maintain the accelerating ability) frequently gives a small advantage in economy in steady high-speed driving over the equivalent manual transmission models.

Overall power train efficiency

The overall efficiency of converting the flow of fuel energy to the engine into power delivered at the wheels for driving the vehicle is the product of the factors we have discussed above. Table 4.2 gives the efficiency of a relatively light and low-powered domestic compact over the Federal Urban Driving Cycle. The second-law efficiency ϵ_t is only 0.08.

TABLE 4.2. Energy flow for a 2500-lb automobile over the Federal Urban Driving Cycle[a].

Fuel available work	100%
Fuel energy (commonly used lower heat of combustion) (× 0.96)[b]	96%
Ideal gas air-cycle Otto engine, $r = 8$ (× 0.56)	54%
Fuel-air Otto-cycle, stoichiometric (× 0.75)	40%
Burning and cylinder wall losses (× 0.8)	32%
Frictional losses (× 0.8)	26%
Partial load factor (estimated) (× 0.75)	19%
Estimated loss due to accessories (including air conditioner)[c]	11%
Automatic transmission (estimated) (× 0.75)	8%
Net output to rear wheels (efficiency)	8%

[a]After Kummer (1974).
[b]See Table 2.6.
[c]Not necessarily expressible as a multiplier.

B. Using the energy

The energy delivered to the rear wheels is used to accelerate the vehicle, to climb grades, and to overcome resistance to motion. This energy must, however, ultimately be dissipated as heat to the road and to the air. The three primary loss mechanisms are the brakes, air resistance, and the rolling resistance of the tires.

Brakes

Kummer (1974) estimates that, of the work done at the rear wheels during level-road driving, both for acceleration and for overcoming resistance to motion, 32 percent is dissipated in the brakes in urban driving, 10 percent in suburban, and 2 percent in highway driving.

The energy that goes into the brakes is, under any set of conditions, a combination of kinetic and gravitational potential energy, and is directly proportional to the vehicle mass. This energy need not be dissipated; in principle, it could be recovered and stored.

Air resistance

An automobile is a rather complex aerodynamic system whose drag is due partly to laminar flow, partly to turbulent, and partly to the air layer between the car and the ground. The classical air-drag formula is a good approximation valid for most normal driving conditions in the absence of wind despite the rather high Reynolds number (~10^7) at most highway speeds:

$$F_d = \frac{C_d A_f v^2}{370}, \qquad (4.5a)$$

where F_d is the drag force in lb, A_f is the projected frontal area in ft^2, and v is the speed in mph. [The factor 370 compensates for the mixed units. In SI units, the formula reads

$$F_d = 0.65 C_d A_f v^2, \quad \text{(SI)} \tag{4.5b}$$

with force in N, area in m^2, and speed in m/sec.] Some typical values for the dimensionless drag coefficient C_d are shown in Table 4.3. A C_d of about 0.3–0.35 is probably near the minimum for a practical automobile, although even lower values are possible in principle.

Rolling resistance

The rolling resistance is due almost entirely to the tires. About 5–10 percent arises from tread slippage on the road, 1–3 percent from air friction with the rotating tire, and the remainder from internal losses in the tire body (Clark, 1971).

Rubber is, strictly speaking, not truly elastic but viscoelastic. The tire deforms as it rolls, and a large fraction (~20 percent) of its deformation energy is converted to heat by internal hysteresis. The two prominent models for rubber are those of Maxwell (a spring in series with a viscous dashpot) and Kelvin (a spring in parallel with a viscous dashpot). Real tires, including the cords, generally behave as some combination of Maxwell and Kelvin elements. In such a model, the losses are partially static, partially velocity-dependent. Below 40–60 mph (18–27 m/sec) the hysteresis losses are linear functions of speed, but above some critical speed nonlinear hysteresis and wavelike deformations of the tire cause additional losses.

An approximate equation for new automobile tires, constructed from many sources (see Clark, 1974), is

$$\text{(British)} \quad F_r = \frac{M}{1000}\left[12 + 0.05v + 0.004(v - 40)^2{}_{v \geq 40}\right] \tag{4.6a}$$

[F_r in lbf, M in lbm, v in mph];

TABLE 4.3. Aerodynamic drag coefficients.

Vehicle	C_d	(Ref.)	Other Shapes	C_d	(Ref.)
Ordinary truck	0.7	(a)	Teardrop (best)	0.03	(d)
Streamlined truck	0.55	(a)	Square flat plate (worst)	1.3	(d)
Jaguar XK-E	0.40	(b)	"Typical" auto shape	0.55	(d)
Porsche 904	0.35	(b)			
1972 Camaro	0.5	(c)			
1972 Dodge Polara Wagon	0.6	(c)			

[a] Smith (1970).
[b] Campbell (1970).
[c] White and Korst (1972).
[d] TEP (1972).

$$(\text{SI})\quad F_r = \frac{M}{1000}\left[118 + 1.1v + 0.2(v-20)^2{}_{v\geq 20}\right] \tag{4.6b}$$

$[F_r$ in N, M in kg, v in m/sec$]$.

As the tire wears down, these coefficients decrease. For a heavy truck with its large, stiff tires and high inflation pressures, the relative deformation of the tire is smaller. A good formula (Smith, 1970) is

$$(\text{British})\quad F_r = \frac{M}{1000}(6.4 + 0.074v) \tag{4.7a}$$

$[F_r$ in lbf, M in lbm, v in mph$]$;

$$(\text{SI})\quad F_r = \frac{M}{1000}(63 + 1.6v) \tag{4.7b}$$

$[F_r$ in N, M in kg, v in m/sec$]$,

about one-half the value for the softer automobile tire at low speeds.

Some composite drag calculations

(A) 2500-lb (1130-kg) compact (after Kummer, 1974): Estimated $C_d \approx 0.5$. Estimated $A_f \approx 22$ ft² (2.0 m²).

Velocity	Rolling Resistance F_r		Air Resistance F_d		Constant-Speed Power $P = v(F_r + F_d)$	
mph (m/sec)	lb	(N)	lb	(N)	hp	(kW)
20 (8.9)	32.5	(145)	12	(53)	2.4	(1.8)
40 (18)	35	(155)	48	(210)	8.9	(6.5)
60 (27)	41.5	(185)	107	(475)	23.8	(17.6)
80 (36)	56	(250)	190	(850)	52.5	(39.3)

Air drag and rolling resistance are equal at 34 mph (15 m/sec).

(B) 5000-lb (2270-kg) "luxury" car: Estimated $C_d \approx 0.52$. $A_f \approx 30$ ft² (2.8 m²).

Velocity	Rolling Resistance F_r		Air Resistance F_d		Constant-Speed Power $P = v(F_r + F_d)$	
mph (m/sec)	lb	(N)	lb	(N)	hp	(kW)
20 (8.9)	65	(290)	17	(75)	4.4	(3.2)
40 (18)	70	(310)	67	(300)	14.6	(10.8)
60 (27)	93	(370)	150	(670)	38.9	(27.8)
80 (36)	112	(500)	270	(1200)	81.5	(60.5)

Air drag and rolling resistance are equal at 41 mph (18 m/sec).

(C) For a 76,000-lb heavy truck with $C_d \approx 0.7$ and $A_f \approx 96$ ft^2 (8.9 m^2), $F_d = F_r$ at 65 mph. The constant-speed power requirement is 170 hp (125 kW) at 60 mph (27 m/sec).

It is difficult to estimate accurately what fraction of the energy used during the Federal Urban Driving Cycle is lost to the rolling resistance. Very little of the energy is used to overcome air resistance during this cycle. Not all the kinetic energy, however, is dissipated in the brakes, as engine drag is frequently used to slow the vehicle down, so that the $\frac{1}{3}$ of the energy estimated to go into the brakes does not represent the total kinetic energy acquired during the cycle. We estimate that somewhere between $\frac{1}{3}$ and $\frac{1}{2}$ of the total energy delivered to the rear wheels is used to overcome rolling resistance in urban driving.

Performance: Acceleration, speed, and hill-climbing

Knowledge of the combined air and rolling resistances is not sufficient for specifying the required engine power, for the necessity to accelerate and climb grades adequately must also be taken into account. In most cases, the minimum engine power is determined by requirements in top gear at highway speeds, and these performance specifications are not independent of one another.

On a grade, the effect of gravity can be treated as another "drag" force,

$$F_g = Mg \sin\theta \cong Mg\theta, \tag{4.8}$$

where Mg is the weight of the vehicle and θ is the angle of the grade. (Setting $\sin\theta \cong \theta$ will usually be an excellent approximation.)

The situation for acceleration is slightly more complicated because of the rotational inertia of the wheels and tires, and of the engine itself. The acceleration of the vehicle will be given by

$$a = (F_T - F_D)/M' \equiv F_a/M', \tag{4.9}$$

where F_T is the total force applied by the engine, F_D is the drag force (including the gravitational term), and M' is the effective mass, which also includes, in addition to the mass of the vehicle, the rotational inertia of the engine and the wheels expressed as a mass. The effect of the wheels and tires is small in an automobile—usually amounting to only 2 or 3 percent of the vehicle mass. The major factor is the rotational inertia of the engine, whose effect is multiplied by the gear ratio of the transmission and rear axle. The effective mass of the engine will be proportional to (rpm/mph)^2, which is greatest in the lowest gear. For a manual-transmission vehicle, the effective mass in low gear can be 1.5 times the vehicle mass (Smith, 1970). With an automatic transmission, where the multiplication of torque requires that the ratio of rpm to mph be very large, the effective mass can be considerably larger, especially when accelerating from a full stop (Burke *et al.*, 1957). In top (high) gear, however, the effective mass will be within 5 or 10 percent of the vehicle mass, and we may approximate the force needed for acceleration in top gear by the elementary formula $F_a = Ma$, where F_a is the net force.

To illustrate how these factors are connected, let us take as an example the 2500-lb (1130-kg) vehicle whose drag forces we analyzed in the preceding subsection. At 60 mph (27 m/sec), the total air and rolling resistance drag is 150 lbf (665 N). If we add the requirement that this car be able to maintain the same speed on a 10-percent grade, an additional 250 lbf (1100 N) will be required. Taking into account an extra 15-percent transmission loss at full load, we find that the engine must be capable of delivering 460 lbf (2050 N) at an engine speed corresponding to 60 mph (27 m/sec) in top gear. On a level road, this reserve power capacity is sufficient to accelerate from 55 to 65 mph (24.5 to 29 m/sec) in only 4.5 sec. These performance standards are more than adequate.

Translating these figures back into power requirements, 24 bhp (18 kW) is required to overcome air and rolling resistance and move the vehicle at a constant 60 mph (27 m/sec) on a level road, and another 46 bhp (35 kW) must be available on demand at this speed. Our hypothetical vehicle must then have an engine capable of delivering 70 bhp (53 kW) at the rear wheels.

The performance in lower gears will follow from the top-gear performance and the choice of gear ratios selected, as well as the torque conversion ratio of the automatic transmission. Taking as some arbitrary, but not unreasonable, figure a torque multiplication factor of 3:1 at 20 mph (9 m/sec) in low gear, our hypothetical vehicle will be able to climb a 50-percent grade (27-deg angle) at this speed, or accelerate at a rate of 8 mph/sec (3.5 m/sec^2) on a level road (for an effective mass ratio of 5:4).

The price in economy which must be paid for specifying this level of performance in an Otto-engine powered vehicle can be deduced from Figure 4.3. Assuming that 60 mph (27 m/sec) corresponds to a piston speed of 2000 ft/min, the level-road, constant-speed load on the engine is only about 30 percent of the maximum available bmep. The specific fuel consumption under these conditions is about 1.4 times the minimum value. At a constant 30 mph (13.5 m/sec) in top gear, the specific fuel consumption will be about twice the minimum value.

The relatively high fuel consumption of high-powered vehicles[80] is seen to be a consequence of the demand for a large amount of reserve power for performance; during most ordinary driving, the engine is operating at very light loads where efficiency is low. In terms of fuel consumption, this power is expensive even when it is not being used. If the engine size were decreased to provide only 400 lbf (1780 N) at 60 mph (27 m/sec), this same 2500 lb (1130 kg) automobile would be able to climb an 8-percent grade at 60 mph (27 m/sec) in top gear, and accelerate from 55 to 65 mph in 5.5 seconds. The 0–60 mph (0–27 m/sec) elapsed time would increase to 16–18 sec, and the top speed would drop from over 100 mph (45 m/sec) to about 90 mph (40 m/sec). This comparatively modest decrease in performance would increase fuel economy by 10–15 percent at a constant 60 mph (27 m/sec), and by 15–20 percent at a constant 30 mph (13.5 m/sec).

Idling

An analysis of the performance of 1973 model year automobiles over the Federal Urban Driving Cycle (DOT, 1974) indicates that, on the average, about 7.5 percent of the total fuel consumption is due to engine idling at stops or during deceleration, when no work is being performed.

---

[80]The model vehicle being considered here can accelerate from 0 to 60 mph in 13–14 sec, which is good but not startling.

Efficiency of current automobiles

A least-squares fit to the EPA data for the fuel consumption of 1971–1973 model year automobiles over the Federal Urban Driving Cycle (EPA, 1972) yields an almost linear dependence of average fuel consumption within each weight class upon the average weight in that class, with a proportionality constant of about 0.28 lb of fuel per ton-mile (0.087 kg/tonne-km).[81] This corresponds to an overall second-law efficiency of about 0.09. Table 4.4 shows how the fuel available work is used, both over the urban cycle and at a constant 60 mph (27 m/sec), for a 3600-lb (1640-kg) domestic automobile.

The observed linear fit for fuel consumption in urban driving upon vehicle mass is not, in retrospect, surprising. Over the urban cycle, very little time is spent at speeds high enough for air drag to have much effect, and both the remaining loss factors and the requirements for reserve power scale directly with the mass. The primary inference of this fit is that, on the average, most domestic vehicles are designed for roughly the same performance. These same data also show that a number of domestic vehicles have a fuel consumption of only $\frac{3}{4}$ of the average in their weight class. Therefore, by simply reducing the average vehicle weight from 3600 lb to 3000 lb and simultaneously improving all vehicles to the level of the current domestic best practice, the average fuel consumption could ultimately be re-

TABLE 4.4. Consumption of available work and calculation of second-law efficiency for an "average" 3600-lb (1640-kg) domestic automobile in two driving patterns (see also Table 4.2). 100×10^6 J (3.0 l) of gasoline will power this vehicle for about 1900 seconds of the Federal Urban Driving Cycle for an average fuel consumption of 13 mpg (5.5 km/l), or for about 700 seconds at a constant 60 mph (27 m/sec) on a level road for an average fuel consumption of 16.5 mpg (7 km/l). [*Notes:* 1. First-law efficiencies η referred to the lower heating value differ little from the second-law efficiencies ϵ calculated here ($\eta \cong 1.04\epsilon$). 2. The effect of accessories is estimated roughly. 3. FUDC time and mpg are calculated using an additional 7.5 percent fuel for idling, which reduces ϵ from 0.093 to 0.086.]

	Energy paths		Net efficiency at successive stages	
Fuel available work	100 MJ		1.0	
Commonly used lower heating value	×0.96		0.96	
Practical fuel-air Otto cycle	×0.33		0.32	
Frictional losses	×0.8		0.26	
	Federal urban driving cycle	Constant 60 mph	FUDC	60 mph
Part-load losses	×0.75	×0.7	0.19	0.18
Accessories	−3.5 kW	−4 kW	0.12	0.15
Automatic transmission losses	×0.75	×0.9	0.09	0.14
Work available at rear wheels	9 MJ	14 MJ	$\epsilon = 0.09$	$\epsilon = 0.14$

[81]We are indebted to M. Fels for her contribution to this calculation.

duced (when current vehicles are totally replaced on the road) to $\frac{2}{3}$ of current averages, saving about 4.5 percent of the total U. S. energy budget and about 10 percent of its petroleum consumption.

C. Increasing power-train efficiency[82]

Emulsified and "novel" fuels

Reduced combustion temperatures and better combustion could allow the Otto engine to run at reduced HC and NO_x emission levels. The use of emulsions of fuel with water or methanol (for instance at a 5:1 ratio of gasoline to emulsifier) might possibly accomplish this, while simultaneously decreasing the fuel-air ratio at which reliable performance is obtainable. The decreased fuel-air ratio would also increase engine efficiency above currently obtainable values. Further research seems worthwhile, both on the use of emulsified fuels and on the possibility of low cost retrofit to older vehicles.

TEP (1972) and NAS (1973) suggest research into the use of other fuels which may become available at lower cost and in greater quantities than petroleum by the end of the century. Ethanol, which requires only 1.4 times the volume of gasoline for an equivalent amount of fuel available work, and methanol, which is slightly less favorable at 1.8 times the volume but may be available in large quantities from garbage conversion, appear to be the most promising candidates at the moment.

Stratified-charge engine

Better combustion in the stratified-charge engine allows it to be run leaner without increased emissions. TEP (1972) predicts 25 to 30 percent savings over comparable future Otto-cycle engines (whose efficiency is expected to be lowered by new emission standards) and 5 to 10 percent over current engines (see Table 4.1).

Shift to diesel

A well-designed diesel engine has a net efficiency at full load which is 20 to 30 percent greater than that of a comparable Otto-cycle engine (NAS, 1973). The greater compression ratio and better fuel-burning efficiency of the diesel more than compensate for the slightly lower air-standard efficiency. At partial loads, the efficiency of the diesel decreases far less than that of the Otto, because the inlet air is not throttled and adding fuel throughout the power stroke reduces heat losses. NATO (1973) reports a 50-percent fuel saving upon substitution of diesel engines in London taxis. TEP (1972) estimates future diesel efficiencies to be 1.5 times present Otto efficiencies and 1.8 times projected 1976 Otto efficiencies (Table 4.1). EPA (1972) reported a 3500-lb diesel-powered vehicle to have a fuel economy 1.75 times the 1972 average for its weight class and 1.4 times bet-

---

[82]In view of the many recent surveys dealing with improved or alternate engines (TEP, 1972; U.S. Senate, 1973; NAS, 1973; Harvey and Menchen, 1974), we treat these only briefly here.

ter than the *best* Otto-cycle vehicle of its weight (some of this economy is due to the low reserve power of this particular vehicle).

Research opportunities include less expensive, less tolerance-critical fuel injection systems; the use of novel fuels (such as emulsions); improved materials for reducing engine weight (currently $\frac{5}{3}$ to $\frac{5}{4}$ that of Otto engines of the same power); further reduction of emissions and noise; and easier starting when cold.

Diesels are considerably more fuel-adaptable than Otto engines, and no fuel supply problems are anticipated other than those due to general petroleum shortages.

Advanced engine cycles[83]

Projected performance standards for the gas turbine and for Rankine- and Stirling-cycle external combustion engines are summarized in Table 4.1. The Brayton-cycle gas turbine, although still under development, does not appear to hold much promise for automobile use at the moment, although the development of appropriate ceramics could change this view (NAS, 1973). The rotary Wankel engine, once considered quite promising because of its high power-to-weight ratio, has suffered some setbacks, first because of poor fuel economy, and then because of difficulties with emission control, but this is still a relatively new technology, and developmental work should be continued to determine its ultimate potential.

External combustion engines, such as the Rankine and Stirling engines, should be thoroughly explored. The advantages of compatibility with a wide range of fuels and readily controllable combustion conditions make them the most promising candidates for technical development in the long run. The Rankine engine (steam cycle) is one with a long history of transportation, and several pilot models exist (U.S. Senate, 1973). Pilot models of the Stirling engine are reported to have an "engine efficiency" (presumably the ratio of brake hp to heat of combustion flow) of 0.4 at full power, dropping only to 0.34 at $\frac{1}{10}$ load. The efficiencies of pilot models, of course, may be considerably greater than that of production ones, but the ability of external-combustion engines to maintain high efficiency and high torque at fractional loads would make them competitive with internal-combustion engines even if their full load efficiency were no higher.

Electric propulsion

The problems and promise of electric vehicle propulsion, whether through rechargeable batteries, fuel cells, or hybrid engine-generator combinations, have also been thoroughly examined recently (TEP, 1972; NAS, 1973; U.S. Senate, 1973) and we have not re-examined them. These reports, as well as NATO (1973), all recommend increased research and development on battery-electric and diesel-electric hybrid systems. One interesting system (mentioned in U.S. Senate, 1973) uses a small, light, internal-combustion engine (such as a Wankel)

---

[83]For a comprehensive discussion of advanced engines, see NAS (1973) and Harvey and Menchen (1974). For extensive discussions of current technology and existing pilot models, see U.S. Senate (1973).

running continuously at very high load factors where the specific fuel consumption is lowest. This is, in turn, connected to a dynamotor (a dc generator which can be reconnected as a motor). The engine is chosen to be just barely adequate to power the vehicle at highway speeds. Under light demand conditions, the excess power is used to charge batteries, and when extra power is needed the batteries are reconnected to the dynamotor to "boost" the engine. Such systems also appear to warrant further investigation.

Recovery and storage of kinetic energy

With present materials, mechanical energy storage systems such as flywheels appear to have only restricted applications as prime movers for vehicles. Post and Post (1973) have suggested the introduction of stronger materials such as high-strength polymers (PRD-49) or fused silica. Using their figures for the energy storage in a 500 to 600-lb package which could replace present engines, we calculate that a fused silica flywheel could power our model 3600-lb vehicle (Table 4.4) for about 6 hr of the urban driving cycle or about 1.5 hr at 60 mph; a PRD-49 flywheel, for about 2.5 hr of the urban cycle or about 0.5 hr at 60 mph; and a high-strength maraging steel flywheel (current technology), for about 20 min of the urban cycle or 5 min at 60 mph.

The figures for the urban cycle would be improved considerably—by as much as a factor of 2—if the vehicle kinetic energy could be recovered and used to repower the flywheel. The use of generators as vehicle brakes has, in fact, some potential. If rolling resistance and air drag could be markedly reduced, as suggested further on in this section, and if the regenerators were used for all decreases in kinetic energy, including that now dissipated by using the engine as a brake, the percentage of energy available for regeneration would increase considerably in urban driving, making the prospect considerably more attractive, whether in concert with flywheels or all-electric propulsion systems.

Even without an adequate system for storing primary drive energy by electrical or mechanical devices, the use of regenerated electricity to replace the present fan, generator, power steering, and air conditioner loads on the engine by means of a small battery or other storage device could have an appreciable effect on fuel economy, provided that the engine size were decreased to take advantage of the reduced engine load.

Better matching of engine to load

Kummer (1974) estimates automatic transmission efficiencies to be only 75 percent over the Federal Urban Driving Cycle. The simple replacement of automatic transmissions with manual ones, however, will not necessarily have a substantial impact on fuel economy in urban driving. The torque-multiplying characteristics of the automatic transmission, and the slippage which allows the engine to operate at relatively high speeds during acceleration, result in operation closer to the regime of higher engine efficiency.

As is shown in Figure 4.3, specific fuel consumption rises very steeply at low engine loads, especially at very low and very high rpm. For any engine there will be a similar "map" of engine efficiency as a function of load and engine speed. Furthermore, although the time rate of fuel consumption is simply the product of specific fuel consumption and brake horsepower output, fuel economy is usually expressed as the inverse of fuel consumption per unit distance travelled—the product of the time rate of fuel consumption and the velocity—which depends on the overall gear ratio between engine and wheels. The choice of gear ratio in turn affects the operating point on the engine map, so that the engine and drive train should be treated as a single system. An infinitely variable transmission with micro-computer control could, in principle, select at any moment the correct gear ratio and engine setting for maximum fuel economy for any combination of vehicle speed, load, and "accelerator pedal" setting. Even a mechanically programmed continuously variable transmission could give marked improvements in fuel economy over present designs (U.S. Senate, 1973), as could a transmission with more gear ratios to select from and better shift-pattern control. A simple re-optimization of present transmissions could increase fuel economy by 10 to 15 percent. A total redesign of present transmissions and reconsideration of power vs. economy tradeoffs could easily result in 20 to 30 percent increases in fuel economy in urban and suburban driving.

The use of smaller engines would also increase mechanical efficiency by increasing the load factor on the engine. A modest decrease in reserve power and performance expectation could increase fuel economy 10 percent. An even smaller engine with a demand "boost"—via superchargers to increase the volume of fuel-air mixture into the engine, injection of exotic fuels to momentarily raise engine power, or a supplementary power system such as a flywheel or electric motor—could increase fuel economy by another 15 percent (TEP, 1972; Rogowski, 1953) even if vehicle weight were unchanged.

D. Redesigning the vehicle

Reducing mass

A reduction in weight reduces kinetic and gravitational potential energy, resistance to acceleration, and rolling resistance all by the same factor. A 20-percent reduction in vehicle weight combined with a 20-percent reduction in engine size and power would leave acceleration and hill-climbing power unaltered, while reducing fuel consumption by 20 percent. A simultaneous reduction in air drag would preserve high-speed performance (if necessary). For instance, given the present figures which indicate that average fuel consumption is proportional to average curb weight with a scaling factor of 0.28 lb/ton-mile (0.087 kg/tonne-km) (Section B), the selection of lighter vehicles from the present mix would decrease average fuel consumption by 2.8 percent for each 100 lb (45 kg) reduction in the average curb weight. Alternatively, consumer preferences for vehicle size could be unchanged and both engine size and curb weight could be decreased, which would have the same effect. In either case, a reduction in average curb weight from 3600 lb (1640 kg) to 2700 lb (1230 kg) would result in fuel savings of about 25 percent, roughly 3 percent of the entire U.S. energy budget.

Reducing air resistance

Effective streamlining can reduce air drag considerably. A reduction from the current U.S. practice of $C_d \approx 0.5$ should be possible with minimal effort (White,

1969). Automobiles with $C_d \approx 0.35$ do exist (Porsche, Citroën), and at least some portion of their reduced drag is due to the fitting of a full belly pan. Beauvais, Tignor, and Turner (1968) report reductions in C_d of about 9 percent on models by fitting a simply enclosed bottom with the rough contours of the actual vehicle, and a further reduction of 7 to 8 percent by fitting a smooth streamlined bottom. Savings for highway trucks could be considerable. The fitting of a streamlined nose can reduce truck drag coefficients from 0.7 to 0.55 (Smith, 1970). Overall savings can amount to as much as 20 to 30 percent of truck and bus fuel (TEP, 1972). Hertz and Ukrainetz (1967) reported a reduction in C_d for a 1965 compact from $C_d = 0.48$ to $C_d = 0.40$ (17 percent) simply by fitting a cardboard "streamlined" nose. They also reported an *increase* of C_d from $C_d = 0.48$ to $C_d = 0.52$ (8 percent) when all windows and vents were opened.

Clearly, much more effort needs to be put into better vehicle streamlining and more efficient ways to introduce fresh air. In particular, the effects of various radiator inlet locations, partial enclosure of the underside, and the effects of varying ground clearance and other ground effects are well worth investigating at passenger vehicle speeds.[84]

Implications for safety

The safety implications of lighter vehicles also need to be considered. Even with a relatively rapid shift to lighter cars, large numbers of 5000+ lb vehicles would still be on the road for a decade, as would buses and trucks. The lighter vehicle must, therefore, be able to protect its occupants during an impact greater than "brick wall," that is, during a head-on collision with a vehicle twice its mass. Projected lighter (but not necessarily smaller) cars need new design and new materials to provide more impact-absorbing "crush" both in the car structure itself and on surfaces which may have to absorb the "second impact" of the occupant with the interior. [*Note:* In the course of this study we accidentally came upon a material which looked very promising for such applications—an aluminum "foam" made by the Ethyl corporation—which has low thermal conductivity, zero rebound, and good structural strength. It could, for instance, be used to make a thermally insulated auto roof or an impact absorbing fender. The pilot plant, unfortunately, has recently been shut down for lack of customers.] This is clearly an area where materials research is badly needed for innovation—imagine an auto made largely of impact-absorbing materials with an unbreakable, inelastic windshield.

E. Tires

About half of the rear-wheel power for the vehicle of Table 4.4 over the urban driving cycle is expended doing work against frictional forces (the other half

---

[84] It is worth noting that vehicles (or models) should be tested in a wind tunnel having a moving belt to accurately simulate vehicle velocity relative to the ground. It is also interesting to note that an automobile-sized sphere moving at 10 mph (4.5 m/sec) has a Reynolds number of $\sim 10^7$, which puts it in the regime where the drag coefficient should be dropping rapidly with velocity (Gray, 1972).

going into kinetic energy, later to be dissipated). Reduced rolling friction offers substantial possibilities for saving.

Hirst (1973) quotes the total mileage breakdown for autos and trucks in 1970 to be the following.

1970 auto miles		1970 truck miles	
Intercity	37%	Intercity freight	33%
Urban	63%	Other	67%

Since the energy spent against rolling resistance is the drag force times the distance traveled, and since we estimate that rolling resistance accounts for $\frac{1}{4}$ to $\frac{1}{3}$ of engine load in urban driving, $\frac{1}{3}$ to $\frac{1}{2}$ in "suburban" driving, and about $\frac{1}{4}$ in intercity driving (a considerably larger fraction for trucks), we estimate that perhaps a third or more of all transportation fuel consumption in the United States is expended solely for the "purpose" of heating the tires (not including heating due to braking effects). This can be reduced in a number of ways.

Increased tire pressures

Increasing the air pressure in the tires decreases the rolling resistance by approximately 5 percent for every 10 percent increase in pressure (Clark, 1971), since there is less deformation of the tire. The average pressure of present tires on present vehicles could easily be increased by about 20 percent without serious erosion of vehicle performance or tire lifetime; this could save between 3 and 4 percent of all automobile fuel.

Radial tires

Current steel-belted radial tires have lower rolling resistance than conventional ones. TEP (1972) estimates of 10- to 15-percent fuel savings seem a bit high for fitting on present vehicles, but 5 to 10 percent appears possible without re-optimizing the drive train, especially with further tire improvements.

Inventing new tires

The tire is a quite complex device, which performs many separate functions, and has enormous safety implications. Tires are made of many different rubber compounds, and contain cords of many different materials. The beneficial aspects of using rubber compounds—high adhesion and high friction—are, at present, not obtainable with low-hysteresis materials. Since the hysteresis produces most of the energy loss at the tire, and appears to be somewhat beneficial only when cornering on wet roads (Kummer, 1966), research into new elastomers and new synthetics with lower losses but similar properties should be strongly encouraged. (Natural rubber, for instance, has 10 to 25 percent less hysteresis loss than synthetic. Why can we not at least match this figure?) Greater use of elastic cords (steel, glass) can markedly reduce cord losses. Radial tires should be examined more carefully to investigate the reasons for reduced rolling resistance, and carefully optimized. Potential reductions of 20 percent or more in rolling resistance below current radial designs seem possible, and every effort should be made to encourage further research on low-loss tire designs, and low-hysteresis polymers.

Also worth examining is a total re-design of the tire itself. A recent example: The Pirelli Company has invented a tire of totally new design with a much smaller quantity of rubber in the actual tire and no cords (Business Week, July 14, 1973). Pirelli claims that the tire runs "cooler" than a radial—which implies reduced rolling resistance. (Possible reductions in fuel consumption are difficult to estimate in the absence of data.)

A worn tire has far less rolling resistance than when it is new (Clark, 1971), and cheaper, thinner tires or replaceable treads might effect large savings. Questions of which materials are the best choice for the tread and which for the carcass, and even the shape of the tire and the distribution of material, should be investigated thoroughly.

Re-inventing the wheel

The tire is used not only for traction, but also for absorbing road irregularities and small fluctuations in the steering. The entire role of the tire, wheel, and suspension, and the apportionment of tasks among them, should be reexamined. The possible use of a light disc wheel with a thin rubber tread, for instance, or other totally new methods of supporting the car through low-hysteresis shock absorbing and irregularity smoothing systems faced with thin pneumatic or rubber layers for adhesion, or the use of elastic, low-loss materials for the non-tread portion of the tires, could reduce rolling resistance by 50 percent—and conceivably come near to eliminating all but the 10 percent or so which keeps the car on the road. A 50-percent reduction in rolling resistance would reduce fuel consumption by at least 15 percent with present car designs, and by 25 percent or more when combined with other improvements.

Implications for safety

Few vehicle components have such profound implications for safety as the tire. There is an extensive body of literature available on analyzing the performance of present tires during braking and cornering, in wet weather and in dry (Clark, 1971; Kummer, 1966). Little of this literature, however, will provide a guide for predicting the performance of innovative wheel and tire designs. Even such simple strategies as mixing radial and conventional tires or inflating tires to significantly different pressures or pressure distributions than the vehicle was designed for is widely known to have potentially disastrous effects. There is need for considerable research and investigation, both experimental and theoretical, in this area.

F. Other areas for improvement

Absorption air conditioning

EPA (1972) estimates up to 9 percent average mileage losses on cars fitted with air conditioning (up to 20 percent in city traffic on hot days). Despite the enormous percentage of fuel energy lost to heat, the use of absorption air conditioners using present designs is held to be impractical (Akerman, 1971). Better, more compact, and more efficient absorption cycles are needed. Alternative heat-powered air-conditioning devices (Chapter 3, Section C) should also be explored. Autos need to be better insulated, and window heat leaks reduced.

Instrumentation

Several instruments which could be retrofitted to existing vehicles to help drivers get better mileage are needed. Some ideas which have occurred to us are gear indicator lights for automatic transmissions, an inexpensive and reliable fuel flow meter coupled to the speedometer so that it can read in mpg, an inexpensive accelerometer, and an engine load monitor interpreted to read specific fuel consumption.

Accessories

The air conditioner has already been discussed. The generator/alternator and power steering could be redesigned without too much difficulty to reduce the power drain. The use of a constant-power drive, for instance, would reduce the excess accessory drive at high engine speeds; the use of a demand-engagement clutch would remove frictional loads when the device is not in use. The fan is usually not needed at high speeds where it has the highest load, or when the engine is cold. A fan clutch is the simplest solution.

G. Summary

Since the ultimate mechanical purpose of the entire engine-transmission system is to do work at the rear wheels, a reduction in air drag or rolling resistance has a direct multiplicative effect *if* engine sizes and/or gear ratios are re-optimized to take advantage of the reduced work necessary to propel the vehicle. Failure to re-optimize the drive train accounts for some conflicting reports on radial tire mileage. A large American auto may (in urban driving) be operating so far into the low-load portion of the engine power curve that specific fuel consumption is almost inversely proportional to load; engine friction then soaks up *all* the savings at the rear wheel. The proper way to obtain proportional fuel savings, whether from reduced mass, reduced air drag, or reduced rolling resistance, is to decrease engine size and/or re-gear the drive train to readjust the system to the reduced load. Fuel savings in urban driving, however, will still be restricted somewhat by the non-multiplicative losses from idling and accessory demands. For the automobile of Table 4.4 on the urban cycle, for instance, $6.4 \times 0.75 = 4.8$ kW are required at the rear wheels, but the engine must put out 10.4 kW, of which 4 kW (38 percent) goes to drive the accessories; even at a constant 60 mph, accessories can use 20 percent of the engine power.

A modest proposal

Not all of the technical improvements listed above can be employed simultaneously. A modest, presently feasible re-design of the private automobile which effects appreciable fuel savings is recorded here primarily to show the effects of altering vehicle construction with a reasonable mix, and assuming *no* transmission improvements or engine innovations.[85]

Property	Net change in property
Average car weight	−20%
Average engine power	−33%
Air drag coefficient	−20%
Rolling resistance coefficient	−20%
Top speed	unchanged
Acceleration	−20%
Average mileage (est.)	+50%
Savings (est.)	33% of present gasoline consumption or 4.4% of total U. S. energy consumption

A less modest proposal

With some of the new devices and techniques described in this chapter, it is possible to construct a hypothetical yet reasonable vehicle for the 1980's and beyond that uses less than half the fuel of present-day automobiles.

Property	Net change in Property
Average car weight	−25%
Average engine power	−33%
Switch to improved diesel, Rankine, or Stirling engine	Efficiency +33%
Better transmission	Efficiency +20%
Air drag coefficient	−25%
Rolling resistance coefficient	−40%
Top speed	+10%
Acceleration (est.)	−10%
Average mileage (est.)	up to 150% (× 2.5)
Fuel savings (est.)	up to 60% of present gasoline consumption or 8% of total U. S. energy consumption

[85]The effects of the various strategies interact, and are not simply additive or multiplicative. For instance, reducing weight by 25 percent and the air drag coefficient by 20 percent is presumed to reduce total air resistance by 40 percent.

5. SAMPLE INDUSTRIAL PROCESSES

The use of energy in industry is complex and varied. Nevertheless, the main components of industrial energy use can be categorized in simple ways—by product, for example, or by quality of the energy, or, as we shall emphasize in this chapter, by the chemical/physical nature of the process. Most studies in the past have employed a standard industrial classification scheme in which industry is subdivided into 20 major groups according to product (see, for example, SRI, 1972). Of these groups the six largest in terms of their energy consumption account for approximately two-thirds of the industrial use. In order of consumption these are

primary metal industries,
chemical and allied products,
petrochemical and related industries,
food and kindred products,
paper and allied products,
stone, clay, glass and concrete products.

In addition to the overview of patterns and historical trends in energy use in these several industrial sectors, more specific analyses of individual products are also of value. The Dow Chemical Company, for example, maintains energy accounting at all steps in its processes so that each kilogram of product can be labelled as to energy expenditure that went into its production from all sources: fuels, feedstocks, steam, electricity, etc. (Decker, 1974). The automobile, a highly complex product, has also been subjected to a detailed energy audit (Berry and Fels, 1973). For other products, systems analyses have been performed that compare energy costs and benefits of alternatives, an example of such a study being that of Hannon (1972), who calculated the energy differences between returnable and non-returnable containers in the bottle beverage industry. As interesting and important as the analysis of the energy content in a product may be in terms of raising concerns and promoting conservation, it does not necessarily indicate where major losses are located. A statement of second-law efficiencies would be far more useful in focusing research attention where it has the opportunity of being the most valuable (Berg, 1974a).

A subdivision perhaps more useful than the standard industrial classification scheme in discussing research priorities is one that focuses on the chemical/physical nature of the process rather than on the final product. In such an approach, processes can be roughly categorized as belonging to one of three general types:

change in chemical state,
change in physical state,
fabrication.

The first of these categories contains processes involving chemical reactions of any kind; obviously, using chemical thermodynamics, these industrial processes can be discussed in terms of available work and second-law efficiency (Riekert, 1974). The category involving changes in physical state covers a wide variety of processes including beneficiation of ores, separation and mixing both at the macroscopic and molecular level, drying, and alloying, to name only a few. Some of these processes can be easily described in terms of efficiencies—purification and the separation of mixtures being good examples—while others cannot. The fabrication subdivision is meant to include machining, assembling, weaving, building and the like, processes for which second-law efficiency is harder to specify because it is so difficult to determine the minimum work required for the process. Although no breakdowns in industrial use have been made in terms of this type of subdivision, it is likely that the categories listed above are in decreasing order of total energy consumption. On the average, they are also in decreasing order of energy requirement per molecule. Whereas a chemical process may typically require 1 eV or more per molecule, the requirement for a change of physical state is more likely to be measured in hundredths of an eV per molecule, and the minimum requirement of energy per molecule for fabrication will in general be even less.

The first two of these categories appear the most probable areas in which scientific research can make major contributions to more efficient energy use. A detailed discussion of all types of industrial processes and an ordering of research priorities associated with them is far beyond the scope of this study. We can only choose to discuss a few selected topics that seem to have some generality with respect to energy use and should therefore be high on a list of priorities for research. Our focus in this chapter is on basic research opportunities in electrochemical processes, in the physical separation of materials, and in heat transfer.

A. Electrochemical processes

The transformation of chemical energy to electric energy is a process of conversion from one form of electromagnetic energy, the energy associated with the charge distribution in molecules, to another form, that of a guided electromagnetic wave. That the economically most attractive methods for producing this transformation have always proceeded through the intermediate step of uncontrolled combustion is in part, at least, a testimony to our lack of understanding of many of the mechanisms occurring in molecular interactions. Eighty years ago Ostwald (1894) advocated the replacement of the heat engine (then only 10-percent efficient) by the fuel cell in which carbon and oxygen could react to produce electricity directly without the loss of available work in going through combustion and a heat engine. We find ourselves today still searching for a way to accomplish this economically.

The problems in the development of practical means for direct conversion of chemical to electric energy have a parallel in the current state of progress in the transformation of solar energy into more immediately useable forms of energy. It presently appears that the conversion of solar energy into useful thermal energy by man is far more likely to have a major economic impact in the next decade or two than the transformation of the electromagnetic energy of solar radiation either to electric energy (direct conversion in photocells) or to chemical energy (photolysis or controlled photosynthesis).

Chemical to electric energy conversion

As discussed in Chapter 2, the first- and second-law efficiencies for a power plant in the conversion of chemical energy to electric energy are about the same. For a modern base-load fossil-fuel-fired power plant, these efficiencies are about 40 percent. The amount of electric energy that could in principle be produced in a fuel cell, operating at the temperature and pressure of the atmosphere, is the difference in Gibbs free energy of the initial and final chemical constituents. The change in Gibbs free energy differs from the change in enthalpy or heat or reaction by an entropy term:

$$-\Delta G = -\Delta H + T\Delta S ,$$

$-\Delta H$ being the enthalpy change and ΔS being the change in entropy from the initial to the final species (it can be either positive or negative). The quantity $-\Delta G$, calculated at atmospheric temperature and pressure, is equal to the available work B (see Chapter 2). For the hydrogen-oxygen reaction at 20 °C and 1 atm, $B = |\Delta G| = 0.83\ |\Delta H|$; under laboratory conditions of very low reaction rates the electric energy output can approach this value very closely. However, it should not be inferred from such an idealized situation that either of the following two propositions is true: (1) that the basic physical mechanisms occurring within a fuel cell are fully understood or (2) that a practical economically competitive device that employs commercially available fuels can be constructed and made to operate reliably for long periods of time. The major impetus for the developments in the technology of fuel cells over the past two decades resulted from their attractiveness for space applications, including high power-to-weight ratio, high efficiencies, and no moving parts; not until recently has there been an economic incentive in terrestrial markets.[86] The development of a practical fuel cell operating on a cheap hydrocarbon fuel for residential, commercial, or utility use would have a profound impact on the pattern of energy consumption in the United States. The economics of electric power distribution, the economies of scale in generation facilities, and the problems of time distribution of demand of electric energy could be fundamentally altered by the widespread introduction of the fuel cell. Such a cell used as a total energy system supplying both electricity and heat to a particular site (as discussed in Chapter 3, Section E) might even require a major restructuring of the present institutions associated with energy supply in the United States.

The subject of electrochemistry is a venerable and well developed discipline. It is on this body of knowledge that the electrochemical industry and in particular the present-day development of the fuel cell rest. However, physical description of

---

[86]It is possible that the major technological spin-off of social benefit from the space program will turn out to be the progress in development of the fuel cell.

phenomena in this field generally has followed industrial application rather than the other way around. As a result, our present understanding tends to be rather empirical when viewed from the discipline of physics. Consider, for example, the single most important phenomenon in a fuel cell—charge transfer. While certain methods are known for improving the process and decreasing the losses in devices, a basic description of the complete process has not been developed.

When two atoms combine to form a molecule, binding energy is made available. Only one method has been suggested for the conversion of this energy into electric energy without an intermediate transformation into thermal energy associated with internal or translational degrees of freedom of the molecule. The technique is to add/remove reversibly an electron to/from an atom, producing an ion that is in electrochemical equilibrium in a solution. Two ions thus formed from the original atomic species at different electrodes can then combine in solution with no release of energy to produce the final molecule. The energy conversion process occurs in the charge transfer between a solid and the atoms. The state of the metal has a higher energy after the transfer than before. Although the charge transfer across the solid surface to or from the atom is central to the energy conversion process, its fundamental nature is not understood. In the early days of quantum mechanics, Gurney (1931) described the phenomenon qualitatively in terms of tunneling of the electron through the potential barrier that exists at the interface between the electrolyte and a conducting solid. However, a microscopic theory has not been developed much beyond this description since then. While the dependence of the charge-transfer process on surface properties is utilized in devices, we do not fully appreciate the mechanisms involved. What is the nature of the electronic wave function in a metal at the electrolyte interface? What is the role of localized states on the surface or the importance of the chemisorption process? How do the various materials that can catalyze the transfer process operate? In what way do other excitations within the system influence the transfer? How can one study possible intermediate steps in the process? These are but a few of the basic questions that surround the transfer process, the answers to which would certainly aid in the design of more efficient and practical devices. The general opinion among experts in the field of fuel-cell technology is that a practical, commercial device will be developed by incremental advances in the field with another decade of hard work. It would appear that physicists have a major opportunity to contribute through basic science to this end.

Solar to chemical conversion

Much of the attention given to the use of solar energy for purposes other than heating has concentrated on the direct conversion to electric energy in photocells. The primary impetus over the past twenty years for the development of solar photovoltaic cells, as with fuel cells, came from the space program. Unfortunately, the present cost of electricity from photocells is prohibitively high in comparison with that from a conventional power plant. Recent progress in the fabrication of large silicon crystals (Hammond, 1974) will lower costs. To make the solid-state photocell economically competitive may require additional scientific or technological breakthroughs. Perhaps the study of more esoteric semiconductors or the further research on such techniques as edge-defined crystal growth will provide a needed breakthrough. Much of the research in this area falls within the domain of solid-state physics.

The possibilities in related but different classes of processes, those which convert solar energy directly to chemical energy, are certainly not as well known to physicists. Studies of photosynthesis in organic systems have begun to unravel the complexities of the energy exchange mechanisms but no serious suggestions have been made on ways to improve on nature. There have been reports in the literature (Heidt, 1951) of the photolysis of water with the aid of an inorganic catalyst but the subject has not been pursued actively until very recently. Fujishima (1973) observed that water can be decomposed into hydrogen and oxygen by shining light on the surface of a semiconductor which forms one of a pair of electrically connected electrodes in an electrolytic solution. The process can be described very briefly as follows: In the surface layer of the semiconductor the photon excites an electron-hole pair, the members of which are spatially separated by the electric field associated with the band bending at the surface. One of the charges moves to the metal electrode; the electron and hole then find themselves at opposite electrodes where, on charge transfer to the electrolyte, hydrogen and oxygen are produced from solution.

This process has advantages over the conventional semiconductor photocell. Problems of recombination may be easier to overcome because one of the carriers may be almost immediately taken up by a surface state where it is ideally located for transfer to the electrolyte. Thus the purity and crystallinity may not be an important consideration for the semiconductor. Work on the physics of this type of process is in its early stages, and an important topic of research is likely to develop in this area between semiconductor physics and electrochemistry.

Electrochemical energy storage

Batteries have long been a subject of both scientific and commercial interest. The rechargeable battery, in particular, has had wide application for energy storage. The upsurge in interest in the technology of electrochemical systems is related in part to problems of the present energy shortage and in part to recent scientific discoveries. There is now hope that new technology of batteries can contribute importantly to more efficient use of energy in society.

The use of batteries to power automobiles is an idea that has appealed to many people. As discussed in Chapter 4, the internal combustion engine and its transmission system transfer to the rear wheels less than 10 percent of the available work of the fuel in urban driving and less than 20 percent in highway driving. An electric drive taking energy from a storage cell might be expected to be 60-percent efficient (Hottel and Howard, 1971). Inclusion of the electric power plant and distribution efficiency of 33 percent in converting primary fuel to electric energy still results in the overall efficiency of the electric-car–power-plant combination exceeding that of the internal combustion engine by a factor of 2 or more in urban driving.[87] (We leave aside any consideration of advantages or disadvantages of a possible shift in primary energy source—e.g., petroleum to coal and uranium—

---

[87]A lesser, but still significant, gain could come from using a battery and electric motor only for peak power requirements in an automobile (see Chapter 4, Section C).

that would accompany the introduction of an electric car.)

One of the main obstacles to the introduction of the battery-powered car has been and continues to be the problem of energy density. Gasoline and other petroleum hydrocarbons have heats of combustion in the range from 40 to 50 MJ/kg, whereas the maximum energy density of a lead/acid storage cell is almost 100 times smaller, 0.6 MJ/kg (see Table 5.1). In addition, a practical battery has an energy-to-weight ratio approximately 6 times less than this figure, so that even with the better efficiency of an electric drive a lead/acid battery is too heavy to be practical for powering automobiles as they are presently constructed. Certainly the technological development of batteries with lightweight electrodes to increase significantly the energy density would have a major influence on the commercial introduction of a practical electric car.

Energy storage has the potential of becoming a very important adjunct for the economic production of electric power. The difficulties that the electric utility companies now face are in part the result of a faster growth in peak demand for electric power than growth in average demand. To meet peak power demands for a few hours a day with a generating unit either very inefficient or having high capital investment costs is undesirable. There is therefore considerable interest in energy storage whereby efficient off-peak generating capacity can be used to store energy which is quickly and cheaply recoverable in electric form. Pumped storage, in which water is transferred between reservoirs at different levels, is one method presently in use but with severe site limitations as the ten-year controversy surrounding the Storm King Mountain project on the Hudson River attests. In part the problem of pumped storage is a consequence of the low energy density of water in the earth's gravitational field. For a practical height difference of 300 m, the energy is only 3 kJ/kg. To provide a reasonable peaking energy, say 10^4 MW hr, requires, with 80 percent conversion efficiency, 1.5×10^{10} kg, or 1.5×10^7 m^3 (12,000 acre-feet) of water. The number of suitable locations in the U. S. where one is able to achieve so massive a transport of water without running into land-use or ecological problems appears to be very limited indeed.

Other energy storage devices are being actively studied; these include compressed gas, flywheels, superconducting magnets, and standard lead-acid batteries. A better and more economical battery with good recharge performance possesses a num-

TABLE 5.1. Energy densities of selected systems.

System	Theoretical energy density MJ/kg	Estimated practical energy density MJ/kg
Water (gravitational potential energy, 300-m height)	0.003	
Gasoline (heat of combustion)	46	
Batteries:		
lead/acid	0.63	0.10
sodium/sulfur	2.9	1.1
lithium/chlorine	7.8	1.4

ber of advantages for use in utility systems. Research in the physics of these electrochemical cells is obviously desirable.

One of the more important scientific developments relating to batteries within the past few years has been the discovery of a new type of solid electrolyte, β alumina (Weber and Kummer, 1967). This sodium polyaluminate, $Na_2O \cdot 11 Al_2O_3$, has sodium ions arranged on planes between blocks of the other constituents. Within these planes the sodium ions are able to move with relatively high mobility at elevated temperatures giving the material very desirable ion-exchange properties. A battery with sodium and sulfur electrodes (Na_2S_3 being the principal reaction product) with β alumina serving as the electrolyte has been shown to be rechargeable, and estimates of the energy density achievable in a practical cell (see Table 5.1 and Sandstede, 1972) are as much as 10 times larger than that of the lead/acid battery. The sodium/sulfur system with β alumina electrolyte is by no means the only electrochemical cell with potential for development. Other alkali/chalcogen as well as alkali/halogen, metal/air, and more esoteric systems have been proposed as possible candidates for batteries with high energy densities [see Sandstede (1972) for a recent review]. However, a better understanding of the basic mechanisms occurring in these systems must be obtained before many of the serious technical problems holding back their development can be overcome. For example, the chemical and phase stability of solid electrolytes, the kinetics of solid electrolyte-electrode interfaces, dendrite formation, and impurity effects are but a few of the many basic phenomena that must be investigated prior to a critical assessment of the technical feasibility of new electrochemical cells for energy storage in automobiles or electric utility systems.

B. Separative work

Much of the recent attention paid to energy use in industrial processes has centered on some of the more energy-intensive processes involving chemical reactions and a change in chemical state of the constituents in the system—the reduction of ore to iron, the electrolysis of aluminum, cement manufacture, etc. In considering improved efficiency for these processes—which do use a significant percentage of the energy consumed by society—it is important to recognize that a large fraction of the energy is employed in the *physical* separation of materials into their chemically nonreacting constituents. Most important industrial plants are a complex mixture of interrelated processes requiring energy for chemical reactions as well as for separative work; an obvious example is the oil refinery, which employs both cracking and distillation.

Although extremely varied, separative processes fall into two broad groups: (1) the separation of macroscopic aggregates such as the filtration of precipitates, the recycling of materials in mixed waste streams; and (2) the separation of substances at the molecular level. Although the first of these groups is a very important subject for energy studies, it is probably an area where improvements in energy use will derive more from technological innovation than from basic science. The analysis of problems of separation at the molecular level, however, can quickly revert to discussions of statistics and mechanics of complex atomic systems, areas in which the present state of knowledge is deficient.

The separation of molecular or ionic species of a multi-component system is very often carried out by a change in phase of the system since the equilibrium concentrations of the components usually depends strongly on state. Fractional distillation and zone refining are two common industrial processes making use of this principle. In a concern for the efficient use of energy the problem is that the minimum work required to separate the chemically non-reacting constituents of a system is $T\Delta S$, where T is the absolute temperature and ΔS is the entropy of mixing, whereas the energy to change the phase of a system involves a latent heat. In many cases, especially in dilute systems, latent heats may be very much larger than the minimum work $T\Delta S$; and although much of the energy required to change the temperature and produce the phase transition may be recoverable with the proper use of heat exchangers, such processes rarely approach efficiencies close to unity. The best solution to many molecular separation problems is often not to employ a phase change but rather to make use of other physical processes such as diffusion, adsorption, osmosis, or ion exchange. The selectivity of these processes to different molecular species makes them very attractive. Notwithstanding the long history of successful use of such processes in industry, better basic physical understanding and improved techniques hold promise of substantially improving their efficiency.

Consider as an example the production of oxygen by its separation from air. Oxygen used in blast furnaces and for other commercial purposes is commonly produced by the liquefaction and subsequent distillation of air. The thermodynamic minimum energy required to separate air at 80 °F and 1 atm into its nitrogen and oxygen components is about 0.19 MJ/kg of oxygen (19 Btu/lb of air). Estimates from various industrial sources indicate that present practice requires an energy expenditure of about 1.5 MJ/kg to provide oxygen by distillation as a gas at STP. Recently an adsorption process has been developed commercially in which nitrogen is selectively adsorbed from air at 40 psi on a molecular sieve that can be regenerated by reducing the pressure. The electric energy required to drive this process is about 2.2 MJ/kg of oxygen, or more than a factor of ten larger than the theoretical minimum, and even somewhat larger than the energy used in the distillation process. This low efficiency is presumably due in part to the lack of complete selectivity in the adsorption and the consequent loss of much of the oxygen, perhaps as much as two-thirds, in the waste stream of nitrogen. (Detailed information is not available since the process is proprietary.) Nonetheless it is interesting to note what can be done in separating two molecules as physically similar as oxygen and nitrogen by an adsorption process.

Desalination of sea water provides a more striking illustration of the differences in energy expenditures by different separation processes. Desalination is an insignificant user of energy at present, but the social and political implications of an inexpensive method for desalting water are staggering to contemplate. The separative work required to convert sea water at 3.5 percent salinity and 20 °C to half fresh water and half brine with twice the salinity is 0.97 kW hr per m^3 of fresh water produced, or 3.5 kJ/kg (see Table 5.2).[88] This number is very much less than the heat of vaporization of water, 540 cal/gm (2300 kJ/kg), or heat of fusion, 80 cal/gm (330 kJ/kg). Present technology using multistage evaporators

---

[88]A simple calculation of the minimum work for desalinating a small fraction of sea water appears in Harte and Socolow (1971).

requires a thermal energy input of the order of 280 kJ/kg while present freezing processes use about 60 kJ(electric)/kg, equivalent to about 180 kJ(thermal)/kg (Reichle, 1965). Systems integration in which electric power generation and water desalination are combined in a single plant can cut the energy requirements assigned to the desalination process but roughly only by a factor of two. In an economic analysis, MacAvoy and Peterson (1969), using arguments based on available work, estimated that 40 percent of the thermal output of a 1250-MW_t nuclear reactor of a dual-purpose plant must be assigned to desalination for an energy expenditure of 110 kJ/kg of desalted water. The dreams of the early 1960's of transforming the deserts of the earth into Gardens of Eden through cheap desalination of sea water have failed to materialize and they are unlikely to materialize with present technology.

Several physical separation techniques which avoid evaporation and freezing have been developed and hold promise of significantly improved efficiency. Electrodialysis is presently used in treating brackish water (<10,000 ppm) but, as indicated in Table 5.2, the technique does not compare favorably either economically or energetically with other processes for sea water (Spiegler, 1966). The process of reverse osmosis in which water is selectively transmitted through a membrane by the application of a pressure difference between the saline and fresh sides has been estimated by Bray (1966) to be 20 kW hr/1000 gal [20 kJ(electric)/kg or 60 kJ(thermal)/kg] with present technology, three times less than the method that employs

TABLE 5.2. Energy use in the desalination of sea water.

A. Comparison Energies	Fuel energy per unit of fresh water (kJ/kg)	Second-law efficiency ϵ [a]
Minimum separative work to desalinate 50 percent of the water	3.5	
Latent heat of evaporation	2300	
Latent heat of fusion	330	
B. Processes		
distillation:		
single purpose plant	280	0.013
combined with electricity generation (1250 MW_t nuclear)	110	0.032
freezing	180	0.019
electrodialysis:		
sea water	450	0.008
brackish water (5000 ppm) [b]	50	0.010
reverse osmosis	60	0.058

[a] The efficiency for sea water is calculated for desalination of 50 percent of the water. It is only approximate because not all processes operate at this level of desalination.

[b] The minimum separative work to desalinate 50 percent of brackish water (5000 ppm) is 0.5 kJ/kg. This is the reference value for calculating efficiency.

freezing. All of these methods, however, fall at least an order of magnitude short of economic feasibility for agricultural irrigation.

[It is interesting to note parenthetically that the natural mixing of fresh water runoff with sea water, the reverse process of desalination, represents the loss of a significant amount of available work. The energy that can be obtained in principle from the mixing of 1 kg of fresh water with sea water is approximately 2.5 kJ, the equivalent of the gravitational potential energy of the same kg of water at a height of 250 m. With an average daily runoff of 10^{12} gallons, this represents an energy of 3.3×10^{15} Btu/yr (1.1×10^{8} kW), roughly 5 percent of the total U. S. energy consumption.]

These examples of separative processes have been discussed in order to illustrate the important gains in energy conservation that can be made by developing materials with better selectivity for different molecular species. Much of the progress made in this area to date owes its origin to an understanding of the basic physical properties of molecules, surfaces, and membranes, but also much of it has resulted from an empirical approach to real problems. The transport of molecules and ions in inhomogeneous systems such as membranes, the distribution and aggregation of molecules in solution, on surfaces and in membranes, the relationship between the nature of an adsorbate and adsorption phenomena, the energetics of biological membranes, and the kinetics of ion exchange are but a few of the areas in which future research can have an important impact on the technological developments in separative processes within industry and therefore on our energy consumption.

C. Research needs in heat transfer

Despite an extensive literature on the phenomena connected with heat transfer,[89] our ability to make predictions in many situations of practical interest is remarkably limited. To be sure, this is in part due to the complexity of "real-life" problems, but our incomplete understanding of many of the basic mechanisms involved also plays a significant role. A case in point, for instance, is heat transfer accompanied by change of phase such as boiling, evaporation, and condensation. Even a cursory look at the figures appearing in two recent articles on boiling heat transfer (Rohsenow, 1971, 1973) is sufficient to demonstrate how large is the scatter of data points around the curves corresponding to the various semi-empirical correlations which embody essentially all our present predictive capability. Aspects of heat transfer phenomena to which physicists could make a significant contribution include basic research in fluid mechanics, surface physics, nucleation theory, optical and infrared emission and absorption, and the computation and measurement of transport coefficients and material properties, particularly under unusual conditions (such as at high temperature or pressure, or near a critical point).

Substantial progress in heat transfer can have an enormous impact on energy efficiency. Basic problems in the design of both nuclear reactors and conventional

[89]Only a few illustrative references are cited in this section. For up-to-date chapters on the various mechanisms and processes of heat transfer, accompanied by a more extensive set of references, see the *Handbook of Heat Transfer* (Rohsenow and Hartnett, 1973). The annual literature surveys by E. R. G. Eckert *et al.* (1972, 1973, 1974) also afford good sources of recent references.

power plants depend on boiling heat transfer. Extensive and economic use of bottoming cycles relies heavily on our ability to produce efficient heat exchangers for small temperature differences. The exploitation of solar energy and the design of all the other devices that involve low-temperature heat require extremely sophisticated heat-transfer equipment. Other areas such as water desalination, control of environmental impact (e.g., thermal pollution), and some manufacturing processes (including paper and food industries and casting of metals), would greatly benefit from advances in heat transfer.

The following discussion (much of it based on Sabersky, 1971) makes no attempt at completeness, but brings out a number of basic research opportunities.

Convection

Fluid mechanics obviously constitutes the basis for a fundamental approach to convective heat transfer. Problems of flow stability and phenomena of both steady and unsteady flow pose challenging questions for the theorist as well as the experimenter. An increased effort in this area can have a very significant impact on the design of heat exchangers and power plants. For instance, there are some fairly common types of single-phase heat exchangers operating in the laminar regime that are overdesigned by as much as 100 percent because of lack of basic information (Sabersky, 1974). Other basic research topics include the effects of temperature and fluid inhomogeneities on the flow and its stability, effects of vibration, surface roughness, and other surface modifications (e.g., fins and fluting). Even some laminar and transition flows in tubes still need research effort; the practical importance here is for the transportation, heating, and cooling of viscous fluids and for the design of oil heat exchangers. The entire subject of turbulence is of enormous importance for convective heat transfer, and a substantial effort is needed in this field.

Convection in liquid metals and cryogenic fluids deserves separate mention because these liquids require substantially different experimental techniques for their analysis. The practical interest in this area is motivated mainly by liquid-metal fast breeder reactors (and possibly by the heat transfer problems that will be encountered in power generation by nuclear fusion) on the one hand, and by the liquid-gas industry on the other. Electric power transmission in superconducting cables may also be significantly affected by such studies.

Free convection problems are of both practical and theoretical interest. The onset of various types of flow patterns and of turbulence, steady and non-steady heat transfer from horizontal and inclined surfaces, free convection in the presence of centrifugal fields and in enclosures—these are only a few examples of problems relevant to space heating, air conditioning, thermal pollution, and meterology.

Finally, one should mention the broad subject of the fluid mechanics and heat-transfer characteristics of non-Newtonian fluids. Here information is needed also at a very fundamental level, such as rheological properties and constitutive relations. In some cases, there are uncertainties on how to characterize these fluids and on what are the basic quantities to measure. Advances in this area would benefit both the chemical and the food processing industries.

Change of phase

Boiling is a very poorly understood process. Nucleation, effects of the heat flux by convection, and transition from nucleate to film boiling are all important phenomena for which our predictive capabilities are surprisingly poor. This is particularly unfortunate in view of the large number of practical applications, such as in the nuclear and chemical industries, and in power generation. The understanding of transient boiling is less developed than that of steady boiling. Perhaps the most important application here is in nuclear reactor safety, where reliable quantitative predictions are necessary to model the events related to various kinds of accidents.

Evaporation and condensation are also processes of major importance that would benefit from an increased research effort. Evaporating and condensing films, for example, are commonly encountered in the chemical industry.

Conduction

Conduction is perhaps the oldest and best understood branch of heat transfer, but it is by no means difficult to find areas of potential interest to the physicist and high usefulness to society. The general problem of conduction with change of phase, such as the solidification of cast metal parts in molds, is a typical case. The uncertainties that one faces in this area are in the thermal resistance of interfaces as well as in the physical properties in the vicinity of the solidification zone. Another broad area in need of fundamental research is that of the measurement and prediction of the physical properties of materials (density, specific heat, heat conductivity, etc.) under unusual conditions such as high temperature and pressure, and also for composite media. The latter area is of considerable importance in food processing.

Two-phase flow

Although not exclusively encountered in heat-transfer problems, two-phase flow should be mentioned here for its importance in power generation and the transport of liquid fuels. In this area essentially all of the existing usable information reduces to some empirical or semi-empirical correlations. Moreover, their validity is often confined to some very special configurations or experimental situations, and no reliable scaling laws are available. In addition, it is impossible in practice to predict the transition between the various flow regimes, and all correlations are of very little use in the vicinity of these transition points. Here a fundamental theoretical approach to the phenomena involved together with accurately controlled experiments may provide the needed insights to reduce the problem to a more tractable form.

Special mention should be made of the study of flows containing droplets and particulates such as one encounters in fluidized beds which are of present interest in coal gasification and oil-shale recovery and appear to be one of the fastest growing areas in heat- and mass-transfer equipment. Some important problems are connected with the basic phenomena of interparticle forces, externally applied forces (gravitational and possibly electromagnetic), and the effect of bounding surfaces. As is discussed at greater length in a companion report (APS, 1974b), the derivation of suitably averaged "macroscopic" equations for the description of such systems (i.e., equations in which the fluid is treated as a continuum) is a very interesting and challenging area.

Heat transfer near the critical point

Heat transfer apparatus operating near critical points (such as power plants operated at supercritical pressures and devices employing cryogenic heat-transfer media) has proliferated in recent years. Here problem areas include the determination of flow fields in the presence of strongly varying properties, and appropriate diagnostic techniques. The derivation of useful equations of state is also a field in which much work is needed; it is a field ideally suited for the physicist.

Radiation

Although the basic mechanisms of radiation are well understood, significant research opportunities in this area still exist. Basic problems include radiative heat transfer measurements and computations in combustion environments (APS, 1974b); determination of physical properties such as surface emissivities and their dependence on temperature, pressure, and direction of incidence; absorption and scattering in liquids, solids, and gases; and radiation properties of vapors, aerosols, and liquid-vapor and gas-solid suspensions. Another interesting application is in the use of special wavelength bands to match specific needs, as in food processing and industrial heating.

APPENDICES

I. SELECTED ENERGY-USE DATA

As noted in Section 1C, the SRI Report, *Patterns of Energy Consumption in the United States* (SRI, 1972), is a standard reference for energy-use data in this country. It covers the period 1960–68. More recent data have been issued by the Department of the Interior (for 1972 and 1973 data, see Bureau of Mines, 1974).

Abbreviated SRI data for 1968 appear in Tables 1.2 and 1.3. More detailed tabulations for 1968, from the same source, are given in Tables A1 and A2. Table A1 (an expanded version of the right side of Table 1.2) classifies energy "consumption" by end use. Table A2 is arranged according to the source of the energy. More extensive tabulations can be found in the original report.

The SRI figures for total national energy consumption are lower (by a few percent) than figures supplied by other sources, such as the Bureau of Mines. As noted in footnote 4, the difference arises from a different treatment of nuclear and hydroelectric power plants.

Here we provide more details on this point in order to display the arbitrariness and uncertainty that (perhaps inevitably) burden the enterprise of energy accounting.

The SRI figures for the generation of electricity in 1968, classified according to the type of power plant, are shown in Table A3. For fossil-fuel power plants, the energy counted is the energy content (heat of combustion) of the fossil fuel these plants reported burning in that year. For the nuclear and hydroelectric plants, the energy counted is only the electric energy generated. This procedure inflates the national average efficiency of producing electricity from 0.315, its value for fossil-fuel plants alone, to 0.364. The latter figure is used throughout the SRI report to calculate the effective fuel energy consumption associated with any electricity consumption for any end use. For an efficiency of 0.364, 9378 Btu are "consumed" to produce 1 kWhr of electricity. For fossil-fuel plants, 10,830 fuel Btu's are needed to produce 1 kWhr. (The units conversion factor is 3414 Btu/kWhr.)

The SRI procedure for handling nuclear and hydroelectric energy is open to question, and it has not been widely copied. Most references follow the Bureau of Mines procedure, which is the following: (1) Hydroelectric power, regarded as a substitute for fossil-fuel power, is assigned a conversion efficiency equal to that of fossil-fuel plants (31.5 percent in 1968, according to SRI, and 32.9 percent

Table A1. Energy equivalent of fuel consumed in the United States in 1968, classified by end use.[a]

Sector and end use	Percent of national total	Energy (10^{15} Btu)	Annual rate of growth, 1960–68
Residential			
Space heating	11.0%	6.675	4.1%
Water heating	2.9%	1.736	5.2%
Cooking	1.1%	0.637	1.7%
Clothes drying	0.3%	0.208	10.6%
Refrigeration	1.1%	0.692	8.2%
Air conditioning	0.7%	0.427	15.6%
Other	2.1%	1.241	5.5%
Total residential	19.2%	11.616	4.8%
Commercial			
Space heating	6.9%	4.182	3.8%
Water heating	1.1%	0.653	2.3%
Cooking	0.2%	0.139	4.5%
Refrigeration	1.1%	0.670	2.9%
Air conditioning	1.8%	1.113	8.6%
Feedstock	1.6%	0.984	3.7%
Other	1.7%	1.025	28.0%
Total commercial	14.4%	8.766	5.4%
Industrial			
Process steam	16.7%	10.132	3.6%
Electric drive	7.9%	4.794	5.3%
Electrolytic processes	1.2%	0.705	4.8%
Direct heat	11.5%	6.929	2.8%
Feedstock	3.6%	2.202	6.1%
Other	0.3%	0.198	6.7%
Total industrial	41.2%	24.960	3.9%
Transportation			
Fuel	24.9%	15.038	4.1%
Raw materials	0.3%	0.146	0.4%
Total transportation	25.2%	15.184	4.1%
National Total	100.0%	60.526	4.3%

[a] Source: SRI (1972).

in 1973, according to the Bureau of Mines, using data of the Federal Power Commission). (2) For nuclear power, the actual average thermal efficiency of nuclear plants is used (32.0 percent in 1973, according to the Bureau of Mines, using data of the Atomic Energy Commission). This method of accounting, applied to the 1968 data, adds 1.92×10^{15} Btu to the total energy account, increasing the total energy used from 60.53 to 62.45 (a 3.2-percent change) and the energy attributed

Table A2. Energy use in the United States in 1968, classified by source.[a]

Sector and source	Energy (10^{15} Btu) non-electric	electric	total	Annual rate of growth, 1960–68
Residential				
Coal	~ 0			...
Dry natural gas	4.606			4.6%
Petroleum products	3.192			2.1%
Total non-electric	7.798			
Electric		1.390		8.2%
Sector total			9.188	4.1%
Commercial				
Coal	0.568			−7.1%
Dry natural gas	1.845			7.2%
Petroleum products	3.389			5.4%
Total non-electric	5.802			
Electric		1.079		9.6%
Sector total			6.881	4.5%
Industrial				
Coal	5.616			1.7%
Dry natural gas	9.258			4.9%
Petroleum products	4.474			3.0%
Total non-electric	19.348			
Electric		2.043		5.8%
Sector total			21.391	3.7%
Transportation				
Coal	0.012			−22.0%
Dry natural gas	0.610			6.9%
Petroleum products	14.514			4.1%
Total non-electric	15.136			
Electric		0.018		...
Sector total			15.154	4.1%
Electric utilities				
Coal	7.130			6.7%
Dry natural gas	3.245			7.8%
Petroleum products	1.181			9.7%
Nuclear energy[b]	0.130			47.0%
Hydroelectric energy[b]	0.757			4.1%
Total consumption	12.443			
Electricity supplied		−4.530		7.3%
Sector net consumption[c]			7.913	7.1%
National fuel consumption				
at points of use	48.084 (79.4%)			3.8%
by electric utilities	12.443 (20.6%)			7.1%
total			60.527	4.3%

[a] Source: SRI (1972).

[b] The nuclear and hydroelectric "fuel" consumption is set equal to the electric energy generated; see the discussion in this appendix and in footnote 4.

[c] The utilities' "net consumption" is primarily the waste heat produced at fossil-fuel plants.

Table A3. Electricity generation in 1968: SRI data and data treatment.

Type of power plant	E_e Electricity generated (10^{15} Btu)	(10^9 kWhr)	E_f Energy counted (10^{15} Btu)	(10^9 kW hr)	E_e/E_f Efficiency ratio (percent)
Fossil-fuel	3.643	(1067)	11.556	(3384)	31.5%
Nuclear	0.130	(38)	0.130	(38)	100.0%
Hydroelectric	0.757	(222)	0.757	(222)	100.0%
All types	4.530	(1327)	12.443	(3644)	36.4%

to power plants from 12.44 to 14.36×10^{15} Btu (a 15-percent change). Correspondingly, the conversion from electric kilowatt hours to effective fuel Btu's is increased by 15 percent from 9378 Btu/kW hr to 10,820 Btu/kW hr.

Finally, we take note of recent trends in energy consumption. In Table A4, electric energy and total energy are extrapolated from 1968 to 1973 assuming continued exponential growth at the average 1960–68 rate, and these extrapolated figures are compared with actual 1973 energy consumption as reported by the

Table A4. Recent trends in energy consumption.

	1968 Energy [a] (10^{15} Btu)	Annual rate of growth 1960–68 [a]	Extrapolated 1973 energy (10^{15} Btu)	Actual 1973 energy [b] (10^{15} Btu)	Change from 1972 to 1973 [b]
Total energy	62.45 [c]	4.3%	77.1	75.6	4.8%
Energy attributable to electricity generation	14.36 [d]	7.1%	20.2	19.8	6.7%
Fraction of total energy attributable to electricity generation	[23.0%] [e]		[26.3%]	[26.2%]	
Electricity distributed	4.53		6.38	6.42	
Average conversion efficiency of all power plants	[31.6%] [f]	g	[31.6%]	[32.4%]	

[a] Source: SRI (1972), modified to reflect Bureau of Mines method of treating hydroelectric and nuclear power.

[b] Source: Bureau of Mines (1974).

[c] Original SRI number, 60.53.

[d] Original SRI number, 12.44.

[e] Original SRI number, 20.6%.

[f] Original SRI number, 36.4%.

[g] Extrapolation assumes negligible change in average power-plant efficiency.

Bureau of Mines (1974). To make possible such a comparison, the SRI figures for 1968 are altered to reflect the Bureau of Mines procedure for reckoning hydroelectric and nuclear energy. Given the inevitable uncertainties in both sets of numbers, the comparison must be viewed with caution. Nevertheless, the table does suggest that the 1973 data lie pretty much on the same exponential curves that pass through the 1960–68 data. These curves are characterized by doubling times of about 10 years for electricity (about 7 percent per year) and about 16 years for total energy (about 4.3 percent per year). Only in the past year (1974) have the gross figures deviated appreciably from these trends.

II. SELECTED DATA AND CONVERSION FACTORS

Temperature

Absolute zero: 0 K = 0 °R = −273.15 °C = −459.67 °F

$T(K) = T(°C) + 273 = T(°R)/1.80$

$T(°R) = T(°F) + 460$

$T(°C) = [T(°F) - 32]/1.80$

Gas constant

$R = 8.314$ J/mole K

= 1.9872 cal/mole K = 1.9872 Btu/lb-mole °R

Thermal energy

($T_0 = 298.15$ K = 25 °C = 77 °F = 536.67 °R)

$RT_0 = 2479$ J/mole = 592.5 cal/mole = 0.5925 kcal/mole

= 1066.6 Btu/lb-mole = 0.02569 eV/molecule

Molecular quantity

1 mole = $N_0 = 6.022 \times 10^{23}$ molecules (= 1 gm-mole)

1 kmole = $10^3 N_0 = 6.022 \times 10^{26}$ molecules

1 lb-mole = $453.59 N_0 = 2.7316 \times 10^{26}$ molecules

$M.W.$ (molecular weight) = $\dfrac{\text{gm}}{\text{mole}} = \dfrac{\text{kg}}{\text{kmole}} = \dfrac{\text{lb}}{\text{lb-mole}}$

Multiply by M.W. to convert from *to*

from	to
Btu/lb	Btu/lb-mole
eV/amu	eV/molecule
J/kg	J/kmole
kcal/kg	kcal/kmole

Multiply by M.W. /1000 to convert from	*to*
kcal/kg	kcal/mole
J/kg	J/mole

Water

Density = 1 gm/cm^3 = 1 kg/liter = 1000 kg/m^3

= 62.43 lb/ft^3 = 8.345 lb/gal

Melting temperature (at 1 atm) = 0 °C = 273.15 K

= 32 °F = 491.67 °R

Boiling temperature (at 1 atm) = 100 °C = 373.15 K

= 212 °F = 671.67 °R

Specific heat = 1 cal/gm °C = 1 kcal/kg K

= 1 Btu/lb °F = 4184 J/kg K

Molar specific heat = 18 cal/mole °C = 75.4 J/mole K

= 18 Btu/lb-mole °F = 7.81×10^{-4} eV/molecule K

Heat of fusion = 79.7 cal/gm = 79.7 kcal/kg

= 143.5 Btu/lb = 3.336×10^5 J/kg

Molar heat of fusion = 1436 cal/mole = 6010 J/mole

= 2585 Btu/lb-mole = 0.0623 eV/molecule

Heat of vaporization = 539.4 cal/gm = 539.4 kcal/kg

= 971 Btu/lb = 2.257×10^6 J/kg

Molar heat of vaporization = 9717 cal/mole = 40, 660 J/mole

= 17, 490 Btu/lb-mole = 0.421 eV/molecule

Air

Typical moist air density (sea level, 20 °C)

= 0.00120 gm/cm^3 = 1.20 kg/m^3 = 0.075 lb/ft^3

Standard dry air density (sea level, 0 °C)

= 0.001293 gm/cm^3 = 1.293 kg/m^3 = 0.0807 lb/ft^3

Composition, by mole fraction (dry air)

N_2, 78.1% O_2, 21.0% Ar, 0.93% CO_2, 0.03%

Mean molecular weight (dry air) = 28.97

Specific heat of standard dry air

$$C_p = 0.240 \text{ cal/gm °C} = 0.240 \text{ kcal/kg K}$$
$$= 0.240 \text{ Btu/lb °F} = 1000 \text{ J/kg K}$$

Molar specific heat of standard dry air

$$C_p' = 6.96 \text{ cal/mole °C} = 29.1 \text{ J/mole K}$$
$$= 6.96 \text{ Btu/lb-mole °F} = 3.503R$$

$$C_p' - C_v' = 1.00R \qquad C_p/C_v = 1.40$$

Water vapor in saturated air

T (°C)	T (°F)	Mass fraction of H_2O in saturated air	Mole fraction of H_2O in saturated air
0	32	0.36%	0.60%
5	41	0.54%	0.86%
10	50	0.76%	1.21%
15	59	1.05%	1.68%
20	68	1.45%	2.30%
25	77	1.96%	3.12%
30	86	2.64%	4.17%
35	95	3.32%	5.23%

CONVERSION FACTORS

Except for fuel economy, all conversions are to SI units. To convert from a non-SI unit to an SI unit, *multiply* by the conversion factor. To convert from an SI unit to a non-SI unit, *divide* by the conversion factor.

A useful reference is Mechtly (1969).

Length

1 in.(= 2.54 cm) = 0.0254 m

1 ft = 0.3048 m

1 mile = 1609 m

Area

1 in.^{2} = 6.452×10^{-4} m^2

1 ft^2 = 0.09290 m^2

1 acre = 4047 m^2

Volume

1 qt $=9.464\times10^{-4}$ m^3

1 liter = 0.001 m^3

1 gal(= 3.785 l) = 0.003785 m^3

1 ft^3 = 0.02832 m^3

1 acre-ft = 1233 m^3

Time

1 hr = 3600 sec

1 day = 86, 400 sec

1 year(= 365.26 days) = 3.156×10^7 sec

Speed

1 km/hr (kph) = 0.2778 m/sec

1 ft/sec (fps) = 0.3048 m/sec

1 mi/hr (mph) = 0.4470 m/sec

Mass

1 lb (lbm) = 0.4536 kg

1 ton = 907.2 kg

1 tonne = 1000 kg

Force

1 dyne = 10^{-5} N

1 lb (lbf) = 4.448 N

Pressure

1 dyne/cm^2 = 0.1 N/m^2

1 lbf/ft^2 = 47.88 N/m^2

1 mm of Hg = 133.3 N/m^2

1 in. of Hg = 3386 N/m^2

1 lbf/in.^{2} (psi) = 6895 N/m^2

1 bar = 10^5 N/m^2

1 atm = 1.013×10^5 N/m^2

Energy

1 eV = 1.602×10^{-19} J

1 erg = 10^{-7} J

1 ft lb = 1.356 J

1 cal = 4.184 J

1 Btu = 1054.4 J

1 kcal(= 1 food calorie) = 4184 J

1 hp hr = 2.686×10^{6} J

1 kW hr(kwh) = 3.60×10^{6} J

1 therm(= 10^5 Btu) = 1.054×10^{8} J

1 kiloton(explosive energy) = 4.20×10^{12} J

1 "quad"(= 10^{15} Btu) = 1.054×10^{18} J

Power

1 Btu/yr = 3.341×10^{-5} W

1 kW hr/yr = 0.114 W

1 Btu/hr = 0.293 W

1 ft lb/sec = 1.356 W

1 hp = 746 W

1 Btu/sec = 1054 W

1 "ton" (cooling) = 3500 W

1 kcal/sec = 4184 W

1 quad/yr = 3.341×10^{10} W

Energy per unit mass

1 Btu/lb = 2324 J/kg

1 kcal/kg (= 1 cal/gm) = 4184 J/kg

1 eV/amu = 9.649×10^{7} J/kg

Energy per mole and per molecule

1 Btu/lb-mole = 2.324 J/mole

1 kcal/mole = 4184 J/mole

1 eV/molecule = 96, 490 J/mole

Molar specific heat, entropy

1 cal/mole °C = 4.184 J/mole K

1 Btu/lb-mole °F = 4.184 J/mole K

Energy flux

1 Btu/ft^2 hr = 3.152 W/m^2

U value

1 Btu/ft^2 hr °F = 5.674 W/m^2 °C

Fuel economy

1 lbm/hp hr = 0.6080 kg/kW hr

1 mile/gal(mpg) = 0.425 km/liter

Fuel heats of combustion (approximate)

1 gal of oil	1.5×10^8 J (1.4×10^5 Btu)
1 barrel (42 gal) of oil	6.1×10^9 J (5.8×10^6 Btu)
1 ft^3 of natural gas	1.1×10^6 J (1030 Btu)
1 ton of coal	2.6×10^{10} J (2.5×10^7 Btu)

Fuel consumption and power (approximate)

10^6 barrels of oil/day	7×10^{10} W (2.1 quad/yr)
10^9 tons of coal/yr	8×10^{11} W (25 quad/yr)

BIBLIOGRAPHY

(For most of the U. S. government publications, the originating or sponsoring agency is cited; most of these publications can be obtained from the Superintendent of Documents, U. S. Government Printing Office.)

APS, *Energy Conservation and Window Systems* (American Physical Society, New York, N. Y.; preliminary version of report available from S. Berman, Stanford Linear Accelerator Center, Stanford, CA 94305, 1974a), final version in this volume (p. 241).

APS, *The Role of Physics in Combustion* (American Physical Society, New York, N. Y.; preliminary version of report available from D. Hartley, Combustion Research Division, Sandia Laboratories, Livermore, CA 94550, 1974b), final version in this volume (p. 151).

ASHRAE, *Fundamentals* (American Society of Heating, Refrigerating, and Air-Conditioning Engineers, New York, NY, 1972).

ASHRAE, *Systems* (American Society of Heating, Refrigerating and Air-Conditioning Engineers, New York, NY, 1973).

J. R. Akerman, "Automotive Air Conditioning Systems with Absorption Refrigeration" (SAE no. 710037), SAE Transactions 80, 132 (1971).

J. Appel and J. J. MacKenzie, "How Much Light Do We Really Need?", Bulletin of the Atomic Scientists (December 1974).

F. N. Beauvais, S. C. Tignor, and T. R. Turner, "Problems of Ground Simulation in Automotive Aerodynamics" (SAE no. 680121), SAE Transactions 77, 451 (1968).

C. A. Berg, "Energy Conservation through Effective Utilization," Science 181, 128 (13 July 1973).

C. A. Berg, "A Technical Basis for Energy Conservation," Technology Review (February 1974a), p. 14.

C. A. Berg, "Conservation in Industry," Science 184, 264 (19 April 1974b).

R. S. Berry and M. Fels, "The Energy Cost of Automobiles," Science and Public Affairs, The Bulletin of the Atomic Scientists (December 1973).

W. B. Boast, *Illumination Engineering* (McGraw-Hill Book Co., New York, NY, 1953).

D. T. Bray, in *Desalinization by Reverse Osmosis*, U. Merten, ed. (MIT Press, Cambridge, MA, 1966), p. 203.

Bureau of Mines, *Minerals Yearbook* (U.S. Department of the Interior, 1969).

Bureau of Mines, press release (U.S. Department of the Interior, 13 March 1974).

C. E. Burke, E. C. Campbell, T. D. Kosier, L. C. Lundstrom, W. A. McConnell, H. L. Welch, and W. E. Zierer, "Where Does All the Power Go?", SAE Transactions 65, 713 (1957).

M. de Cachard, note T.T. 457, Centre d'Études Nucléaires de Grenoble (38041 Grenoble-Cedex, France, 1974).

C. Campbell, *The Sports Car* (Robert Bentley Press, Cambridge, MA, 1970).

Citizens for Clean Air, *Electricity Demand in the Service Territory of Consolidated Edison Company of New York, Inc.* (Citizens for Clean Air, Inc., New York, NY, October 1973).

S. K. Clark, ed., *Mechanics of Pneumatic Tires*, NBS Monograph 122 (U.S. National Bureau of Standards, 1971).

S. K. Clark, R. N. Dodge, R. J. Ganter, and J. R. Luchini, *Rolling Resistance of Pneumatic Tires*, report DOT-TSC-74-2 (U.S. Department of Transportation, Transportation Systems Center, Cambridge, MA 02142, July 1974).

DOT, *Analysis of 1973 Automobiles and Integration of Automobile Components Relevant to Fuel Consumption* (Transportation Systems Center, U.S. Department of Transportation, September 1974).

Decision Sciences Corporation, *Economic Evaluation of Total Energy*, a report to HUD (U.S. Department of Housing and Urban Development, July 1973).

G. L. Decker (Dow Chemical Co.), private communication (June 1974).

R. M. E. Diamant, *Total Energy* (Pergamon Press, New York, NY, 1970).

Dubin-Mindell-Bloome Associates, *Energy Conservation Guidelines for Office Buildings* (U.S. General Services Administration, January 1974).

M. Dumont, note T.T. 451, Centre d'Études Nucléaires de Grenoble (38041 Grenoble-Cedex, France, 1974).

EPA, *Fuel Economy and Emission Control*, monograph (U.S. Environmental Protection Agency, 1972).

EPA, *Energy Conservation Strategies*, EPA-R5-73-021 (Office of Research and Monitoring, U.S. Environmental Protection Agency, July 1973).

E. R. G. Eckert *et al.*, bibliographies of heat-transfer literature, Int. J. Heat Mass Transfer 15, 1969 (1972); 16, 1969 (1973); 17, 351 and 615 (1974).

Federal Council on Science and Technology, *Total Energy Systems, Urban Energy Systems, Residential Energy Consumption*, a report to HUD (U.S. Department of Housing and Urban Development, October 1972).

Federal Power Commission, *The 1970 National Power Survey, Part II* (U.S. Federal Power Commission, 1970), p. 2-12.

Federal Power Commission, *The 1970 National Power Survey, Part I* (U.S. Federal Power Commission, 1971), p. 3-15.

Federal Power Commission, *A Technical Basis for Energy Conservation*, staff report FPC/OCE/2 (U.S. Federal Power Commission, April 1974).

Ford Foundation, *Exploring Energy Choices*, a preliminary report of the Ford Foundation's Energy Policy Project (Ford Foundation, New York, NY, 1974).

J. Fox, *Energy Consumption for Residential Space Heating, a Case Study*, CES Report no. 4 (Center for Environmental Studies, Princeton University, Princeton, NJ 08540, 1973a).

J. Fox, H. Fraker, R. Grot, D. Harrje, E. Schorske, and R. Socolow, *Energy Conservation in Housing: First Year Progress Report*, CES Report no. 6 (Center for Environmental Studies, Princeton University, Princeton, NJ 08540, 1973b).

J. Fox and R. Socolow, *Conservation Research Results from a Statistical Analysis of Monthly Gas and Electric Consumption at Twin Rivers, N.J.*, CES Working Paper W-11 (Center for Environmental Studies, Princeton University, Princeton, NJ 08540, 1974).

A. P. Fraas and M. N. Ozisik, *Heat Exchange Design* (John Wiley & Sons, Inc., New York, NY, 1965).

A. Fujishima and K. Honda, "Electrochemical Photolysis of Water at a Semiconductor Electrode," Nature 238, 38 (1973).

M. G. Gamze, *Total Energy System Primer, Part I* (Gamze-Korobkin-Calogen, Inc., Chicago, IL 60606, 1972).

General Electric, *Solar Heating and Cooling of Buildings, Phase 0*, prepared for the National Science Foundation Public Technology Projects Office (Space Division, General Electric Co., Philadelphia, PA 19101, May 1974).

J. W. Gibbs, *The Collected Works of J. Willard Gibbs* (Yale University Press, New Haven, CT, 1948), vol. 1, p. 77.

P. W. Gill, J. H. Smith, Jr., and E. J. Ziurys, *Fundamentals of Internal Combustion Engines* (U.S. Naval Institute, Annapolis, MD, 1954).

D. E. Gray, ed., *American Institute of Physics Handbook* (McGraw-Hill Book Co., New York, NY, 1972), third edition, pp. 2-266–268.

R. W. Gurney, "The Quantum Mechanics of Electrolysis," Proc. Roy. Soc. A 134, 137 (1931).

W. Häfele, "Energy Choices That Europe Faces: A European View of Energy," Science 184, 360 (19 April 1974).

A. L. Hammond, "Solar Power: Promising New Developments" (Research News), Science 184, 1359 (1974).

B. Hannon, *System Energy and Recycling: A Study of the Beverage Industry* (Center for Advanced Computation, University of Illinois, Champaign, IL, 1972).

B. Hannon, *Options for Energy Conservation,* report CAC-79 (Center for Advanced Computation, University of Illinois, Urbana, IL, 1973).

H. Harboe, "Economical Aspects of Air Storage Power," Proceedings of the American Power Conference, vol. 33 (1971).

J. Harte and R. Socolow, *Patient Earth* (Holt, Rinehart, and Winston, Inc., New York, NY, 1971), pp. 273–275.

D. G. Harvey and W. R. Menchen, *The Automobile—Energy and the Environment* (Hittman Associates, Columbia, MD, March 1974).

H. R. Hay and J. I. Yellott, "Natural Air Conditioning with Roof Ponds and Movable Insulation," ASHRAE Transactions 75, 165 (1969).

L. J. Heidt, "Non-Biological Photosynthesis," Proc. Amer. Acad. Arts and Sciences 79, 228 (1951).

R. A. Hein, "Superconductivity: Large-Scale Applications," Science 185, 211 (19 July 1974).

R. A. Herendeen, *An Energy Input-Output Matrix for the United States, 1963: Users Guide,* report CAC-69 (Center for Advanced Computation, University of Illinois, Urbana, IL, 1973a).

R. A. Herendeen, *Use of Input-Output Analysis to Determine the Energy Cost of Goods and Services* (Center for Advanced Computation, University of Illinois, Urbana, IL, 1973b).

P. B. Hertz and P. R. Ukrainetz, "Auto-Aerodynamic Drag-Force Analysis," Experimental Mechanics 7 (no. 3), 19A (March 1967).

E. Hirst, "Transportation Energy Use and Conservation Potential," Science and Public Affairs, The Bulletin of the Atomic Scientists (November 1973).

E. Hirst, *Direct and Indirect Energy Requirements for Automobiles,* ORNL-NSF-EP-64 report (Oak Ridge National Laboratory, Oak Ridge, TN, February 1974).

Hittman Associates, *Residential Energy Consumption* (Federal Council on Science and Technology, October 1972; Hittman Associates, Columbia, MD, March 1973).

H. C. Hottel and J. B. Howard, *New Energy Technology: Some Facts and Assessments* (MIT Press, Cambridge, MA, 1971).

R. Hudson, Jr., *Infrared System Engineering* (Wiley-Interscience, New York, NY, 1969).

W. Kauzman, *Thermodynamics and Statistics* (W. A. Benjamin, New York, NY, 1967).

J. H. Keenan, *Thermodynamics* (John Wiley & Sons, Inc., New York, NY, 1948).

J. H. Keenan, E. P. Gyftopoulos, and G. H. Hatsopoulos, "The Fuel Shortage and Thermodynamics," in *Proceedings of the MIT Energy Conference* (MIT Press, Cambridge, MA, 1973).

H. W. Kummer, *Unified Theory of Rubber and Tire Friction,* Engineering Research Bulletin B-94 (College of Engineering, Pennsylvania State University, University Park, PA 16802, July 1966).

J. T. Kummer, *The Automobile and the Energy Crisis,* report (Ford Motor Company, Dearborn, MI, 1974).

N. Kurtz, quoted in "Cutting the Energy Costs in a Home is a Snap," *New York Times,*

Section 8, p. 1 (7 April 1974a).

N. Kurtz (Flack and Kurtz, New York, NY), private communication (July 1974b).

D. O. Lee and W. H. McCulloch, "A New Parameter for Evaluating Energy Systems," *8th Intersociety Energy Conversion Engineering Conference Proceedings* (American Institute of Aeronautics and Astronautics, New York, NY, 1973).

L. Lees and M. P. Lo, *Time Factors in Slowing Down the Rate of Growth of Demand for Primary Energy in the United States*, EQL Report no. 7 (Environmental Quality Laboratory, California Institute of Technology, Pasadena, CA 91109, June 1973).

G. S. Leighton, "HUD Takes Total Energy Concept One Step Beyond," Air Conditioning and Refrigeration Business (January 1973), p. 81.

H. Lorsh, ed., *Conservation and Better Utilization of Electric Power by Means of Thermal Energy Storage and Solar Heating* (University of Pennsylvania, Philadelphia, PA 19104, July 1973).

W. S. Lusby, "Total Energy Systems," in *Report of Technical Advisory Commission on Conservation of Energy for the National Power Survey of the Federal Power Commission* (U.S. Federal Power Commission, 1974).

P. W. MacAvoy and D. F. Peterson, *Large-Scale Desalting* (Praeger Publishers, New York, NY, 1969).

A. Makhijani and A. Lichtenberg, *An Assessment of Residential Energy Utilization in the USA* (Electronics Research Laboratory, College of Engineering, University of California, Berkeley, CA 94720, January 1973).

S. Marsden (Stanford University), private communication (1974).

E. A. Mechtly, *The International System of Units: Physical Constants and Conversion Factors*, NASA report SP-7012 (Office of Technology Utilization, National Aeronautics and Space Administration, Washington, DC, 1969).

Mitre Corporation, *An Agenda for Research and Development on End Use Energy Conservation*, two volumes, report no. 6577 (Mitre Corporation, Bedford, MA, 1973).

J. C. Moyers, *The Value of Thermal Insulation in Residential Construction: Economics and the Conservation of Energy*, ORNL-NSF-EP-9 report (Oak Ridge National Laboratory, Oak Ridge, TN, December 1971).

J. J. Mutch, *Transportation Energy in the United States: a Statistical History, 1955–1971*, report no. R-1391-NSF (Rand Corporation, Santa Monica, CA 90406, December 1973).

J. J. Mutch, *Residential Water Heating: Fuel Conservation, Economics, and Public Policy*, report no. R-1498-NSF (Rand Corporation, Santa Monica, CA 90406, May 1974).

NAS, *Physics in Perspective: Recommendations and Program Emphases* (National Academy of Sciences, Washington, DC, 1972).

NAS, *An Evaluation of Alternative Power Sources for Low-Emission Automobiles* (National Academy of Sciences, Washington, DC, 1973).

NAS, *Materials and Man's Needs, Summary Report of the Committee on the Survey of Materials Science and Engineering (COSMAT)* (National Academy of Sciences, Washington, DC, 1974).

NATO, *Technology of Efficient Energy Utilization*, E. G. Kovach, ed., report of a NATO Science Committee Conference held at Les Arcs, France, 8–12 October 1973 (Scientific Affairs Division, NATO, 1110 Brussels, Belgium, 1973). [Also available from Pergamon Press, New York, NY, 1975.]

NEDO, *Energy Conservation in the United Kingdom* (National Economic Development Office, Millbank Tower, Millbank, London SW1P 4QX, December 1974).

New Jersey, *Energy*, a report to the Governor by the New Jersey Task Force on Energy (New Jersey State Energy Office, Trenton, NJ, May 1974).

OEP, *The Potential for Energy Conservation*, part 1, October 1972, and part 2, January 1973 (U.S. Office of Emergency Preparedness, 1972, 1973).

V. Olgyay, *Design with Climate* (Princeton University Press, Princeton, NJ, 1963).

W. Ostwald, "Die Wissenschaftliche Elektrochemie der Gegenwart und die Technische der Zukunft," Z. Elektrochem. 1, 122 (1894).

B. Peavy, F. Powell, and D. Burch, *Dynamic Thermal Performance of an Experimental*

Masonry Building, Building Science Series no. 45 (U.S. National Bureau of Standards, July 1973).

H. Perry, *Conservation of Energy*, a report of the Senate Committee on Interior and Insular Affairs (U.S. Government Printing Office, 1972).

P. Petherbridge, "Transmission Characteristics of Window Glasses and Sun Controls," in *Sunlight in Buildings: Proceedings of the CIE Intersessional Conference, April 1965* (Bouwcentrum International, Rotterdam, Netherlands, 1967).

G. F. Pinder (Water Resources Program, Princeton University), private communication (July 1974).

R. F. Post and S. F. Post, "Flywheels," Scientific American (December 1973).

Rand Corporation, *California's Electric Quandary*, report no. R-1116-NSF/CSA, by R. D. Doctor *et al.* (Rand Corporation, Santa Monica, CA 90406, September 1972).

Regional Plan Association and Resources for the Future, *Patterns of Energy Consumption in the Greater New York City Area* (Regional Plan Association, New York, NY, and Resources for the Future, Washington, DC, 1973).

L. F. Reichle, "Evaluation of All Potential Sources of Energy for Desalting," in *Proceedings of the First International Symposium on Water Desalination* (U.S. Department of the Interior, 1965), p. 605.

L. Riekert, "The Efficiency of Energy-Utilization in Chemical Processes," Chem. Eng. Sci. 29, 1613 (1974).

A. R. Rogowski, *Elements of Internal Combustion Engines* (McGraw-Hill Book Co., New York, NY, 1953).

W. M. Rohsenow, "Boiling," Ann. Rev. Fluid Mech. 3, 211 (1971).

W. M. Rohsenow, "Boiling," in Rohsenow and Hartnett (1973), Section 13 (see following reference).

W. M. Rohsenow and J. P. Hartnett, eds., *Handbook of Heat Transfer* (McGraw-Hill Book Co., New York, NY, 1973).

W. F. Rush, *Statement Relative to HR10952* (Institute of Gas Technology, Illinois Institute of Technology, Chicago, IL 60616, 14 November 1973).

SRI, *Patterns of Energy Consumption in the United States*, prepared by the Stanford Research Institute in 1970 using 1960–68 data (U.S. Office of Science and Technology, 1972).

R. H. Sabersky, "Heat Transfer in the Seventies," Int. J. Heat Mass Transfer 14, 1927 (1971).

R. H. Sabersky, "Areas of Heat Transfer Research," lecture, AIAA/ASME Thermophysics and Heat Transfer Conference, 17 July 1974 (available from the author, California Institute of Technology, Pasadena, CA 91109, 1974).

G. Sandstede, ed., *From Electrocatalysis to Fuel Cells* (University of Washington Press, Seattle, WA, 1972).

G. L. Smith, *Commercial Vehicle Performance and Fuel Economy*, 16th L. Ray Buckendale Lecture (Society of Automotive Engineers, New York, NY, 1970).

K. S. Spiegler, *Principles of Desalination* (Academic Press, New York, NY, 1966).

R. G. Stein, "Spotlight on the Energy Crisis: How Architects Can Help," AIA Journal (June 1972).

W. R. Stevens, *Building Physics: Lighting* (Pergamon Press, New York, NY, 1969).

C. M. Summers, "Conversion of Energy," Scientific American (September 1971).

TEP, *Research and Development Opportunities for Improved Transportation Energy Usage*, summary report of the Transportation Energy Panel, report no. DOT-TSC-OST-73-14 (U.S. Department of Transportation, September 1972).

TRW, *Solar Heating and Cooling of Buildings, Phase 0*, prepared for the National Science Foundation Public Technology Projects Office (TRW Systems Corp., Redondo Beach, CA, May 1974).

C. F. Taylor and E. S. Taylor, *The Internal Combustion Engine*, second edition (International Textbook Company, Scranton, PA, 1961).

Thermo Electron Corporation, *Potential Fuel Efficiencies in Industry,* a report to the Energy Policy Project of the Ford Foundation (Ballinger Publishing Co., Cambridge, MA, 1974).

U. S. Senate, *Automotive Research and Development and Fuel Economy,* Hearings before the Committee on Commerce, serial no. 93-41 (U.S. Government Printing Office, 1973).

G. J. Van Wylen, *Thermodynamics* (John Wiley & Sons, Inc., New York, NY, 1959).

W. Vielstich, *Fuel Cells* (Wiley-Interscience, New York, NY, 1970).

N. Weber and J. T. Kummer, "Sodium-Sulfur Secondary Battery," in *Proceedings of the 21st Annual Power Sources Conference* (PSC Publishing Company, Red Bank, NJ, 1967).

Westinghouse, *Solar Heating and Cooling of Buildings, Phase 0,* prepared for the National Science Foundation Public Technology Projects Office (Westinghouse Electric Corp., Special Systems, Baltimore, MD 21203, May 1974).

R. G. S. White, "A Method of Estimating Automobile Drag Coefficients" (SAE no. 690189), SAE Transactions 78, 829 (1969).

R. A. White and H. H. Korst, "The Determination of Vehicular Drag Coefficients from Coast-Down Road Tests," SAE Transactions 81(1), 354 (1972).

W. C. Yee, "Thermal Aquaculture: Engineering and Economics," Environmental Science and Technology 6, 273 (1972).

EFFICIENT USE OF ENERGY

PART II

THE ROLE OF PHYSICS IN COMBUSTION

Edited by

D. L. Hartley

D. R. Hardesty

M. Lapp

J. Dooher

F. Dryer

PART II

PREFACE AND ACKNOWLEDGMENTS

This report represents one of the first attempts by the American Physical Society to identify priority physics research areas in specific energy-related engineering disciplines. Since efficiency in combustion systems is recognized as a major factor in energy conservation, it was felt that the role physics could play in helping overcome the inherently scientific limitations of combustion systems would be a significant contribution to the "Technology of Efficient Energy Utilization."

The basic philosophy behind this APS summer study is to relate physics research to applied energy problems; in this case, more efficient combustion devices. To do this it is necessary to assimilate the recommendations of other panel studies in order to recognize the practical systems of interest, define the technical limitations of those systems, and then detail the promising approaches, both analytical and experimental, for overcoming those limitations.

Fortunately, a great deal of effort has already gone into recognizing technology areas which would have the greatest impact on energy conservation, and into addressing the problem of assigning priority categories for energy-related research and development. To illustrate a path of high priority recommendations of these earlier panels, consider the following example of evolution toward specifics: Energy consumption → transportation → diesels and stratified charge engines → combustion research → better diagnostics. It is one of the aims of this study to take the next step, i.e., to define specifically what options to explore next and to define the areas of physical research which can contribute to this exploration.

We chose to concentrate on three specific areas of combustion research within a broad spectrum of combustion science problems. These were the general areas of combustion diagnostics (particularly laser-oriented schemes), combustion modeling, and the specific area of emulsified fuels. The latter was chosen because of recent interest of the combustion community and because such virgin territory might offer challenge for unconventional and hopefully constructive input from the physics community.

To address our goals we assembled nineteen physicists and engineers actively involved in research in areas relevant to the study topics. These scientists were asked to establish the relevance of each research area, to delineate the technical limitations, and to suggest specific physics research areas that might lead to overcoming these limitations. These topics were discussed interactively for four weeks, with undoubtedly as much benefit for the participants themselves as for the audience of this report. We also concluded that significant new work could be done now within the limitations of our current research capabilities, and we attempted to define those specific research areas.

The primary benefactor of this study might be thought to be the academic physicist who awaits identification of "relevancy" for his work, or perhaps the applied physicist who utilizes his unique capabilities to address the practical problems identified herein. Rather, the ultimate benefactor will hopefully be the combustion community for whom new tools and new insight might be offered from a fresh and unconventional input from the physics community at large.

D. L. Hartley,
chairman and
principal organizer

ACKNOWLEDGMENTS

We are grateful to Dr. D. Hardesty of Sandia and to Dr. M. Lapp of General Electric for coordinating and preparing the sections on modeling and diagnostics, respectively, and to Professor J. Dooher of Adelphi and Dr. F. Dryer of Princeton for cooperatively preparing the section on emulsions.

We also acknowledge the many valuable contributions by the participants and briefers listed in the front of this book. Additional valuable conversations were held with Professor W. Sirignano of Princeton, Professor A. Roshko of Caltech, and Dr. E. Storm of Lawrence Livermore Laboratories.

The support and sponsorship of Dr. Paul Craig of the National Science Foundation, Dr. Kurt Riegel of the Federal Energy Administration, and Dr. Craig Smith of the Electric Power Research Institute is appreciated.

PART II
THE ROLE OF PHYSICS IN COMBUSTION

INTRODUCTION

The opportunities

Virtually all our energy derived from fossil fuels comes through some form of combustion. Therefore, any technical advance which can result in more efficient combustion systems without sacrificing environmental quality would be a major contribution to energy conservation. As in nearly every practical engineering system, however, combustion devices are faced with performance limitations of an inherently scientific nature. The environmental crisis emphatically demonstrated to the technical community the inadequacy of our basic scientific understanding of the phenomena involved in the technical concepts for utilizing our energy-consuming devices. Initial attempts to rectify the environmental problems by after-treatment of the products of combustion rather than by scientific modification of the combustion processes themselves have not solved the problem, but have compounded it.

With the recognition of the energy crisis, both regulatory and scientific pressure is being brought to bear on a wide range of energy-related studies, with added emphasis on sound technical solutions rather than temporary stop-gap measures. Now, perhaps more than ever before, the limitations of our engineering design capabilities are being strained for the tools and insight of a broader and more innovative technical base of fundamental physics. It is therefore the purpose of this report to delineate priority physics research areas which can have impact on efficient utilization of our energy resources in combustion systems.

The practical devices which plead for better understanding of combustion processes include the spark ignition and diesel internal combustion engines, which consume about 65 percent of our petroleum resources (Figure 1); the aircraft gas turbine, which is projected to be as energy-consumptive as autos by 1990 (Figure 2); furnaces for utility boilers, which consume half our coal and 18 percent of our natural gas; and residential, commercial and industrial furnaces, which utilize 75 percent of our natural gas, 30 percent of our coal and 30 percent of our petroleum.[1] Also strikingly evident from Figure 1 is the inherently low efficiency of energy conversion for these systems. In fact, half of the energy available is rejected as waste. The potential for increased efficiency in these systems exists and should be developed into practical use.

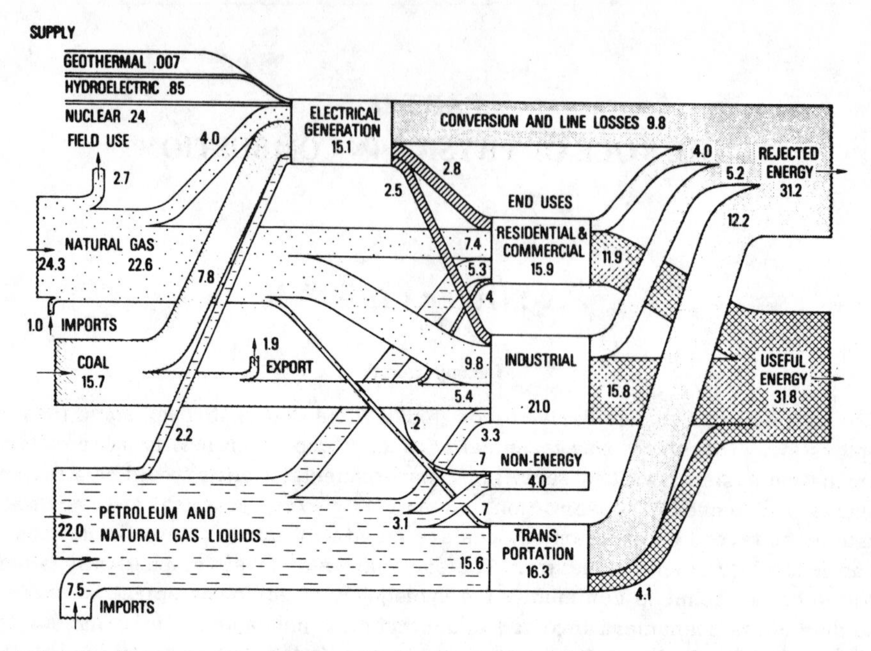

Figure 1. U. S. energy flow patterns, 1970. (Note: All values are $\times 10^{15}$ Btu. Total production = 71.6×10^{15} Btu.)

The largest immediate payoff can come from automobile engine improvement because it is the major fuel consumer and is the least efficient in energy conversion. Recent studies[3, 4, 5] have indicated that development of either light weight diesels or stratified charge engines can result in 25–30 percent improvement in efficiency. These systems, however, are based on complex combustion processes which are not well understood. The fundamental aspects of these combustion systems are discussed in more detail in this report and are used to illustrate technical shortcomings.

Longer range payoff will come from fundamental studies of combustion systems in gas turbines because of the projected usage of aircraft for passenger transport from the 1980's on. Aircraft propulsion systems have experienced significant technical advance in the past decade, largely due to government involvement in research and development funding, but efficiency and pollution abatement gains are still to be made through more sophisticated design.

The utility boiler, though designed to operate at 85–90 percent efficiency, does not always operate at maximum efficiency and, for large systems, requires combustion modification for pollution control. The market for utility boilers responds rapidly to even small efficiency increases, in contrast to the automobile market, and therefore economic incentives exist today for improved combustion techniques for these systems. In particular, new concepts for coal-fire/power generating systems offer considerable challenge to the combustion scientist and to the physicist. Fluidized bed steam generators, with integrated schemes for the catalytic removal of oxides of sulfur (SO_x), are poorly understood and would benefit greatly from improved combustion modeling capabilities

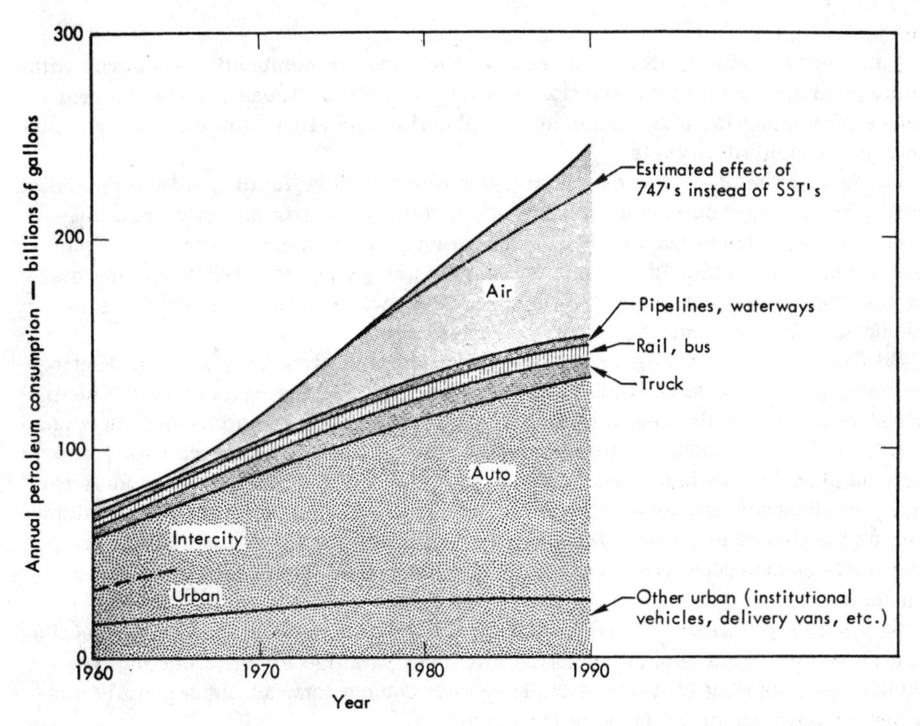

Figure 2. U. S. petroleum consumption for transportation. Data source: W. K. Fraize and J. K. Dukowicz, "Transportation Energy and Environmental Issues," Report M72-25, Mitre Corporation, McLean, Va., February, 1972.

and new diagnostic techniques.

The approach

The National Science Foundation recently sponsored a Workshop on Basic Combustion Research.[6] The attending combustion scientists were invited to provide guidance and direction to the Foundation so that funding for basic research in this area could have greatest impact. One of the conclusions of that workshop is that more research is needed in the area of turbulent combustion, with the strong suggestion that better experimental diagnostics are needed before the field can advance significantly. Other priority research areas suggested by the workshop chairman[7] include the surface physics of catalysis, condensation and nucleation in soot formation, and, with particular aim at the physicist, innovation.

In the American Physical Society Topical Conference on Energy, Heywood[8] recommended research areas for physicists in combustion modeling applied to automotive engines. He stressed the need for more accurate knowledge of fuel distribution in diesels and stratified charge engines, turbulent flow, soot formation, flame quenching, and spray combustion in diesels.

Similar assessments and recommendations of a more general nature will be found in the present report. It is hoped, however, that by considering selected areas in more depth, specific suggestions of a more substantive nature can be made.

This report concentrates on three specific areas of combustion research within a broad spectrum of combustion science problems. These are the general areas of combustion diagnostics and combustion modeling, and the more specific area of emulsified fuels.

(1) *Experimental diagnostics* is an area which is very familiar to the physicist. With increasing requirements in the interpretation of data obtained from measurement techniques such as optical scattering, more input from modern physics is required. Many of the new techniques are on the verge of supplying diagnostic information for which the theorists have waited years. Hopefully this report will supply some direction to that effort.

(2) *Combustion modeling* includes mathematical representation of interacting physical processes such as chemical kinetics, kinetic theory, and turbulence, all of which are in the realm of classical physics. With the addition of new analytic and computational capabilities, new critical diagnostic information, and new insights into turbulent phenomena, the generation of predictive models for many combustion processes may be an accelerated science where the physicist can play a meaningful role. Modeling and its relationship to diagnostics are put into some perspective for the physicist in this report and needed research is outlined.

(3) *Emulsified fuels for combustion systems* are discussed in the final section of this report. Emulsion combustion may hold promise for efficient improvements. As a new field it offers challenge for unconventional and hopefully constructive input from the physics community.

The amplification of these three topics is not meant to imply that they are the only combustion research areas to which physicists can contribute, but merely represent a pre-defined limited scope for this study. Some other important combustion physics problems, such as catalytic combustors, are discussed in cursory fashion.

The technology which has gone into the development of practical combustion systems has evolved over several decades and yet is still in its infancy in scientific detail. The opportunity for technical breakthroughs and innovation still exists and as such forms the impetus for this study on the involvement of physics in combustion science.

REFERENCES

[1]Austin, A. L., Rubin, B., and Werth, G. C., "Energy: Uses, Sources, Issues," Lawrence Livermore Laboratory Report UCRL-51221, May 1972.

[2]Fraize, W. K., and Dubowicz, J. K., "Transportation Energy and Environmental Issues," Report M72-25, Mitre Corp., McLean, Va., February 1972.

[3]Harvey, D. G. and Menchen, W. R., *The Automobile — Energy and the Environment* (Hittman Associates, Inc., Columbia, Maryland, March 1974).

[4]"Research and Development Opportunities for Improved Transportation Energy Usage," Summary Technical Report for the Transportation Energy Panel to the FCST Energy R & D Goals Committee, July 1972.

[5]"Energy R & D Programs for Transportation Systems," Department of the Army Study,

November 1973.
[6]Glassman, I. and Sirignano, W. A., "Summary Report of the Workshop on Energy-Related Basic Combustion Research," sponsored by NSF, Princeton Univ. Dept. of Aerospace and Mechanical Sciences Report No. 1177 (1974).
[7]Glassman, I., private communication.
[8]Heywood, J. B., "Combustion Modeling in Automobile Engines," in AIP Conference Proceedings No. 19, *Physics and the Energy Problem — 1974*, edited by M. D. Fiske and W. W. Havens, Jr. (American Institute of Physics, New York, 1974).

1. MODELING OF COMBUSTION PROCESSES

Discussions relating to the study and modeling of combustion processes have been carried out with a rather broad focus. This derives from the basic importance of and strong coupling between experiment and analysis as they relate to the understanding of combustion phenomena which range from the fundamental laboratory scale to the extreme complexity of practical devices such as engines and stationary power plants. A common thread to our discussions has been the effort to delineate those areas of study which have special promise of making a significant impact upon a wide range of practically important combustion problems. Within the selected areas the potential impact of new effort by the physics community has been assessed. A conscious attempt has been made to underscore those aspects of combustion analysis and modeling which will benefit significantly by the input of information obtained by the application of new diagnostic techniques (e.g., laser Raman spectroscopy, laser fluorescence, and laser Doppler velocimetry) to the *in situ* evaluation of local properties in combustion flow fields of fundamental scientific as well as practical interest. The complementary impact of predictive modeling on the selection and design of diagnostic techniques, experiments and practical devices has been a major consideration.

I. INTRODUCTION

In the broadest context, combustion analysis may be defined as the formal application of the scientific method to the study of simple and complex chemical reactions in systems of scientific and practical interest. That aspect of combustion analysis designated as *combustion modeling* has lately assumed prominence in the literature because of current usage of the term. In a recent discussion of the application of physics to new technology, Heywood[1] defined combustion modeling in the context of automotive engine research as "the prediction of performance and emissions characteristics of automotive engines using computational procedures developed from a fundamental analysis of the combustion process." Similarly, Osgerby[2] used the term combustor modeling to identify the "quantitative calculation of performance characteristics and emission levels" for gas turbine combustors by means of "complex analytical models."

As we shall see in the following sections, practical combustion devices may involve the interacting complexities of multiphase, compressible, turbulent, chemically reacting fluid dynamics under the constraints of unforgiving geometries and initial conditions. Historically, the term modeling has been applied in the narrowest sense to the mathematical description of individual phenomena or collections of phenomena called processes (e.g., droplet burning). In the remainder of this report we will use the term *combustion modeling* in both the limited and broader senses.

The resurgence of interest in combustion modeling in the broadest sense is attributable to several causes. Firstly, increased concern exists for predicting and restricting combustion-generated pollutant emissions (including noise) while simultaneously maintaining and improving thermodynamic efficiencies and fuel economies of fossil fuel combustion devices. Previously, trial and error development of combustion devices was based on, at best, relatively crude overall system models (or indeed on none at all). Nevertheless, satisfactory systems in the context of earlier economic and societal demands were achieved with only rudimentary understanding of the details of the combustion process. As emphasized by Chiu[3] and depicted in Figure 1, the present capacity to predict characteristics of combustion processes is inversely related to the sensitivity of those characteristics to the details of the combustion process. Accordingly, there was little necessity to formulate comprehensive models of combustion system details in order to predict for example, horsepower, fuel consumption, or rates of steam generation.

While ultimate system optimization will continue to require significant empirical testing, the importance of combustion modeling to the improvement and development of new, efficient, low emission combustion systems is evident. Indeed, positive reinforcement for the application of combustion modeling in this broad sense is already at hand. Relatively simple models of the combustion process in liquid and solid propellant rockets and gas turbine combustors have been partially successful in the prediction of performance and exhaust gas composition. Design modifications based on predictive models of gas turbine combustors have successfully reduced pollutant emissions (see, for example, Osgerby[2]). To a lesser degree combustion modeling has influenced operational and design changes in present day internal combustion engines. Heywood[1] emphasizes, however, the importance of combustion details by observing that small empirical modifications in the combustion process and fuel composition in conventional spark ignition engines have produced significant changes in emission characteristics. Similarly, emissions and performance of large scale utility boilers have responded to empirical adjustment and predictions based on new, yet relatively crude combustion models.[4-7] There is great promise for further improvement through detailed combustion modeling in these and other conventional devices and, perhaps more importantly, in proposed advanced systems (for example, the stratified charge engine[8]). The evolution of competitive, cleaner, more efficient, novel combustion techniques including fluidized and porous beds, catalytic combustors, and electrically augmented systems may hinge on the application of effective modeling to reduce design costs and maximize the utility of information to be recovered from basic and pilot studies.

A second, perhaps equally important contribution to increased interest in combustion modeling, is the rapidly growing availability of large, scientific computers and efficient numerical techniques which have had tremendous impact on the entire combustion field and, in particular, on modeling. Launder and Spalding[9] refer to the digital computer as the "agent of revolution" (through combustion modeling) in the design of combustion equipment. They envision the virtual replacement of physical experiment by computer experimentation by 1980. While this may be regarded as unbridled optimism on the part of practitioners of the arts of numerical methods, there can be no argument with the fact that the formulation and execution of detailed combustion models is directly coupled to the ongoing explosion in computer facilities. This aspect of combustion modeling is treated in some detail in Part III.C of this Section.

A final factor to be considered in the emergence of detailed combustion modeling is the recent appearance of several new tools which may be applied by the experimentalist and diagnostician to the study of basic combustion phenomena and combustion systems. Many of these new tools involve the application of current

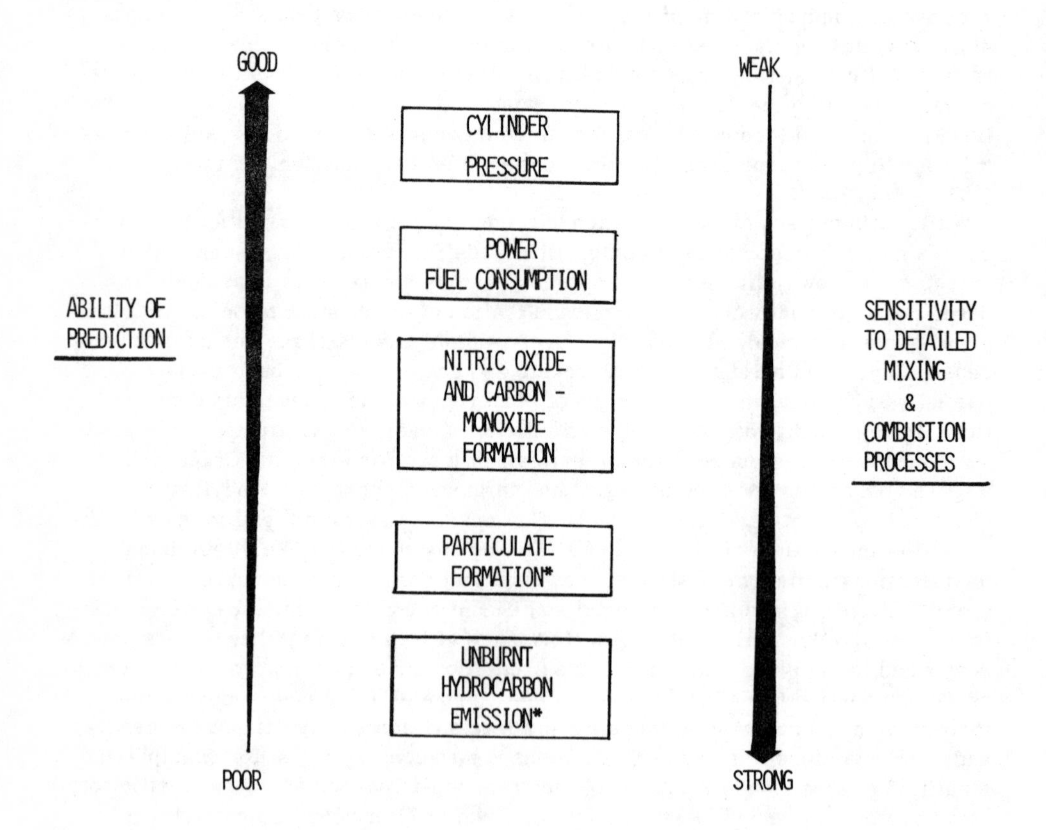

Figure 1.

laser technology in optical diagnostic probe techniques such as laser Raman spectroscopy, laser fluorescence, and laser Doppler velocimetry. Their importance to combustion modeling is manifold and is elaborated upon in Section 2 of this report.

Finally, it is appropriate to stress the fact that combustion science necessitates an integrated effort in experimental diagnostics and analytical modeling. It is recognized that a theoretical model is no more and no less than a mathematical representation of a set of interacting physical processes. Consequently, it is only as real and meaningful in describing a physical system as is our understanding and representation of these elementary processes.

A model is sometimes hampered by the difficulty in determining the accuracy of its solution whether numerical or analytical. Due to the complexity of a physical model such as that of an internal combustion engine, the model accuracy must be validated by measurement. The most direct validation of the theoretical model would be the experimental verification of modeling results and predictions. Through detailed analysis of the theoretical model we can determine the physical parameters that characterize the system, and the measurement of these parameters will then provide a definitive evaluation of the theory represented by the model. In fact, advances in diagnostic techniques can be of central importance in the formulation of a model. It is not always clear what mathematical variables are the "natural" representation of a given system. Given several alternatives, diagnostic compatibility may be quite useful in selecting one or the other.

On the other hand, the model-identified characteristic parameters will provide the motivation for the development or extension of particular diagnostic tools or techniques. Since diagnostics is by definition a quantitative process, we must verify the measurability of any variable with respect to a given technique before the implementation of such diagnostic techniques. A physically realistic model will often help in this process of decision making.

II. SCOPE AND FRAMEWORK OF COMBUSTION MODELING

In the preceding sub-section we have attempted to put into perspective combustion modeling as it relates to combustion science and technology. We have adopted the convention that combustion modeling includes the conceptual formulation and mathematical representation of combustion systems (e.g., engines and boilers) as well as elemental combustion phenomena (e.g., droplet burning). In this section, we will outline the scope of combustion modeling and emphasize the requirement for what we have chosen to call subscale modeling of the elements of combustion systems.

Figure 2 provides a comprehensive framework for our discussion of combustion modeling and is particularly useful in delineating interrelationships among various elements of combustion systems and "levels" of the modeling process. At least a cursory discussion of certain aspects of the framework is necessary at this point in order to mitigate against a knee jerk reaction—to recoil in dismay at the seeming irresolvable complications of combustion phenomena.

The objective in presenting the schematic is to depict the commonality among various combustion processes concerning their dependence upon the underlying physics and chemistry. Accordingly, the analysis and modeling of diverse com-

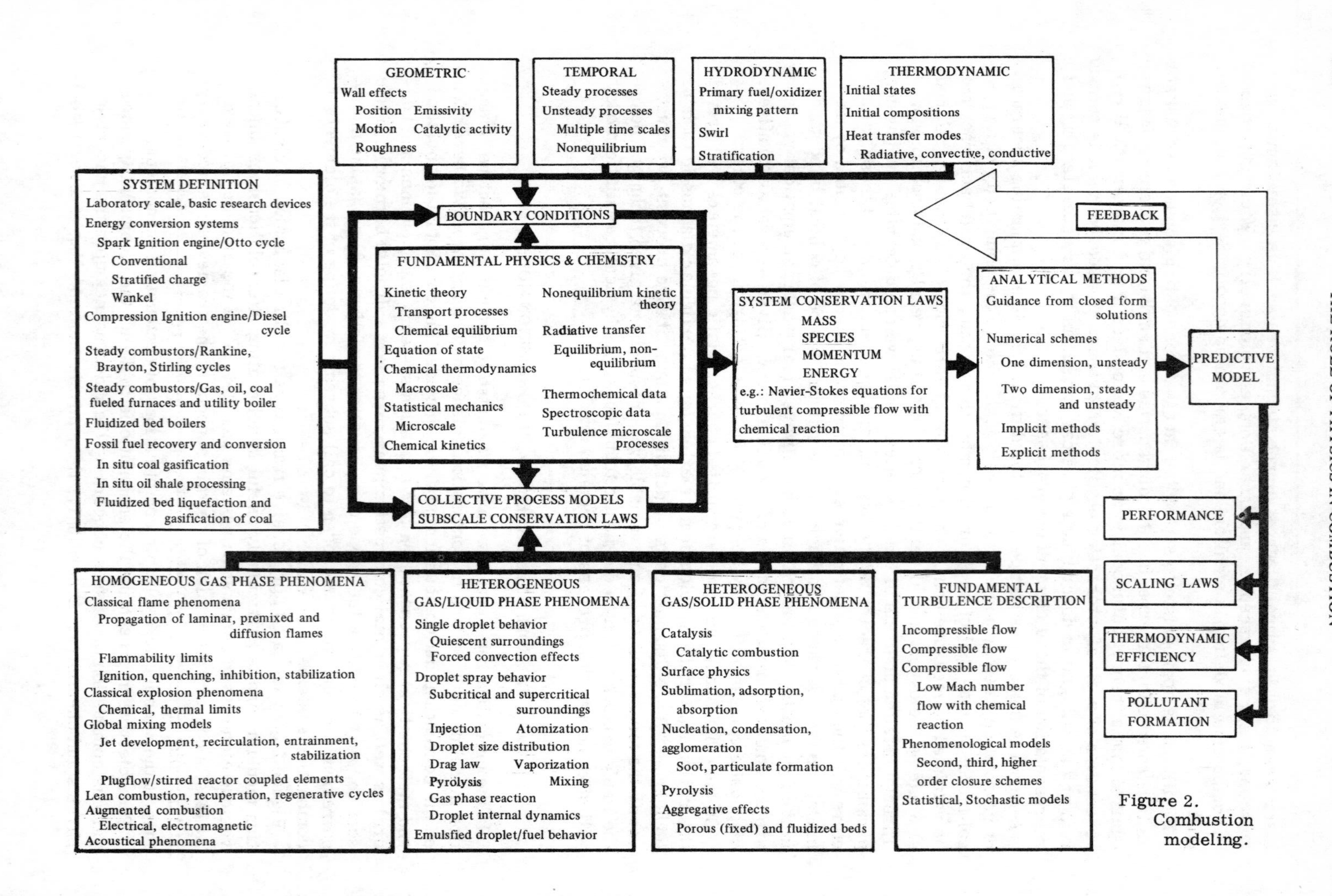

Figure 2. Combustion modeling.

bustion systems (such as the stratified charge spark ignition engine and a prevaporized liquid-fueled utility boiler) follow similar paths and involve the input of smaller models of fundamental phenomena common to both (e.g., turbulent mixing, chemical kinetics, and soot formation). The principal drawback to such an outline (besides being incomplete) is that the cataloguing imposes the requirement that lines be drawn and boxes created, where in fact there is significant interaction and feedback between component processes at all levels. Furthermore, not shown in Figure 2 is the important impact to all levels of modeling of process monitoring, basic experimentation, and diagnostics. The prospect for significant, immediate impact of new diagnostic techniques on the measurement of local thermodynamic and hydrodynamic properties within the combustion zones of practical devices is unclear. Hence, verification of comprehensive combustion models by measurement of local properties within devices may not occur as rapidly as desired. However, a conclusion of this study is that the prospects are good for the near term application of laser diagnostic techniques (discussed in Section 2) to the improvement and verification of models of component combustion processes.

A. System definition

The first step in any modeling process is the specification of the physical system to be described. Here we recognize small scale, laboratory systems as a special class along with various types of practical devices. We define practical combustion devices as those devices whose large scale industrial feasibility has already been proven. It is assumed that such a device has the capability of producing or utilizing a non-negligible fraction of the total fossil fuel energy resource in the near future. By this definition we include internal combustion engines (following both Otto and Diesel cycles) of conventional and advanced design,[10] steady state combustors employed in Rankine and Stirling cycle systems, gas turbine combustors, gas, oil, and coal fueled boilers and furnaces, fluidized bed boilers, and porous bed and catalytic combustors. We may also add at least three fossil fuel recovery or conversion systems, namely, *in situ* coal gasification, *in situ* oil shale recovery and fluidized bed liquification and gasification of coal. There are, of course, other schemes which may find application in several systems, such as the so-called excess enthalpy burners[11] and electrically or laser augmented combustors.[12,13]

B. Objectives of system modeling

The five principal objectives of overall combustion system modeling are:

1. the prediction of combustor performance, including, for example, rate of fuel consumption, power output, and the principal properties (temperature, density, major constituents) of the exhaust gases;
2. the determination of scaling laws to allow extrapolation of results and predictions to other similar systems;
3. the prediction of thermodynamic efficiency of the system;
4. the prediction of pollutant emissions (including noise) from the system;
5. the determination of the sensitivity of the model predictions to changes in all major variables and fundamental processes.

C. Components of system modeling

In meeting the above objectives, the output of combustion modeling is expected to be a predictive model (e.g., a "user-oriented" computer program) which reflects the modeler's knowledge of the fundamental physics and chemistry involved. The fundamentals will be summarized and generalized in mathematical formulations in accordance with their influence on the minimum number of important smaller scale processes and boundary conditions which comprise the overall system. Thus, the second major step in the development of a combustion system model involves the identification and modeling of the *essential* collective or subscale processes and the system boundary conditions. This may involve at some stage the writing and solving of subscale conservation laws to describe each collective process. For example in order to model combustion in a diesel engine it may be necessary to specify the functional dependence of the burning rate of a single liquid fuel droplet in a hot turbulent, convective, oxidizing atmosphere in terms of the local thermodynamic environment. Ideally, a quasi-steady model of the behavior of liquid fuel droplets in similar surroundings but in the absence of turbulent fluctuations would allow the droplet burning rate law to be written in terms of local properties. A separately obtained turbulence model could then be applied to formulate a coupled subscale model to be input to the diesel engine model.

An important part of Figure 2 is the section denoted by subscale models, and one of the most crucial of these models to combustion phenomena is hydrodynamic turbulence. *All practical combustion systems have turbulent flow* in them and many times employ turbulence to reduce mixing and combustion times, eliminate local hot spots and improve heat transfer. A basic requirement of any overall combustion model is the need to describe quantitatively the interaction between turbulent fluctuations and each subscale process and the influence of turbulence on the overall system. Because of this fundamental importance of turbulence in all practical systems we have devoted special attention in this study to current treatment of turbulent combustion. We are reinforced in this by the real prospect for significant improvement in our understanding of turbulent combustion through simultaneous modeling and diagnostics applied to relatively simple laboratory scale systems. Our discussions in this area are summarized in Parts III and IV of this section and in Section 2.

Other subscale processes are conveniently grouped in Figure 2 according to the physical state of the species involved. Clearly there is overlap. For example, combustion of fuel droplets injected into an oil fired furnace may involve eventual homogeneous gas phase mixing and reaction of the volatilized fuel and air, requiring a knowledge of the detailed chemical kinetics of high temperature hydrocarbon oxidation. Examples of modeling of many of the subscale combustion processes (excluding turbulence) cited in Figure 2 are given by Williams,[14] Fristrom and Westenberg,[15] Lewis and von Elbe,[16] and Spalding.[17] In all circumstances the development of appropriate subscale models relies on a strong interaction with experimental diagnostics. It is anticipated that new diagnostic techniques discussed in Section 2 will be applied to improve our understanding of most of these subscale processes.

D. Derivation of system conservation equations

With the exception of fluidized bed, porous bed, and catalytic combustors, the governing nonsteady equations for conservation of mass, species, momentum and energy may be written (for a gaseous medium) for an arbitrary coordinate system. The set of equations constitutes the Navier-Stokes equations for which no general closed form solutions exist. Conservation equations for nonreactive idealized fluidized beds have been suggested by Murray[18] and Jackson.[19] General fundamental equations for fluidized beds with chemical reaction have not been obtained.[20] Recently Kuo[21] presented equations for reactive porous media. The case of catalytic combustion is relatively new and has not been formalized.

For most practical and laboratory scale systems the logical "fundamental" approach to modeling is to attempt a solution of the governing Navier-Stokes conservation equations. This is a formidable task requiring the solution of a set of coupled, nonlinear, elliptic, partial differential equations. While the equations are valid for *both* laminar and turbulent flows, all practical combustion systems actually involve turbulent flows. For these flows the conservation equations for the time-averaged mean flow properties contain terms which depend on fluctuating quantities (due to turbulence) and therefore do not form a complete set of equations. Additional equations or approximations (called "closure" schemes in the trade) are needed to account for the turbulent fluctuations, and this problem of closure is a classic unsolved one in the theory of turbulent flows. Recently, Mellor and Herring[22] reviewed closure models applied to calculate mean turbulent flow fields for incompressible flows. The closure problem for chemically reacting, variable density (low Mach number, however) flow is potentially much more difficult because many new fluctuation terms appear in the reacting flow equations which are not present for incompressible flows.

Further complicating the formulation of the conservation equations for many practical systems is the presence of liquid droplets or solid particles which occupy a small fraction of the total volume. This is the case for liquid or solid fuel particles in air or gaseous combustion products and for soot particles formed *in situ* in the combustion process. Correct mathematical expressions for the source and sink terms in the conservation equations due to the presence of condensed fuel or agglomerating, radiating particulates must be obtained from subscale modeling studies.

Finally, in the case of turbulent flows it should be noted that whereas the mean flow field may be steady (e.g., in gas turbine combustors and furnaces) and essentially one dimensional or axisymmetric, the turbulent properties are always nonsteady, rotational and three dimensional. Proper modeling must account for this unhappy reality. From the foregoing it is clear why few attempts have been made to solve the governing equations in complex systems. Indeed, it is not obvious that a large scale attempt at solving these equations in a fundamental manner should be undertaken until there are clear indications that the subscale modeling of the physico-chemical processes, including turbulence, is at least approximately correct.

Most current models of combustion systems rely on semi-empiricism and *ad hoc* approximations of the chemically reacting, turbulent flow processes in order to simplify and make tractable the governing equations. More recent examples of such efforts include those reviewed by Osgerby[2] (gas turbine combustors), and the studies by Bellan and Sirignano[23] (stratified charge engines), Bracco[25] (Wankel engine), Anasoulis and McDonald[26] and Swithenbank *et al.*[27] (furnaces). The work of Swithenbank *et al.* is perhaps the most advanced of the so-called "tube and tank" models of combustors which are prevalent today. Such models represent the combustion process in terms of a number of stirred and plug-flow reactors connected in series and parallel. They have been applied to the modeling of gas turbine combustors and furnaces and have proven useful in evaluating system performance and predicting the effects of slight design and operating changes in specific applications. However, these models cannot be used to predict the effects of large changes or to design completely new combustion systems. Furthermore, although most of these models can handle complex chemical reaction mechanisms consisting of many elementary steps, gross assumptions about hydrodynamic processes are required. Many of the shortcomings of such models can be attributed to these approximations. However, no clear way currently exists to improve on these "tube and tank" models.

It has been observed that turbulence remains a major unsolved problem in physics today.[28] In view of its fundamental importance in all practical combustion systems and (we believe) in view of the possibility for significant improvement in our understanding of turbulent phenomena through the coupling of detailed modeling and diagnostics, a separate discussion of the most promising approaches to modeling compressible, reacting, turbulent flows is included in Part III.B. The numerical solution of turbulent flows is also discussed in Part III.C. In both cases we have attempted to underscore areas where the physics community may contribute.

E. Numerical solutions of the governing conservation equations

In view of the limited applicability of most present combustion modeling techniques it has been suggested that the only practical means of solving the governing equations in their most fundamental form is through the use of efficient numerical schemes on large digital computers. Numerical techniques and machine capacity have both developed to the point where their application in this area seems appropriate and feasible. Numerous examples of complex flow calculations by machine are now available in the literature (see Part III.C). However, the additional complexity of the equations for flows with chemical reaction has prevented similar calculations for combustion systems. In addition to the problems posed by turbulence, there are numerical problems arising from the appearance of strongly temperature sensitive chemical kinetic rates. It is questionable whether detailed chemical kinetic information need be input to first generation efforts in this area.

In Part III.C, the numerical aspects of combustion modeling are treated in detail. The point to note here is that numerical capabilities are reasonably well developed and computer capacities are large and growing. These would not seem to be the limiting factors in fundamental combustion modeling. In fact, the pioneering efforts of Spalding's group at Imperial College have already produced useful results in the treatment of two dimensional, steady boundary layer flows[29] and recirculating flows with and without chemical reaction,[30] and more recently in three dimensional recirculating flows with chemical reaction.[31] While Spalding is perhaps over-optimistic concerning the relative roles of numerical computation and experimental work in the near future, even he cautions against overstating the case for detailed modeling.[9, 31] Much more than a large computer and a clever numerical analyst is required. Correct representation of the essential physics and chemistry of the elementary processes composing the combustion system is demanded. Well designed diagnostic experimentation by the physicist, combustion scientist, and engineer will continue to occupy a prominent position in the formulation of predictive models.

III. SELECTED PROBLEMS FOR COMBUSTION MODELING

We will now focus attention on a few important problem areas where improvement in modeling is required. These include:

(a) elemental processes occurring in practical combustion devices;

(b) turbulent combustion;

(c) application of numerical techniques and current computer facilities to modeling of steady and nonsteady, multidimensional, turbulent combustion;

(d) advanced or novel combustion techniques including very lean combustion, catalytic combustion, fluidized and porous bed combustion, and electrically and laser augmented combustion. (The burning of emulsified fuels is treated in Section 3.)

A. Elemental processes occurring in practical devices

A principal goal of the recently concluded Workshop on Basic Combustion Research[32] was the identification of specific fundamental problems associated with practical combustion devices. In the context of the present discussion, it is the modeling of such "problems" that constitutes the development of subscale conservation laws for the elemental processes which comprise each practical device. To identify the problems, the definition mentioned earlier of *practical device* was suggested, namely, a combustion device which is producing or utilizing a non-negligible fraction of the fossil fuel energy resource or is likely to do so in the future. For the purpose of assigning priorities to elemental problems, three groups of combustion systems were designated according to the type of medium which prevails in their combustion chambers or reaction zones. It is convenient here to use the same format in order to list some of the problems where improved models are required. *Both* experimental and analytical work are needed on each problem.

1. First group: The medium is mostly gaseous with liquid or solid particles occupying a small fraction of the volume. This is the case in gas, oil and coal dust furnaces and in internal and external combustion engines. Here, specific fundamental problems include:

(a) better characterization of the turbulence (see III.B below);

(b) better characterization of the liquid droplet and solid particle vaporization, pyrolysis and combustion, particularly of droplets and particles within sprays and dusts;

(c) models of *in situ* nucleation, condensation and agglomeration of particles such as soot where the role of electrically charged species (ions, electrons, as well as microscopic particles) remains controversial; the influence of quenching environments on hydrocarbon formation in well mixed combustion;

(d) chemical kinetics of combustion and pollutant formation reactions, including elementary reaction processes as well as engineering oriented global kinetics allowing estimations of overall volumetric heat release rates and

—the need for an approximate but unified chemical kinetic theory of the oxidation of families of hydrocarbons, and in particular, the higher molecular weight hydrocarbons such as the aromatics which may see added use as octane number boosters in future fuels,

—the further development and careful evaluation of transition state theory and its predictions of thermochemical rates (see, for example, the work of Benson[33]),

—the development of the theory of high temperature homogeneous catalysis and the kinetic role of additives such as water in combustion processes,

—the kinetics of fuel rich combustion and the influence of carbonaceous species such as hydrogen cyanide and fuel-bound nitrogen on nitric oxide formation;

(e) the influence of gas turbulence on the motion and combustion or agglomeration of droplets and particles in sprays and dusts;

(f) the influence of nonsteady phenomena on all finite rate processes contributing to heat release and pollutant formation and their interaction with turbulent time scales;

(g) heat transfer partitioning among various modes (radiative, conductive, convective) and species (gas phase and particulate).

2. Second group: The medium in the combustion chamber or reaction zone consists predominantly of moving solid particles with the rest being gas or liquid This is the case, for example, in fluidized bed combustors for steam raising, fluidized beds for coal gasification and liquification. See III.D below.

3. Third group: The medium in the combustion chamber or reaction zone consists predominantly of a fixed, structured solid and the rest is gas or liquid. Packed beds, porous propellants, incinerators and *in situ* coal gasification are examples of systems in this group. Sec. III.D below.

B. Turbulent combustion

In Part II of this section it was observed that turbulence is one of the important unsolved problems in physics and, furthermore, that turbulent flows are common (and, in fact, essential) to most practical combustion devices. It is consequently not difficult to argue the case for developing a fundamental understanding of compressible turbulent flows with substantial heat release from chemical reaction. The promise in pursuing such studies lies in the reality that many incompressible and inert compressible turbulent flows in a variety of configurations are observed to have much in common. The hope is that through coupled diagnostic experimentation and modeling, fundamental laws will emerge to allow the description of whole groups of turbulent flow including various compressible flows with chemical reaction and heat release.

Even for simple incompressible flows no fundamental approach to the solution of turbulent flows has led to a model suitable for making quantitative estimates.[28] The difficulties involved in attempting to model more complicated turbulent flows were mentioned earlier. Here it is appropriate to comment on some of these difficulties in more detail, particularly as they apply to new treatments of turbulent combustion.

All fundamental turbulent flow models begin with the basic macroscopic time-dependent conservation laws.[14] In principle, these Navier-Stokes equations could be solved on a super computer having sufficient storage capacity. (For the requirements of such a device, see Launder and Spalding,[9] page 2.) In actual practice, however, only the time-averaged information at each point in the flow-field is required. The process of replacing each dependent variable in the governing equations by its time-averaged mean value and a fluctuating component which is small relative to the mean value and then taking the time-average of the equations is referred to as Reynolds averaging. This process produces the time-averaged Reynolds equations for the mean mass, concentration, momentum, and energy balances. Most of the difficulties inherent in turbulent flow theory arise due to the appearance in these equations of various second order correlation terms which are non-zero time averages (barred) of products of fluctuating (primed) quantities, e.g., $\overline{\rho' T'}$, $\overline{\rho' u_i'}$, $\overline{u_i' T'}$ and higher order terms (where u_i, ρ and T represent velocity in the ith direction, density and temperature variables).

Obtaining expressions for the correlation terms that appear in the mean flow conservation equations is the classic problem of turbulent flow theory. In principle, dynamic equations for the unknown correlations may be obtained by taking moments of the governing equations (which consists of multiplying the Navier-Stokes equations by a dependent variable, taking the time average of all terms and subtracting the mean flow equations). The drawback to this approach is that triple and higher order correlations are involved. Furthermore, combustion phenomena introduce compressibility, temperature and heat release fluctuations. Thus, many additional correlations appear, the relative importance of which are completely unknown. Since it is almost hopeless to obtain equations for higher than second order correlations a closure mechanism must be imposed.

Historically, the development of turbulent flow closure schemes has proceeded along rather separate and independent paths. The first is the classical statistical theory for analysis of the structure of turbulence. This dynamic theory is based on the mechanics of turbulence and usually consists of postulating a wave mixing or wave transfer function (e.g., the Heisenberg transfer function) for the energy transfer between low and high frequency eddies. Because of the extreme complexities of the theory, such analyses have been limited to very idealized homogeneous fields, and no serious attempt has been made to apply the theory to an analysis of trubulent flow fields of practical interest.

The second category of studies consists of the so-called turbulent viscosity models and originates with Boussinesq in 1877. A rather complete review of these theories (and their limitations) which utilize Prandtl's mixing-length proposal and various modifications is given by Launder and Spalding.[9] Although Prandtl's theory contains many short-comings, solutions of most engineering problems have almost exclusively depended upon this theory in the past. There are, however, certain problems which Prandtl's theory is intrinsically unable to describe correctly. The problem of chemically reacting turbulent flow (particularly, turbulent combustion) is one of these.

A third category of closure schemes is the so-called phenomenological theories which attempt to remove the unknown correlation terms by relating them to the Reynold's stresses $(\overline{u_i' u_j'})$ through physical hypotheses based on experimental measurements. There are many such theories existing today. The review by Mellor and Herring[22] summarizes successful approaches for the case of incompressible flow, while Launder and Spalding[9] treat the so-called two-equation models which have been applied to compressible and reacting flows. There is great hope that phenomenological models complemented by detailed information on the correlation terms obtained through new diagnostic experiments (as described, for example, in Section 2) will lead to closure models appropriate for compressible and reacting flows.

Diagnostic information of the type required for turbulent combustion modeling is presently unavailable. In its absence the present policy is to extrapolate to reacting flows information obtained from studies (e.g., using hot wires) of simpler turbulent flows. Perhaps present Reynolds stress models can be extended directly to recirculating flows. Clearly, however, the overall problem of closure for turbulent, reacting flows will be much more difficult to solve than for incompressible flow due to the large number of correlation terms.

A fourth approach to the analysis of turbulent flows has emerged in recent years. Designed to overcome the most critical shortcomings of the mixing-length, turbulent-viscosity theories for turbulent shear flows and combustion, several new statistical theories have been proposed. (See, for instance, references 34–39.) These theories are fundamentally different from the classical statistical turbulence theories both in their approach and objective. They are kinetic theories whose objective is to evaluate the correlation terms of the turbulent shear flow fields with and without chemical reaction. These theories are not primarily concerned with the structure of the turbulence field. In fact, the kinetic theory particularly designed for turbulent combustion[37–40] accepts and exploits certain known descriptions of the structure of the turbulence field given in the classical statistical turbulence theory. The kinetic theory usually involves determining and solving an equation for the probability density functions (e.g., a Boltzmann-type equation) of the fluid elements and chemical species contained in the fluid elements. All important correlation functions can then be readily constructed in a self-consistent manner from the probability density functions.

The major shortcomings of the kinetic theories as they are presently formulated are two-fold. First, the effect of chemical reaction on the structure of the turbulence field itself is being neglected. The theory analyzes the turbulent flow field initially with the density considered constant. Chemical reactions taking place in the turbulent field are then described. Second, the solution of the kinetic equations requires substantially greater computer effort than is necessary for mixing-length equations for realistic complicated flow configurations.

Recently it has been proposed[40,41] that certain assumed forms of the probability density functions of the chemical species be employed to evaluate the average chemical reaction rates in conjunction with the mixing-length (Prandtl) equation to alleviate the second shortcoming of the kinetic theory. The flow-field description given by Prandtl's equation, which has been shown to be usually quite accurate, would be used in the solution of the kinetic equations governing the probability density functions of the chemical species. This use of the Prandtl equation should substantially reduce the amount of difficulty involved in the solution of the kinetic equation for engineering problems of practical interest. By employing the kinetic equations, however, even in conjunction with the Prandtl equation, one would obtain the probability density functions which are consistent with the overall statistical behavior of the flow field. This should be a more acceptable approach to the chemically reacting flow problem than that of simply assuming the probability density functions in the evaluation of the average chemical reaction rate term.

One of the first steps in solving the closure problem inherent in both the phenomenological and new statistical theories of turbulence should be the complete and systematic development of the governing mean flow equations from the Navier-Stokes equations and an evaluation of the correlation terms that appear in these equations to determine which are the important ones. Obviously the Reynolds stresses and equivalent scalar transport terms in the species and energy conservation equations will be important, as will the impact of turbulent fluctuations on mean volumetric chemical reaction rates. Not so obvious is the possible appearance of new turbulence production terms due to the presence of combustion or flame-generated turbulence. Flame-generated turbulence has long been a subject of controversy which must be resolved if the governing equations are to be properly posed and solved. In discussing flame-generated turbulence, it is important to distinguish between changes arising from alteration of the mean flow pattern and changes resulting from some new mechanism of interaction which further complicates the closure problem. (Combustion instability is an important example of coupling between chemical reactions and fluid dynamics. Whether or not such a coupling is important in turbulent combustion is not clear. However, it should be noted that turbulence is a much more random, "high entropy" process than the acoustic and non-linear wave phenomenon associated with instability.) If combustion introduces new mechanisms for converting mean flow energy to turbulence energy or for converting chemical energy to turbulence, then these additional interactions must be modeled and the closure problem becomes that much more difficult. Recent laser Doppler velocimeter measurements indicate that flame-generated turbulence may be important.[44] More such work is needed to fully and completely resolve the question.

Of the obvious closure problems, handling the chemical reaction rate terms will probably be the most difficult one, especially if details of the chemical composition are desired. In general, the elementary chemical rate expressions are nonlinear in temperature and composition. These nonlinearities are strong enough in some cases (usually due to the temperature dependence) to introduce into the equations very high order correlation terms (if not the entire distribution function) for temperature and reactant composition.[41] Such detailed information will generally not be available, and therefore it is highly doubtful that any combustion model for turbulent flow will be able to predict chemical composition to a high degree of accuracy. On the other hand, the mean volumetric heat release rate which is the most important effect of chemical reaction will probably be easier to obtain since it is an average over many different reactions and not sensitive to the role of any one particular step. It may be possible to obtain heat release rates by a semi-empirical model, such as an advanced version of Spalding's recently proposed models.[31,43]

An alternative approach to the problem involves solving for the necessary temperature-composition distribution function required by the statistical methods. The pay-off from such an approach would be considerable. However, the applicability of the statistical approach to complicated flows is not clear. There are doubts concerning the universality of any probability density function. The influence of flow geometry on the probability density function is not known.

An extensive experimental program in support of these modeling efforts is essential. A significant amount of empiricism will most likely be involved in any turbulence closure model, and thus experimental data are needed to develop and test the models. Quantities to be measured include mean velocity, temperature and composition as well as instantaneous values of these quantities. The correlation terms involving these quantities which appear in the governing conservation equations can be obtained by continuous measurements or by discrete sampling. If discrete sampling is employed the data must first be used to develop distribution functions which may then be used to obtain correlation terms. Measurements of the distribution functions are also needed to test the statistical theories. Many of the new diagnostic techniques discussed at length in Section 2 have the capability of obtaining these data.

Even if not directly applicable to turbulent combustion flows, the results of studies on simple flow systems will be of considerable qualitative value in helping to develop more precise approaches to the more complex problems. Existing theories apply to nonreactive flows. They may be extended to the solution of compressible reactive flows provided that the turbulence structure is not fundamentally altered by combustion. Much information concerning this subject could be obtained simply by measuring the change (if any) of turbulence scales, turbulent energy, and Reynolds stress produced by combustion. Many of the fundamental aspects of the turbulent combustion theories can be checked without having to produce and probe turbulent streams with complex high-temperature combustion. Some of the more important types of measurements and techniques are discussed in Section 2.

C. On the application of numerical methods to the solution of combustion models

1. General observations

It is not without some trepidation that we offer the following discussion on the application of the methods of computational fluid dynamics to the evaluation of combustion models of varying degrees of complexity. To the casual observer there would appear to be a polarization of attitudes on this subject which was in fact reflected in the differing assessments of its merits by various participants in this workshop. The controversy, which invariably surfaces during discussion periods at meetings of the combustion community, can rudely be stated as one between the engineer and the physical scientist. It is a classic one rooted in the very nature of applied science. It is also one in which the numerical fluid dynamicist, much to the dismay of the physicist, has formed a comfortable liason with the engineer. The objective of the liason is to obtain, by the optimum means available, the most useful prediction of combustion system behavior and performance. The optimum procedure is, of course, that which is the cheapest and the least time consuming. The most useful predictions are naturally those which provide a reliable statement of the real system and which allow accurate inferences about "what to do next"—for example, with regard to exercising control of a particularly noxious pollutant without decreasing fuel economy. Heretofore, the optimum procedure has usually been to resort to empiricism, perhaps with expensive and tedious testing of a scaled version of the real system or of the full scale system itself. It is claimed by the numerical fluid dynamicist that computational facilities now admit the solution of arrays of coupled, nonlinear, partial differential equations and therefore permit the direct evaluation of detailed two and three-dimensional models of complex combustion phenomena (see, for example, Gosman and Lockwood[45] and Patankar and Spalding[31]). The implication, of course, is that the numerical exercise will be cheaper and provide just as reliable results as the now widely used experimental testing programs. While these are admirable goals, and certainly widely shared by the scientific community as a whole, the controversy stems from the means exerted by the numerical analyst to bring about the desired result—which is, the ability (a computer code) to predict. The dispute is not diminished by the concomitant enthusiasm exhibited by the analyst and, which is more common, by the users of his product, with respect to the validity and utility of the results. For, in the void created by our lack of understanding of most of the important features of turbulent, reacting, recirculating flows, the analyst has pursued what Gosman *et al.* (Ref. 30, page 60) have euphemistically championed as "the controlled neglect of reality." In order to close the system of equations governing a two- or three-dimensional, turbulent, reacting flow field a truly staggering number of assumptions and arbitrarily specified constants, functional dependencies and equation forms are required (see Gosman *et al.*[30]). The physical scientist may rightly, and, in fact, does, question the use of the word *controlled*. To him, *complete* might seem more appropriate. He argues that until the correct subroutines can be written to describe the fundamental subscale processes which constitute combustion phenomena it makes little sense to claim improved accuracy and predictive capability by virtue of detailed numerical modeling. He knows what is being done to squeeze quantitative information from the governing equations and is equally comfortable with his position that much more basic research into the various fundamental processes (some of which are listed in part III.A), using techniques such as those described in the following section, is required in order to pose the subroutines in even an approximately correct fashion.

Such are the superficial ingredients of the dispute. Each side satisfies its constituency with the claim of realism. The goal is the accurate prediction of complex combustion phenomena. The combustion scientist cautions that the goal cannot be met without a much improved knowledge of the basic subscale processes and lends his support to basic research studies where theoretical models and assumptions can be thoroughly examined and improved as experimental information is obtained. The engineer knows that the goal must be met, and met as quickly as possible. The numerical analyst abets the engineer in doing what is necessary to obtain an immediate predictive capacity.

In actual fact, as Launder and Spalding[9] and, more recently, Patankar and Spalding[31] have emphasized, there is ample need for serious effort from both camps. Furthermore, a strong case can be made for *simultaneous,* and relatively uncoupled, developments on each front. Fundamental studies will of course, involve the integrated comparison of theoretical models and experimental data. On the other hand the code developer's principal contribution is that of improving our ability to handle, by means of accurate and efficient numerical schemes, large numbers of coupled, nonlinear, parabolic and elliptic, partial differential equations. Certainly this capability will be required whenever the physical scientist ultimately proves his mettle by unlocking the mysteries of three-dimensional, turbulent, reacting flows. Indeed it is the development of this capability that provides the certain justification for pursuing the fundamental studies. However, until such time as the physical scientist meets his objectives, the validity of the quantitative outputs from such codes remains highly dubious. Many would argue that even the qualitative predictions must remain suspect. Only by continued field testing of actual or closely simulated equipment can the arbitrary constants, functional relationships and simplifying assumptions, which are required as inputs to the codes, be obtained in order to "calibrate" the systems of equations. Once this is accomplished the codes may indeed have utility in the engineering sense of allowing small extrapolation in design and tenuous predictions as to the effect of changes in operating variables. However, in the absence of a foundation in physical understanding, it is a dubious claim that our *understanding* of combustion phenomena and systems will be greatly advanced by such a process, nor is it justifiable to speculate on the imminent demise of the physical experimentalist or the field test engineer. In fact, it can be argued that equally valid performance predictions can already be obtained, at much less expense, from the extended "tube and tank" models such as that proposed by Swithenbank *et al.*[27]

We shall have occasion to return to some of these observations later in our remarks on the importance of combustion model validation in connection with the use of the digital computer to solve the model equations. First, however, it is helpful to comment briefly on the types of models which have been formulated to describe basic combustion processes as well as overall combustion systems. Without exception, each model has benefited from the application of numerical techniques to the solution of the governing coupled, nonlinear, algebraic and differential equations. In closing this section a few suggestions are offered as to logical extensions of numerical models of combustion processes.

2. *A classification of combustion models*

At various points in our commentary, we have remarked on the significant role played by the digital computer in the formulation and analysis of combustion models. At many large laboratories and universities access to a machine of the quality and computational speed of a CDC 7600 can be obtained to some limited degree. This has had far reaching implications. With a machine of this capacity it is *possible* to obtain solutions to complex models of combustion phenomena in some detail.

The hierarchy of the classes of combustion models listed below is an outgrowth of that proposed by Gosman *et al.*[30] and follows chronologically the course of expansion in computational facility and numerical capability. Specific examples of models are given for each class. Some of these examples and many others are described in greater detail in the review by Caretto[46] where a different ordering scheme is applied. Each of the models reflects to some degree a compromise between the modeler's knowledge of the basic subscale processes involved, the numerical and computational facilities available to the modeler, and his inclination to obtain a "useful" result. As a consequence, as our knowledge, capabilities, and inclinations are upgraded, each model will be found to be wanting and new models will be required.

We have noted in the preceding discussion that the rapid growth in computational facility already makes possible the consideration of combustion models of much greater complexity, although of somewhat dubious validity. The physicist is advised to familiarize himself with the details of existing models of interest and to bear in mind the tenuous foundations of all existing models of steady and unsteady, turbulent reacting flows.

(a). Zero-dimensional models.

(i). Steady flows. The most common examples of this simplest of combustion models are the so-called stirred and plug flow reactor models of turbulent combustion zones. The experimental and conceptual framework of the well-stirred reactor was originally provided by Longwell and Weiss[47] while the plug flow approximation has long been a mainstay of chemical reactor theory. In their simplest forms these models neglect the details of turbulent fluid dynamics and deal with spatially averaged properties of the combustion process. In their grandest forms, coupled volume arrays of partially or perfectly stirred reactors and plug flow zones are analyzed in order to model the mean properties of nominally steady combustion in tunnel burners, utility boilers and gas turbine combustor cans. Despite a gross treatment of the fluid dynamics, these models usually provide for a detailed treatment of the chemical kinetics of the pollutant formation reactions and occasionally of the dominant hydrocarbon oxidation reactions. Heuristic integrated forms of the conservation equations are obtained as a system of coupled, nonlinear, algebraic equations. The degree of spatial information obtained depends upon the number and location of the discretized volume elements. As we have noted earlier, such models have proved useful in obtaining performance predictions and in estimating certain pollutant emissions where mixing times are sufficiently short that finite chemical kinetic rates become important.

Examples of this class of models include the work of Breen *et al.*,[48] Swithenbank *et al.*,[27] Fletcher and Heywood,[44] Pratt *et al.*,[50] Hammond and Mellor[51] and others reviewed by Osgerby[2] and Mellor.[52]

(ii). Unsteady flows. Simple models of the time-dependent processes in conventional and stratified-charge, spark ignition engines and diesel engines have been obtained by similar gross approximations of the unsteady, turbulent combustion processes. The chemical composition of the combustible fuel/air charge is assumed either to be completely uniform or to exist in discrete lumps of arbitrary mixedness. Spatially averaged, equilibrium thermodynamic properties are assumed, with isentropic compression and expansion. The dynamics of the

injection and exhaust processes are essentially disregarded and empirical burning rate laws are assumed based on fits to measured cylinder pressure-versus-time curves. A simple energy balance relation is used to derive the nonlinear, ordinary differential equation governing the time-dependent, spatially averaged, combustion chamber properties. Numerical solutions to such models have been obtained, for example, by Newhall,[53] Blumberg,[24] Lavoie *et al.*,[54] and Chiu.[56]

(b). One-dimensional models.

(i). Steady flows. One-dimensional models of steady combustion phenomena are typical of the earliest studies of laminar and turbulent premixed and diffusion flames. In many cases, sufficiently idealized formulations of the conservation equations have made possible closed-form solutions to the governing first or second order ordinary differential equation in one space dimension.[14] Bracco's model[57] of nitric oxide formation in a laminar droplet diffusion flame in the absence of buoyancy and forced convection is a recent example of a numerical study of a complex, but still highly idealized, steady one-dimensional problem.

(ii). Unsteady flows. Perhaps the earliest time-dependent, one-dimensional models of combustion phenomena were derived from the intensive effort to describe the process of unstable combustion in liquid and solid propellant rocket engines. Typically, crude models of the quasisteady or unsteady combustion processes were grafted to elaborate treatments of the unsteady gas dynamics. The governing, nonlinear, hyperbolic partial differential equations were occasionally simplified by perturbation analyses in order to obtain ordinary, integro-differential equations which described the periodic, shock wave phenomena. Subsequently, unsteady two-dimensional analyses of the gas dynamics were obtained. Examples of the combustion instability work have been summarized recently by Harrje and Reardon.[58] The critique of modeling efforts in this field by Summerfield and Krier[59] underscores some of the difficulties encountered.

The most recent one-dimensional, time-dependent combustion models are those due to Sirignano[23] and Bracco.[25] In these, the unsteady combustion process in conventional and Wankel spark ignition engines is modeled in similar fashion. The models are proposed as more realistic representations of the real phenomena than the unsteady, zero-dimensional models mentioned earlier, by virtue of their accounting for a mechanism of spread of a plane or cylindrical flame front through the uniform combustible mixture. An *ad hoc* model of the spatially homogeneous but time-dependent, turbulent flow is used. An eddy diffusivity is assumed to depend upon "piston-generated turbulence" and an "intake-generated turbulence" which decays exponentially in time. The effects of turbulence and heat release rate are implicitly coupled through the appearance of the temperature and species concentration in the rate law. Laminar flow equations are applied with the turbulent eddy diffusivity substituted for the laminar transport coefficient. The one-dimensional, time-dependent, mass, energy and species conservation equations are combined to describe the spatially averaged combustion chamber properties. A numerical study of the governing coupled, nonlinear, parabolic, partial differential equations is performed.

(c). Two and three dimensional flows—direct numerical solution.

All combustion processes have a common basis in that they can be modeled to satisfy some form of the conservation laws. In their general form these equations involve both time and space variables for mass, momentum, energy and species concentrations. In the previously discussed models heuristic forms of the equations have been solved to obtain spatially averaged combustion zone properties. Recently, direct numerical solutions to the conservation law equations have been obtained, for *nonreacting* flows. These have involved detailed two- and three-dimensional, unsteady flow solutions to the full Navier-Stokes equations and their various asymptotic forms (see, for example, Roach[60]). While the majority of these solutions have been for laminar flow, considerable progress has been made in solving the Navier-Stokes equations for steady, two dimensional, turbulent flows.[9,22] It has recently been demonstrated that a two-dimensional, time-dependent, compressible turbulent flow can efficiently be calculated in the high-Reynolds-number regime.[61,62] The turbulent flow calculations consist of solving the time-averaged mean Navier-Stokes equations and other mean conservation equations, along with a set of equations for the turbulence quantities appearing in the mean flow equations. The turbulent flows are essentially treated with the laminar flow equations, but with turbulent transport coefficients which depend upon the properties of the flow rather than upon the fluid material properties. The required turbulence closure information has been postulated arbitrarily or extrapolated from experimental measurements. There has been some success in obtaining universal closure relations which are independent of the flow system used in the measurements.[9,22]

It is this general progress in solving nonreacting, two-dimensional, turbulent flows that offers the promise of eventual improvement in the prediction of practical combustion processes by means of direct numerical solutions to the governing Navier-Stokes equations. However, as we have cautioned in our introductory remarks to this section, the turbulent flow fields within all practical combustion devices involve a myriad of basic physical and chemical processes which interact with the three-dimensional turbulent flow field properties in, as yet, a poorly understood fashion. The closure conditions for turbulent reacting flows are simply not known. In most situations the added complication of the presence of regions of reverse flow—or recirculation—requires that a coupled set of nonlinear elliptic equations be treated. Thus while there is promise in this newest of modeling techniques, there is the realization of a measured return for some time to come.

In spite of these difficulties and realities some progress in handling the numerics for such flows has been made, principally by Spalding and his co-workers at Imperial College. Patankar and Spalding[29] developed a code to treat the parabolic set of coupled partial differential equations governing plane and axisymmetric, two-dimensional, steady, turbulent, boundary layer flows. Turbulent transport coefficients were obtained from experimental data and a simple Prandtl mixing length model of turbulence was assumed. While agreement between code predictions and experiment was reported for nonreacting flows, there is no evidence that the scheme successfully predicts reacting boundary layers. Spalding's[63] attempt to use the code to predict flame spread in the wake of a stabilizing baffle in a tube was not successful.

In 1969, Gosman *et al.*[30] published one of the first direct numerical treatments of a general two-dimensional, turbulent flow field. The original Gosman code is a relatively efficient, unified ("user oriented") treatment of the elliptic form of the two-dimensional Navier-Stokes equations with an arbitrary two-equation turbulence closure model. Turbulent flow is modeled by the time-averaged laminar flow equations of conservation with assigned effective transport coefficients. Turbulent dissipation and convection is treated by assuming that the mixing length and turbulent kinetic energy obey elliptic partial differential equations of a general form. The numerous constants and functional forms in the turbulent transport relations, boundary conditions and the two-equation closure model (e.g., the source and sink terms for the mixing length and turbulent kinetic energy) are specified by intuition and extrapolation of experimental data.

Initially Gosman reported success in predicting nonreacting, plane and axisymmetric, turbulent flow fields with recirculation. By incorporating an idealized, one-step, overall reaction rate model, the code was extended to treat reacting flows. Species and heat transport coefficients were specified and instantaneous reaction was assumed. The idealization of the reaction chemistry was argued as being consistent with the requirement to treat the turbulence properties in an *ad hoc* fashion.

It is not surprising that the original Gosman code did not accurately predict details of turbulent reacting flows. To be fair, however, it should be emphasized that the code developers did not intend the original models of the turbulence, heat transfer, and reaction rate properties to be the final word. Rather it was suggested that the unified treatment of the two-dimensional Navier-Stokes equations would provide the framework for more exact treatments of practical combustion flow fields as improved information on the basic subscale processes became available. Subsequent refinements and extensions of the code have been made at Imperial College and elsewhere (see, for example, the development by Anasoulis and McDonald[26]). A major effort has been to expand the original adiabatic flow model to allow the modeling of combustion in furnaces where convective and radiant heat transfer from the bulk gases to the walls becomes important. Gosman and Lockwood[45] reported results obtained with a four-flux radiation model added to the code. While the numerical scheme was successfully modified, the details of the temperature and radiation field of a simple furnace were poorly predicted. Elghobashi and Pun[64] have reported preliminary results of a recent extension of the code to include a three-equation turbulence model. By writing a third equation which governs the concentration fluctuations it is felt than an imporved treatment of the "unmixedness" of combustion flows can be made. Perhaps the grandest extension to date is the development of a three-dimensional computer model, Mammoth If (mathematical model of heat transfer in furnaces), by Patankar and Spalding.[31] The eventual goal of this work is the direct numerical solution of the three-dimensional, time-averaged, Navier-Stokes equations with allowance for unsteady flow, turbulence, recirculation, chemical reaction, heat transfer to walls, buoyancy and two-phase flow.

3. Closing remarks and suggestions for further work

There is agreement between the combustion engineer and scientist that the accurate prediction of the details of practical combustion flow fields will require the direct numerical solution of the governing equations. We have remarked on the differences of opinion as to what can or should be done *now* toward meeting this common goal. Computer capacities are large and growing and numerical capabilities have advanced to the stage where it is indeed possible to "turn on" coupled arrays of nonlinear, elliptic, partial differential equations and see what happens. The question remains as to what is to be gained by turning on the wrong equations.

The benefits to be gained by extensive numerical computation of nominally *steady* combustion flows (e.g., gas turbine combustors, tunnel burners, furnaces), given the present state of physical uncertainty, are of two types. First, it is possible that some useful qualitative information, properly validated by direct comparison with diagnostic information obtained from testing of the full scale system, will be obtained. This has yet to be demonstrated. Many physical scientists in the combustion field are dubious of obtaining this minimum achievement, even for homogeneous flows, in light of the difficulties of properly treating the turbulence closure problems—particularly the chemical closure aspects. It is not known, for example, whether second order closure information will be sufficient. The numerical complications and uncertainties of physical data, in regard to the treatment of numerous simultaneously appearing chemical species and reactions, compound the problem and reinforce the skeptics.

Given the rather pervasive arbitrariness that must now be imposed on the governing equations in order to treat the unknown character of the turbulence and its interaction with the chemical rate terms, it is worth commenting on the important role that model validation will play in the future efforts at predicting complex combustion phenomena. A predictive model can predict not only the characteristic behavior of the physical system under consideration but also the sensitivity of the model results to input physical parameters. The usefulness of such a model as a design tool has long been recognized, but it has been most useful only in areas where experimental data are available as an aid in the interpretation of model results. These experimental data in the past have been used to tell the designer the extent of validity (i.e., direct interpretability) of the model. In the absence of such data one must be extremely careful in accepting model results at face value, even those derived from the most carefully constructed model. The essence of the predictive process is the projection into the unknown. Consequently the model sensitivity to uncertainties in all input parameters must be analyzed. Such analysis will provide a basis for the proper interpretation of model results. Sensitivity analysis has been used widely in control theory, but not in other fields. At the present it is not clear how various analysis techniques may apply to combustion modeling. A preliminary estimate indicates that direct application of the standard procedure will not be economical, but there may be viable alternatives.

In any case when one develops hierarchies of increasingly detailed, yet imperfect, physical models of complex combustion phenomena, the desirability of systematic model validation cannot be overemphasized. This is especially true with regard to the inclusion of detailed chemical kinetics involving numerous chemical species. In practice, only limited numbers of chemical species will be feasible in the more complex models, even using the largest imaginable computers. High priority is accordingly attached to deducing compact kinetic mechanisms that reliably simulate the truly extensive kinetics obtained physically. Even given success in this area, caution is advised in accepting results where the numerical difficulties inherent in solving stiff nonlinear differential equations with time dependent reaction rate coefficients are recognized.

This brings us to the second and possibly more significant benefit assigned to the present development of direct numerical solutions to models of combustion phenomena. Efficient, stable and convergent numerical schemes will be required to handle combustion flows when the mysteries of turbulent flow and chemical reaction are solved by physical experimentation and theory. While we have documented some of the advances in this area of computational fluid dynamics, much remains to be done. Problems in numerical analysis arise due to specific features of chemically reacting flows. Some of the problems are yet to be discovered and may well depend upon the schemes used to handle particular flow situations.

As Roach[60] has pointed out, the mathematical theory for numerical solution of nonlinear, elliptic, partial differential equations remains inadequate. There are no rigorous stability analyses, error estimates or convergence proofs. Physical intuition, heuristic arguments, experimental data, trial and error testing, and analysis based on idealized simple models are an integral part of the numerical solution of nonlinear partial differential equations. For combustion problems where recirculatory flows are possible, the equations are elliptic and iteration is required. Iterative solutions introduce hidden variables which themselves may strongly influence the numerical solution (see, for example, the discussion by Anasoulis and McDonald, Ref. 26, page 49, of the need to judiciously choose the iteration acceleration parameter in their ADI algorithm).

Aside from the problems of turbulent closure, the major new difficulties due to physical phenomena which will appear in numerical solutions to most combustion phenomena will be associated with the use of chemical species equations and strongly temperature-sensitive reaction rates. For the numerical solution of species equations with chemical reaction a considerable amount of information is available for two-dimensional, laminar boundary-layer flows (see, for example, Blottner[65]), but detailed treatment of these reactions will considerably increase the total amount of computer time necessary for a solution. This circumstance occurs because the time scales spanned in specific applications are usually not sufficiently long to allow all important system components to relax to asymptotic states. Such components are fluid velocity distributions, chemical species concentrations, and internal atomic and molecular states. Thus, at physically interesting times the nonsteady evolution of specific system components from earlier conditions must be evaluated in considerable detail.

Specific physical complexities of a combustion system derive from the circumstance that chemical kinetic reactions provide the significant energy sources, which in turn drive the fluid dynamic properties of the system. The loop is completed by the fact that the dynamic properties of the reactive components, in turn, determine the extent of possible kinetic reactions for driving the system. The fact that the dynamic and kinetic coupling in combustion systems is nonlinear gives rise to many time scales that are generally disparate and variable. As we mentioned earlier, completely satisfactory degrees of detail in modeling the evolution of all degrees of freedom that are likely to be important in combustion systems is presently out of range, even with our largest computers. Nevertheless, approximate and somewhat incomplete models of chemical kinetic schemes can serve useful functions for the present.

It is logical to assume that our ability to handle the complex numerics required in such direct solutions will improve as more people spend more time experimenting with them.

In light of these considerations the choice of practical combustion systems appropriate for numerical simulation probably depends more on the source of funding for the study rather than on any valid claim of the likelihood for an accurate result. Certainly the analysis should be simpler for nominally steady combustion systems found in gas turbines and furnaces. The fact that nonadiabatic effects due to radiant and convective heat transfer must be modeled in the furnace situation perhaps makes it the more difficult problem. In both, the turbulence model should allow for nonisotropy of the turbulence field which may be a consequence of the use of high levels of swirl in the inlet flows.

The most difficult practical problem is that of the time-dependent flow and flame spread in spark ignition and diesel engines. Given our present level of understanding of combustion in these systems it is questionable whether even qualitatively accurate information, beyond the type of results that Sirignano and Bracco have obtained with much simpler schemes, can be obtained. The natural extension of Sirignano's model would be a time dependent two-dimensional model of flame propagation with a more exact treatment of the intake flow. The requirement to unravel the multiple time scales composed by the piston motion, the turbulent inlet and exhaust flows, and the reactant and pollutant chemistries is formidable. The numerics are difficult, but probably not insurmountable. However, the treatment of the details of the fluid dynamics, the physics and the chemistry would require an even greater number of *ad hoc* assumptions than in the simpler one-dimensional models.

Considering all this it seems reasonable to accord highest priority to the application of direct numerical techniques to the study of a variety of carefully constructed laboratory scale turbulent flows, with and without chemical reaction. The recommendation carries special weight in light of the recent advances made in the development of laser diagnostic probes which should allow *in situ* measurement of many of the properties of turbulent flows required by various closure schemes. By comparing detailed predictions and diagnostic measurements of relatively simple turbulent reacting flows there is the hope that the relative importance of the various correlation terms in the chemical and mechanical closure schemes can be determined and modeled in a systematic fashion. Examples of such simple "model experiments" have been suggested by Launder and Spalding.[9] Another is discussed in some detail at the end of the Section II of this report.

D. Advanced or novel combustion techniques

In the preceding discussions, we have considered areas of combustion research where a substantial modeling effort is ongoing. Much of the physics and chemistry underlying the models in these areas remains to be correctly treated, but at least the points of uncertainty are rather well defined. It is appropriate to conclude with a few comments regarding several advanced or novel combustion systems. It is in these areas that innovative contributions including modeling by the physics community may be most needed.

We restrict our attention to three specific new combustion techniques: (1) very lean combustion, (2) fluidized and fixed bed combustion, and (3) augmented combustion.

Very lean combustion (including catalytic combustion, and the use of noncombustible porous media in the combustion zone). Interest in the combustion of lean mixtures (i.e., air/fuel mixtures for which the amount of air is in excess of that required to burn the fuel completely) at modest combustion temperatures has increased recently. In part, this is due to concern about the emission of those pollutants (principally nitric oxide) which increase in concentration at high combustion temperatures. The desirability of utilizing all fossil fuel resources including, for example, lean waste gases from mines (which are presently flared or dumped into the atmosphere) has also contributed to this interest. Two techniques have been suggested for the burning of very lean air/fuel mixtures.[11,66] Both rely on substantial preheating of the unburned air/fuel mixture by means of heat recirculation from hot product gases. As the amount of air in the inlet mixture is increased beyond that required by stoichiometry, the final product flame temperature decreases. Eventually, with sufficient excess air (and an initially unheated mixture), the flame temperature is reduced below that required to sustain the heat feedback to the incoming mixture needed to produce a stable ignition process. The flammability limit thus appears as a fundamental limitation to the burning of lean mixtures. However, the lean flammability limit decreases linearly (in terms of percent fuel by volume at the limit) with increasing temperature of the inlet mixture for a wide range of hydrocarbon fuels.[11] Thus, merely by providing efficient heat recirculation between products and reactants without direct mixing, the advantages of lean combustion may be realized. The consequences and potential rewards of this combustion technique in terms of power generation from gases of low heat content, fuel saving, pollutant emission, and flame stability (flammability limits and mass throughput rates) are discussed by Hardesty and Weinberg.[11]

The question of flame stability is a particularly important one from a practical standpoint. In order to improve or assure flame stability it may be necessary to confine the lean combustion process within a fixed porous medium (such as alumina) having a high surface area/volume ratio which, when temperature equilibration is reached, floats at a relatively high temperature, but below that of the combustion gases. Heat recirculation may then be arranged to occur within the porous bed (for example, by immersing inlet tubes in the bed). In lieu of creating an extremely efficient (rapid) heat transfer process in order to accommodate high mass throughputs, it has recently been suggested[66] that the inert porous substrate may be used to support a catalytic material which affords a bypass to the thermochemical flammability limit and catalyzes the burning of very lean mixtures at lower temperatures with somewhat less preheat. Blazowski and Bresowar[66] have completed an initial experimental "proof of principal" study with such a catalytic combustor designed to yield combustion product gases at a temperature amenable to direct injection into a gas turbine. The relative importance of the hot porous bed as a thermal ignition source and a "catalytic convertor" is unfortunately not clear on the basis of their work.

This entire area is relatively virgin to exploration, particularly if inert or catalytic porous beds are used to define the combustion zone volume. While the application of heat recirculation principles introduces no particular theoretical stumbling blocks, the conservation equations for the combustion of air/fuel mixtures in inert or catalytic beds are not well developed. Theories of high temperature (greater than 1000 °C) heterogeneous catalysis are neither quantitative nor predictive.[67] Indeed, while catalysis is a classical subject for chemists, like turbulence it remains one of the principal unsolved problems in the physical sciences. There is obvious need for serious consideration (beyond the catalyst manufacturers) of lean combustion techniques both from an experimental and modeling standpoint. The principle of low temperature catalytic combustion is intriguing, possibly of revolutionary impact, yet remains unproven.

Fluidized and fixed bed combustion. Fluidized beds comprise the second group of practical combustion systems discussed in Part III.A and have long been used by the petrochemical industry in catalytic refining processes. More recently it has been suggested that they be applied to the gasification and liquification of coal (see, for example, the review by Squires[68]). Of more relevance to this report, however, is the recent resurgence of interest in direct power generation by means of the fluidized bed combustion of high sulfur content coal to allow the *in situ* removal of SO_2 formed in the combustion process. A review of earlier work in Great Britain and elsewhere is given by Skinner.[69]

The Environmental Protection Agency presently sponsors a modest research and development program to investigate the practical aspects of fluidized-bed combustion, with an immediate goal being the construction of a 630 kw continuous fluidized-bed combustion pilot plant, capable of operating over a wide range of conditions of interest, including pressure.[70] Present emphasis has turned toward the development of high pressure (10 to 20 atmospheres) fluidized-bed boilers.[71]

The fluidized-bed coal combustor consists primarily of a moving bed of solid dolomite or limestone particles (principally, $CaCO_3$ or CaO) with a small fraction of sized coal particles and the remainder being pressurized air or oxygen-enriched air. By virtue of the extremely high heat transfer rates to water-carrying tubes immersed in the bed, a relatively low and uniform bed temperature with low nitric oxide production can be maintained with relatively high rates of steam generation. Sulfur dioxide produced in the combustion process is absorbed with high efficiency on the solid dolomite-limestone particles due to the relatively long gas residence times in the bed. Regeneration of the sulfated limestone particles and a bottoming cycle using the hot scrubbed product gases to power a gas turbine generator complete the system.

The development of fluidized-beds and especially fluidized-bed combustors has proceeded in a cut-and-try fashion. This has been necessitated by the meager detailed knowledge of the dynamic physical and chemical processes occurring within the bed. The prediction of the behavior of a new system from the known behavior of a working one of somewhat similar design is anything but a sure matter. Indeed, detailed conservation equations for the reactive, high temperature fluidized-beds have not been derived.[20] As mentioned earlier, the most advanced theoretical work on non-reactive fluidized-beds is that of Murray[18] and Jackson.[19] Jackson presents mass and momentum conservation equations for constant density beds with solid pellets of equal and constant size and no reactions. The extension of this work to the case of fluidized-bed coal combustion where the fraction of coal particles in total solids is small would seem appropriate. The development and solution of conservation equations in a unified fluidized-bed model is required in order to identify the important system variables.

Reactive porous beds constitute the third category of practical combustion systems discussed in Part III.A. With respect to modeling, the situation here is only slightly more advanced than in the case of fluidized-beds. The most developed model of porous media combustion is due to Kuo.[21] The model applies to an unsteady, one-dimensional porous medium composed of reactive spherical particles of equal size. The particles are postulated to release combustion products at a rate which is a function of the pressure alone. Accordingly, Kuo's model does not include the effects of finite-rate chemistry. An extension of Kuo's model which includes some kinetic effects has been developed by Ohlemiller.[72]

Areas of application of porous media combustion include incinerators, gun powder combustion, packed beds, cigarette burning, and *in situ* coal gasification. Clearly, many types of porous media combustion exist and it is unlikely that a single model will apply to all. Modeling of these systems requires the determination of thermodynamic and transport properties of the material involved as well as the identification of the types and rates of controlling reactions. Scaled laboratory experiments and matched computer modeling efforts are needed, particularly concerning the study of *in situ* gasification techniques.

Augmented combustion. The subject of electrically augmented combustion has been mainly an academic curiosity for many years. (See, for examples, Lawton and Weinberg.[12]) Interest remains in the principle of using electrical discharges to stabilize (through the generation of large numbers of hot radicals) high mass throughputs of combustible material[12,13] or to initiate reaction of fuel rich mixtures to provide desired reducing environments (for example, in cutting or welding torches[73]). An intriguing possibility of low temperature flame stabilization was suggested recently by Fontaine.[74] The concept is based on recent observations indicating the important influence of nonequilibrium vibrational energy distributions in chemical reactions. By placing a small amount of energy into selected vibrational modes, for example, by focused electromagnetic radiation, rather than into translational modes, a much greater influence on the reaction rate is observed. By this means it may be possible to speed up combustion reactions by selecting slow steps in the kinetic process and accelerating them through low power inputs of laser energy.

IV. CLOSING REMARKS

We have noted that a discussion of combustion modeling should reflect the strong influence of experimental diagnostics. The following Section will serve to emphasize and elaborate upon this point. It is appropriate, however, to close this Section with a further assessment of the potential for impact by the physics community (as distinguished from the already vital combustion science community) in selected areas of combustion modeling. For this purpose we may identify three areas of combustion science and technology, namely, the traditional areas, the developing areas and, finally, the novel or innovative areas.

In the *traditional areas* of combustion research, which include the study of "subscale processes" such as classical flame phenomena, droplet and spray combustion, and chemical kinetics of combustion and pollutant formation reactions, the physicist will find numerous talented workers. Many views are well-established, although not all are well-founded, principally due to inadequate diagnostic information about fundamental processes. The interloper must be cautious in order to sift fact from dogma. Much of both is to be found in the traditional sources of combustion literature, e.g., the 14 biennial Symposia (International) on Combustion published by the Combustion Institute, and various journals including Combustion and Flame and Combustion Science and Technology. A conscientious exercise in homework to grasp the fundamentals is recommended. Perhaps the key inputs by physicists in the traditional areas will be in the development of advanced diagnostic and monitoring tools to be applied to laboratory scale studies and practical system evaluation.

In the *developing areas* of combustion research, we place such efforts as turbulent flow modeling and the development and extension of numerical techniques to reacting compressible flows. Here, promise exists for substantial contribution in the field by the physicist. Physicists are encouraged to take a fresh, non-conventional look at the problems in chemically reacting turbulent flows. The classical statistical theories and Prandtl's mixing-length theory contributed much toward the present understanding of the problem. Both of these theories, however, have more or less reached their limit as far as advancing the technology of turbulent combustion is concerned. Physicists with a good background in other concepts, such as quantum mechanics, but free from the prejudices of classical statistical theories and Prandtl's theory may be able to produce a viable and tractable approach to the combustion problem. Here also a powerful contribution to modeling may be made in the pursuit of the development and application of new diagnostic techniques to make local measurements (in simple systems) required by the new phenomenological and statistical theories and to provide verification of assumptions and predictions of detailed numerical models.

Finally, in the *advanced or novel areas* of combustion science, including lean and catalytic combustion, fluidized-bed and reactive porous bed combustion, and electrically or electromagnetically augmented combustion, we find the greatest probability for significant impact by the physicist. This is so because it is in these areas where he may exercise perhaps his greatest asset—namely, innovative thinking based on a firm foundation in basic physical and chemical principles. It is also true because in all of these areas, in some sense, there is the vacuum created by the nonexistence of unified theories and governing conservation equations.

V. REFERENCES

[1]Heywood, J. B., "Combustion Modeling in Automotive Engines," *Physics and the Energy Problem — 1974*, AIP Conference Proceedings, No. 19, edited by M. D. Fiske and W. W. Havens, Jr. (American Institute of Physics, New York, 1974).

[2]Osgerby, I. T., "Literature Review of Turbine Combustor Modeling and Emissions," AIAA Journal 12, 743 (1974).

[3]Chiu, W., "Combustion in Diesel Engines," Informal presentation to APS Summer Study on Technical Aspects of Efficient Energy Utilization, Princeton, July, 1974.
[4]Breen, B. P., Presentation to NSF Workshop in Basic Combustion Research, Princeton, June 1974.
[5]Bagwell, F. A., Rosenthal, K. E., Teixera, D. P., Breen, B. P., de Volo, N. B., and Kerho, S., "Utility Boiler Operating Modes for Reduced Nitric Oxide Emissions, J.A.P.C.A. 21, 11 (1971).
[6]Krippene, B. C., "Burner and Boiler Alterations for NO_x Control, Paper presented at Central States Combustion Section Meeting, Madison, March 1974.
[7]"Power Plants and Clean Air: The State of the Art," Special Publication No. 6, Power Engineering Society, IEEE 1973 Winter Meeting, Jan. 1973.
[8]*Stratified Charge Engines*, Special Issue of *Combustion Science and Technology*, 8 (1973).
[9]Launder, B. E. and Spalding, D. B., *Mathematical Models of Turbulence* (Academic Press, New York, 1972).
[10]"An Evaluation of Alternative Power Sources for Low-Emission Automobiles," NAS, Committee on Motor Vehicle Emissions, Alternate Power Sources panel, April 1973.
[11]Hardesty, D. R. and Weinberg, F. J., "Burners Producing Large Excess Enthalpies," Comb. Sci. and Tech. 8, 201 (1974).
[12]Lawton, J. and Weinberg, F. J., *Electrical Aspects of Combustion* (Oxford University Press, London, 1969).
[13]Harrison, A. J. and Weinberg, F. J., "Flame Stabilization by Plasma Jets," Proc. Roy. Soc. A321, 95 (1971).
[14]Williams, F. A., *Combustion Theory* (Addison-Wesley, Reading, 1965).
[15]Fristrom, R. F. and Westenberg, A. A., *Flame Structure* (McGraw-Hill, New York, 1965).
[16]Lewis, B. and von Elbe, G., *Combustion, Flames and Explosions of Gases* (Academic Press, New York, 1961).
[17]Spalding, D. B., *Some Fundamentals of Combustion* (Academic Press, London, 1955).
[18]Murray, J., Fluid Mech. 21, 465 (1965).
[19]Jackson, R., "Fluid Mechanical Theory," in *Fluidization*, edited by J. F. Davidson and D. Harrison (Academic Press, New York, 1971).
[20]Jackson, R., private communication.
[21]Kuo, K. K., Vichnevetsky, R., and Summerfield, M., "Theory of Flame Front Propagation in Porous Propellant Charges Under Confinement," AIAA J. 11, 444 (1973).
[22]Mellor, F. L. and Herring, H. J., "A Survey of the Mean Turbulent Field Closure Models," AIAA J. 11, 590 (1973).
[23]Bellan, J. R. and Sirignano, W. A., "A Theory of Turbulent Flame Development and Nitric Oxide Formation in Stratified Charge Internal Combustion Engines," Comb. Sci. and Tech. 8, 51 (1973).
[24]Blumberg, P. N., "Nitric Oxide Emissions from Stratified Charge Engines: Prediction and Control," Comb. Sci. and Tech. 8, 5 (1973).
[25]Bracco, F. V., "Theoretical Analysis of Stratified, Two-Phase Wankel Engine Combustion," Comb. Sci. and Tech. 8, 69 (1973).
[26]Anasoulis, R. F. and McDonald, H., "A Study of Combustor Flow Computations and Comparison with Experiment," EPA-65012-73-045, December 1973.
[27]Swithenbank, J., Poll, I., Vincent, M. W., "Combustion Design Fundamentals," *Fourteenth Symposium (International) on Combustion* (Combustion Institute, 1973), p. 627.
[28]Tennekes, H. and Lumley, J. L., *A First Course in Turbulence*, (MIT Press, Cambridge, 1972).
[29]Patankar, S. V. and Spalding, D. B., *Heat and Mass Transfer in Boundary Layers* (Intertest Books, London, 1970).
[30]Gosman, A. D., Pun, W. H., Rundral, A. K., Spalding, D. B., and Wolfshteir, M., *Heat and Mass Transfer in Recirculating Flows* (Academic Press, London, 1969).
[31]Patankar, S. V. and Spalding, D. B., "A Computer Model for Three-Dimensional Flow in Furnaces," *Fourteenth Symposium (International) On Combustion* (Combustion Institute, 1973), p. 605.

[32]Glassman, I. and Sirignano, W. A., "Summary Report of the Workshop on Energy-Related Basic Combustion Research," sponsored by the NSF, Princeton Univ. Dept. of Aerospace and Mechanical Sciences Report No. 1177 (1974).

[33]Benson, S. W., *Thermochemical Kinetics* (John Wiley & Sons, New York, 1968).

[34]Lundgren, T. S., "Distribution Functions in the Statistical Theory of Turbulence," The Physics of Fluids 10, 969 (1967).

[35]Fox, R. J., "Solution for the Correlation Functions in a Homogeneous Isotropic Incompressible Turbulent Field," Sandia Laboratories, Research Report SC-RR-70-912, February 1971. (Also to be published in The Physics of Fluids.)

[36]Lundgren, T. S., "Model Equation for Nonhomogeneous Turbulence," The Physics of Fluids 12, 485 (1969).

[37]Chung, P. M., "A Simplified Statistical Model of Turbulent, Chemically Reacting Shear Flows," AIAA J. 7, 1982 (1969).

[38]Chung, P. M., "Diffusion Flame in Homologous Turbulent Shear Flows, The Physics of Fluids 15, 1735 (1972).

[39]Chung, P. M. and Shu, H. T., "HF Chemical Laser Amplification Properties of a Uniform Turbulent Mixing Layer," Acta Astronautica 1, 835 (1974).

[40]Donaldson, C. duP., Lecture delivered at the NSF Workshop on Energy-Related Basic Combustion Research, Princeton, 1974.

[41]Gouldin, F. C., "Role of Turbulent Fluctuations in NO Formation," Comb. Sci. and Tech. 9, 17 (1974).

[42]Bywater, R. J., Aerospace Corporation, Los Angeles, California, private communication.

[43]Spalding, D. B., "Mixing and Chemical Reaction in Steady Confined Turbulent Flames," *Thirteenth Symposium (International) On Combustion* (Combustion Institute, 1971), p. 649.

[44]Chigier, N. A. and Dvorak, K., "Laser Anemometer Measurements in Flames with Swirl," paper presented at the Fifteenth International Symposium on Combustion, Tokyo, August 1974.

[45]Gosman, A. D. and Lockwood, F. C., "Incorporation of a Flux Model for Radiation into a Finite-Difference Procedure for Furnace Calculations," *Fourteenth Symposium (International) on Combustion* (Combustion Institute, 1973), p. 661.

[46]Caretto, L. S., "Modeling Pollutant Formation in Combustion Processes," *Fourteenth Symposium (International) on Combustion* (Combustion Institute, 1973), p. 803.

[47]Longwell, J. P. and Weiss, M. A., "High Temperature Reaction Rates in Hydrocarbon Combustion," Ind. Eng. Chem. 47, 1634 (1955).

[48]Breen, B. P., Bell, A. W., de Volo, N. B., Bagwell, F. A., and Rosenthal, K., "Combustion Control for Elimination of Nitric Oxide Emissions From Fossil-Fuel Power Plants," *Thirteenth Symposium (International) on Combustion* (Combustion Institute, 1971), p. 391.

[49]Fletcher, R. S., and Heywood, J. B., "A Model for Nitric Oxide Emissions from Aircraft Gas Turbines," Paper 71-123 at AIAA Ninth Aerospace Sciences Meeting, New York, 1971.

[50]Pratt, D. T., Bowman, B. R., Crowe, C. T., and Sonnichsen, T. C., "Prediction of Nitric Oxide Formation in Turbojet Engines by PSR Analysis," AIAA Paper No. 71-713, 1971.

[51]Hammond, D. C. and Mellor, A. M., "Analytical Calculations for the Performance and Pollutant Emissions of Gas-Turbine Combustors, Combustion Science and Technology 4, 101 (1971).

[52]Mellor, A. M., "Current Kinetic Modelling Techniques for Continuous Flow Combustors," in *Emissions from Continuous Combustion Systems*, edited by Cornelius, W. and Agnew, W. G. (Plenum, New York, 1972).

[53]Newhall, H. K., "Kinetics of Engine-Generated Nitrogen Oxides and Carbon Monoxide," *Twelfth Symposium (International) on Combustion* (Combustion Institute, 1969), p. 603.

[54]Lavoie, G. A., Heywood, J. B., and Keck, J. C., "Experimental and Theoretical Study of Nitric Oxide Formation in Internal Combustion Engines," Comb. Sci. and Tech. 1, 313 (1970).

[55]Bastress, E. K., Chng, K. M., and Dix, D. M., "Models of Combustion and Nitric Oxide Formation in Direct and Indirect Injection Compression-Ignition Engines," SAE Paper No. 719053, 1971.

[56]Chiu, W., Cummins Diesel, private communication.

[57]Bracco, F. V., "Nitric Oxide Formation in Droplet Diffusion Flames," *Fourteenth Symposium (International) on Combustion* (Combustion Institute, 1973), p. 831.

[58]Harrje, D. T. and Reardon, F. H., eds. "Liquid Propellant Rocket Combustion Instability," NASA SP-194, 1972.

[59]Summerfield, M. and Krier, H., "Errors in Nonsteady Combustion Theory in the Past Decade," AIAA Paper No. 69-178, 1969.

[60]Roach, P., *Computational Fluid Dynamics* (Hermosa, Albuquerque, 1972).

[61]Baldwin, B. S. and MacCormack, R. W., "Numerical Solution of the Interaction of a Strong Shock Wave with a Hypersonic Turbulent Boundary Layer," AIAA Paper No. 74-558, 1974.

[62]DeWiert, G. S., "Numerical Simulation of High Reynolds Number Transonic Flows," AIAA Paper No. 74-603, 1974.

[63]Spalding, D. B., "Mixing and Chemical Reaction in Steady Confined Turbulent Flames," *Thirteenth Symposium (International) Combustion* (Combustion Institute, 1971), p. 649.

[64]Elghobashi, S. E. and Pun, W. M., "A Theoretical and Experimental Study of Turbulent Diffusion Flames in Cylindrical Furnaces," paper presented at the Fifteenth International Symposium on Combustion, Tokyo, August 1974.

[65]Blottner, F. G. and Johnson, M., "Chemical Reacting Viscous Flow Program for Multi-Component Gas Mixtures," SC-RR-70734, August 1971, Sandia Labs, Albuquerque, New Mexico.

[66]Blazowski, W. S. and Bresowar, G. E., "Preliminary Study of the Catalytic Combustor Concept as Applied to Aircraft Gas Turbines," Air Force Aero Propulsion Laboratory Rept. AFAPL-TR-74-32, Wright-Patterson AFB, May 1974.

[67]Langley, R., Englehard Industries, private communication.

[68]Squires, A. M., "Clean Fuels from Coal Gasification," Science 184, 340 (1974).

[69]Skinner, D. G., *The Fluidized Combustion of Coal* (National Coal Board, London, 1970).

[70]Turner, P. P., "EPA-CSL Program to Control Pollution from Stationary Sources," in *Proceedings of Third International Conference on Fluidized-Bed Combustion*, EPA-65012-73-053, December 1973, p. 0-2-1.

[71]"Evaluation of the Fluidized-Bed Combustion Process," Vol. I–IV, EPA-650/2-73-048a, b, c, d, December 1973.

[72]Ohlemiller, T., private communication.

[73]Weinberg, F. J., "Combustion Temperatures: The Future?" Nature 233, 239 (1971).

[74]Fontaine, A., comments at NSF Workshop on Energy-Related Basic Combustion Research, Princeton Univ. (1974).

2. DIAGNOSTICS FOR EXPERIMENTAL COMBUSTION RESEARCH

I. INTRODUCTION

Successful development of comprehensive mathematical models for both basic combustion phenomena and practical systems (turbine combustors, reciprocating engines, furnaces, etc.) requires local information on gas-phase chemical composition, physical parameters, and particulate properties. In fact, the modeling procedures and the diagnostic techniques interact so strongly that it is practically impossible to consider one without the other. Modeling would be an empty art without significant effort spent in diagnostics, since the basic aim of modeling is the prediction of new phenomena from a necessarily simplified view of a physico-chemical combustion system, the validity of which can only be judged from experimental data. Similarly, diagnostics would be greatly diminished in importance without conceptual models, since *practically every* set of potentially useful experimental conditions would have to be subjected to measurement procedures in that case. Thus, the development of modeling and diagnostics must go hand in hand, and those who perform combustion modeling (usually theoreticians) and those who perform combustion diagnostics (usually experimentalists, who may or may not have significant combustion experience) must enter into close dialogue in order to determine (1) what should be measured, (2) what can be measured, (3) how the measurements are to be performed, and (4) how the resultant data are to be interpreted and fed into the framework of theory.

These comments are, on one level, obvious, and clearly apply to science in general. But often we do not consciously think of these considerations because we deal in much of our work with fields with which we become relatively familiar. Thus, for example, an experimentalist doing Langmuir probe measurements in an electrical discharge rapidly unites his knowledge of probe methodology with the science and technology of gas discharges. However, in this APS study, we are viewing fields in which the physicist has *not* conventionally contributed to a substantial extent. This is, in fact, increasingly true as both diagnostic methods and modeling procedures become increasingly specialized and demanding of highly focused effort.

For the purposes of our discussion of combustion diagnostics, we choose to divide our consideration of probes into three areas: (1) non-perturbing, (2) perturbing (non-purposeful), and (3) perturbing (purposeful). The major focus of our discussion will be on the first area—particularly with respect to newly developed and emerging light scattering probes. To a lesser extent, other non-perturbing and purposely perturbing probes will also be discussed. [By "purposely perturbing," we imply that the probed system is given a minor perturbation (as, for example, a taut wire is plucked) and the resultant behavior of the system (e.g., the oscillation of the wire) is then observed. By "non-purposely perturbing," we imply that the probed system is altered (to a greater or lesser degree, depending upon both the test system and probe geometries) by insertion of the probe.]

In many instances, perturbing probes only slightly alter the measurement environment, and so are entirely acceptable. For example, a small thermocouple carefully inserted into a large furnace at moderate temperatures is capable of producing dependable results. However, the same thermocouple inserted into a smaller test volume at higher, rapidly fluctuating temperatures would be likely to produce far less reliable results, and, eventually, temperatures and temporal fluctuations could be achieved which would totally prohibit the use of such solid probes.

An enormous amount of useful data for combustion science and engineering has been supplied in the past by perturbing probes, such as thermocouples, hot wires, etc., and undoubtedly use of these probes will continue unabated in the future. However, strenuous environments of new combustion systems dictate that new probes be devised in order to provide data for conditions for which solid probes cannot even survive, much less be expected to give accurate results. Thus, our attention has been placed upon optical probes and, in order to exploit their characteristics fully, we have chosen to discuss in detail those techniques based upon light scattering: laser Doppler velocimetry (LDV) measurements of velocity; Raman scattering, Rayleigh scattering and fluorescence measurements of temperature, density and composition; and particle scattering (often termed Mie scattering) measurements of parameters relating to particle and droplet densities and size distributions.

In order to provide a physical basis for establishing the relative intensities of the various scattering processes associated with the diagnostics described here, we present typical values of scattering cross sections for these processes in Table 1. Here, the scattering cross section is defined by the relation

$$Q_s(\text{joules}) = Q_0(\text{joules})N\left(\frac{\text{molecules}}{\text{cm}^3}\right)\sigma\left(\frac{\text{cm}^2}{\text{sr}}\right)l(\text{cm})\Omega(\text{sr}),$$

where Q_s is the light energy scattered into solid angle Ω (and into a given polarization state and spectral region), Q_0 is the incident light energy, N is the number density of scattering molecules, and l is the length along the incident beam from which scattering is observed.

It is evident from Table 1 that very small cross sections are associated with some of the scattering processes considered here. While we do not minimize the experimental difficulties that these small magnitudes introduce into the diagnostic methods, modern experimental apparatus and signal processing techniques have made great strides in bringing these methods to their current advanced stage—which we believe to be the point of applicability to practical engineering systems.

The advantages and disadvantages of using light scattering diagnostics will be

Table 1

Scatterer	Type of scattering	Cross section cm^2/sr
0.1 μ diam. particles	Mie	10^{-13}
10 μ diam. particles	Mie	10^{-7}
Atoms	Fluorescence	10^{-16}–10^{-13}
Molecules	Fluorescence	10^{-19}–10^{-22}
N_2	Rayleigh	9×10^{-28}
N_2	Rotational Raman (strong line)	5×10^{-30}
N_2	Vibrational Raman	5×10^{-31}

discussed in succeeding sections, but here we point out that most of the non-perturbing optical diagnostics done in combustion has been either emission or absorption studies.[1] While these techniques have and will continue to have a strong impact upon combustion research, they are relatively well-developed, and so need little further comment here. We do wish to point out that they have one substantial disadvantage, viz., interpretation of the resultant data is often complicated due to the necessary deconvolution of the effect of the nonuniform test material conditions from the "line-of-sight" type of data that is obtained. On the other hand, many instances exist for which transmission, absorption, or emission studies provide either the only possible source of non-perturbing data, or provide significant additional data in experiments relying upon other techniques (such as light scattering).

II. NON-PERTURBING LIGHT-SCATTERING PROBES

Within the last decade, a significant impetus has been given to optical diagnostics by the introduction of the laser into techniques which had previously depended upon much less intense, incoherent sources. Over this period, the development of laser technology has progressed significantly, and the commercial availability of a wide variety of highly useful and specialized laser light sources has increased substantially. Along with this rapid development of light source technology, signal discrimination and processing equipment has kept pace—so much so that experimentalists often find their equipment outdated in terms of optimal performance after relatively short periods of time. Although sometimes frustrating to active workers, this state of affairs represents a healthy condition for a fast-growing field.

Nowhere in diagnostic measurement technology has new equipment been of more profound impact than in the development of light scattering probes. The basic ideas behind these probes—whether, for example, the fundamentals of electromagnetic signal heterodyning (as in LDV) or the quantum mechanical principles of Raman scattering (for species state property measurements)—have been well-known for many years. However, the application of these ideas to homogeneous gas phase and heterogeneous gas/liquid and gas/solid phase studies on a *diagnostic* basis is new, and is strongly related to a combination of the availability of advanced equipment, as has just been described, and the needs of science and engineering for new system probes of high capability.

To be sure, some of these techniques have been widely used in laboratory analyses in the past. (For example, Raman scattering has been strongly utilized in basic laboratory studies for identification of chemical species, usually in the solid or liquid state, and for elucidation of chemical bonding schemes.) However, these uses have been for either the exploration of molecular structure in a fundamental fashion or for the routine analysis of material samples. What is specifically new is the application of these techniques to *in situ* gas phase (as well as gas-liquid and gas-solid phase) diagnostics, particularly as applied to highly stressed environments. Because of the general weakness of most light-scattering processes, they were not serious contenders for these diagnostics before the advent of the laser, but now, they are clearly within the realm of technical reality. This is not to say that the experiments are necessarily simple to perform, or in fact, that the weakness of the scattered signature—particularly for Raman scattering—is not a serious obstacle to be overcome in many instances. But laboratory scale experiments have already demonstrated the utility of laser light-scattering diagnostics for state property measurements, so that a logical step in the evolution of diagnostic probes is to develop prototype instruments applicable to larger-scale, engineering test systems.

The advantages of light-scattering probes for utilization with combustion systems include a variety of features common to many of the schemes, and some specific to the various individual techniques. In general, these techniques have the advantage of providing data over a wide sensitivity range without probe interference. Local measurements can be performed by moving the point of intersection of the laser beam and the line of observation so that different positions within a combustion zone can be probed remotely. Spatial resolution approaching the diffraction limit of light is possible with collimated laser beams. Sensitive detectors and electronic signal-processing systems with subnanosecond time responses are also readily available for transient combustion measurements, since several of these techniques provide essentially "instantaneous" time response. The capability exists, therefore, for real time, on line diagnostics —fed directly to a computer which can then provide feedback to the experiment, if desired.

These various scattering techniques, once developed and employed together with other diagnostic methods in an integrated fashion, should lead to significant advances in our knowledge of practical combustion processes and help to provide the necessary basis for model development, which in turn should provide a rationale in combustion processes for the optimization of fuel utilization with a balance between maximizing energy extraction and minimizing pollutant formation.

A. Laser velocimetry

The form of laser velocimetry currently most discussed is termed laser Doppler velocimetry, because of an analysis based upon the Doppler effect of the particular technique used. This analysis can also be performed in an alternative fashion and, in addition, other basic configurations exist for laser velocimetry, such as time-of-flight and flow visualization techniques, which are of use in specific experimental situations. Here, we will confine ourselves to consideration of LDV, which has great promise for worthwhile application to combustion studies.

A great variety of LDV systems using different principles, optical configurations and signal processing electronics have been devised and applied to many flow situations. The best type of system for any given experiment is quite dependent on the flow situation, i.e., there is no universal instrument. Here, after we describe briefly the basic methods and options, we discuss the application of LDV to combustion.

Several hundred papers have been written on LDV since its inception, approximately ten years ago. A review of laser velocimetry has recently been published by Trolinger.[2] Reference may also be made to the proceedings of the International Workshops on Laser Velocimetry held at Purdue University in 1972 and 1974. Technological development is still being carried out, but experimentation with instrument design has slowed and largely stabilized now on a single basic type of instrument with variations. LDV apparatus is now being used to make measurements with quite impressive accuracy and spatial resolution. Among the flows studied are laminar and turbulent jets and channel flows in liquids and gases, such as Mach 2-3 cold wind tunnel flows (including boundary layers effects), gas turbine compressor flows, free air turbulence, aircraft wake studies, gas turbine exhausts, pre-mixed gas combustion flows, combustion MHD channel flows and solid fuel rocket flows. LDV has even been applied to fluid mechanics studies in arteries and veins and in tiny blood vessels in the retina. However, relatively little work has been done in hot and/or reacting flows. The problems generally increase with gas velocity, temperature, and the level of turbulence. Furthermore, confined flows and boundary layers are more difficult to handle experimentally due to surface scattering, window fouling, etc.

1. *Basic methods*

Detailed discussions of LDV theory have been presented (see Ref. 2 for sources), so that we only wish to outline some of the fundamental ideas here for purposes of clarity.

We consider first particle scattering characteristics. For macroscopic particles, where the ratio d/λ of particle diameter d to scattered wavelength λ is ≥ 1 we find the small angle forward scattering intensity $\gg$backscatter (often $100X$). Thus, we find that single particle scattering is strongly anisotropic. For spherical dielectric particles, the solution to this problem has been worked out mathematically, and is usually termed Mie scattering.[3] For $d/\lambda \ll 1$, the phenomenon approaches essentially elastic atomic or molecular scattering, termed Rayleigh scattering.

In laser Doppler velocimetry there are two basic schemes, single scatter (or reference beam mode) and differential or dual scatter (also known as the fringe method).

In *single scatter,* light scattered from some position (usually a focus) in the incident beam is collected at some scattering angle ψ to the incident direction, and its frequency compared with that of a reference beam from the same laser. The frequency shift is proportional to the particles' velocity component along the bisector of the supplement of the scattering angle ψ, and is proportional to $\sin(\psi/2)$. Thus for backscatter ($\psi \sim \pi$) the shift is a maximum and the system is sensitive to velocities along the line of sight, while for forward scatter ($\psi \ll 1$), the shift is much smaller and the system is sensitive to velocities perpendicular to the line of sight. The reference beam, whose intensity is much less than that of the incident beam but somewhat larger than that of the scattered beam, may either pass through the measurement point, or follow a path external to the flow system. The Doppler frequency shift may be determined either by homodyning (beating) the scattered and reference signals on a photodetector to give a radio-frequency signal, or by direct comparison of the optical frequencies using an interferometer[4] (Fabry-Perot). The latter method is useful for high velocities and backscatter, when the Doppler frequency may be too high ($\gtrsim$100 MHz) for light beating with available detectors, unless very small scattering angles are used.

In the *differential or dual scatter* (fringe) scheme, which is now widely employed, two equal intensity beams from a laser are made to intersect at a common focus, and light scattered from this control volume is collected at some angle onto a photodetector. An explanation can be given in terms of the beating of light scattered (with equal intensity) from the two incident beams, and it may be noted that the spatial coherence requirements for beating are automatically satisfied since the two scattered signals originate at the same point. The beat frequency is proportional to the velocity component normal to the bisector of the incident beams and is proportional to $\sin(\psi/2)$ where ψ is now the angle between the beams. The beat frequency is independent of the collection angle, which allows large aperture collection optics to be used without aperture broadening, i.e., loss of frequency (velocity) resolution, inherent to the single scatter scheme. An alternative explanation of the differential scatter scheme, which is equivalent and more illuminating, may be given in terms of the formation of a real fringe system at the intersection of the incident beams, with spacing $\lambda_0/2\sin(\psi/2)$. A particle passing through the fringes scatters a Doppler burst light signal which is intensity-modulated at a frequency proportional to its velocity component normal to the plane of the fringes.

Either the single scatter or the differential scatter (fringe) method may be used in forward scatter or backscatter. Forward scatter has the advantage of a larger scattered signal, allowing the use of lower power lasers or smaller particles, but requires access to the flow system from both sides, and accurate alignment of the incident and collection optics. The latter is difficult when scanning in large unmovable systems, when the incident and collection optics must be ganged together. Furthermore, in hot turbulent gas flows, refractive index inhomogeneity may cause the illuminated and viewed volumes to wander independently, leading to signal drop-out and errors. For these reasons a backscatter geometry, in particular a co-axial backscatter system, may be preferable even though the signal level is smaller and one may encounter problems with excessive light scattered from surfaces. Even in backscatter, the differential scatter (fringe) mode may be subject to refractive index inhomogeneity causing lack of intersection of the two beams. A single scatter, coaxial backscatter system with external reference beam[4] is free of this problem because the incident and collected beams occupy the same space, though of course the probed volume may wander due to refractive index effects. However, unlike the dual scatter system it can measure only one velocity component, along the line of sight, and gives high frequency shifts which may require the use of a Fabry-Perot rather than light-beating for frequency determination, unless the velocity is low.

2. Seeding

The basis of LDV depends upon the sensing of the position of particles as a function of time as they pass through a test volume. The natural density of small particles in the flow may be adequate for this purpose in some instances. In many others, small light-scattering particles are purposely seeded into the flow. A substantial effort has been spent on determining the required characteristics of seed materials and their ability to accurately follow flow configurations. The basic desire is to have the particles respond to an average over the molecular velocity distribution function, i.e., be large enough not to exhibit Brownian motion, but small enough not to slip relative to the fluid. Usually, particles whose diameter is some fraction of a micron up to about one micron are considered to be well-suited for LDV gas flow studies, except for strongly accelerated flows and shocks.

The problem of adequately seeding LDV experiments has not been properly solved yet, in the sense that workers in the field often feel that this technology is really an art. Methods for seeding include a variety of basically mechanical methods, as well as fluidized-bed systems. A particular need exists for very high temperature seed materials, i.e., stable solid materials with good scattering properties at ca. 3000 °K and higher.

With heavily seeded flows and low spatial resolution one can have high duty ratio, ≥ 100 percent, or almost a continuous signal. With low seeding and high spatial resolution one has very low duty ratio. Low duty ratios are adequate for stationary flows but require long measurement times and hence result in time-average moments only.

3. Signal processing

Single particles passing through the sample volume generate Doppler bursts at random time intervals. Provided the intensity is such that there is a statistically large number of photodetections per Doppler cycle, the electrical output from the detector is a smooth replica of the scattered light intensity with negligible quantum noise. The problem is then to measure the frequency of the train of bursts (high duty cycle) or individual bursts (low duty cycle). A number of methods for doing this are possible. These include the following.

(1) *A single channel R. F. spectrum analyzer* is effectively a tunable filter but has a low information rate. It suffers, however, from Doppler "ambiguity" because it effectively Fourier analyzes a finite wave train. The natural width has to be subtracted from the signal width to get the true velocity spread due to turbulence.

(2) *The multi-channel spectrum analyzer* is a filter bank capable of achieving full information rate. It also suffers from Doppler ambiguity.

(3) *The frequency tracker* was developed by Harwell and Imperial College but is only applicable for high duty cycles ($\geq 20\%$).

(4) *Individual realization* requires hand processing of data recorded on tape,

and is generally good only for low velocity studies.

(5) *The digital counter,* developed by General Electric and Arnold Research Organization, times the zero crossings of individual bursts and in principle eliminates Doppler ambiguity. It is very good for low duty ratios when the signal to noise is good.

(6) *A high speed digital photon correlator* has been used for detecting very low signal intensities such as from natural seed in wind tunnels. Such systems can measure down to one photodetection per 500 Doppler cycles, at the expense of long observation times.

(7) *Direct optical frequency determination* is used by Self[4] with a Fabry-Perot interferometer as an optical spectrum analyzer. In this system the rate of signal pulses is counted as the interferometer is swept in frequency. The display gives the time-average velocity probability function in real time. However, the information rate is low and the seed concentration must be steady.

LDV systems have become very sophisticated and several special applications of the technique have recently been performed. These include simultaneous multi-component velocity measurements, high resolution boundary layer measurements, and frequency offset measurements to determine the sense of velocity in recirculating flows.

4. *Applications to combustion*

Relatively little work has been done in combustion flows using LDV systems. Mossey at General Electric,[5] Whitelaw *et al.*[6] at Imperial College, Baker *et al.*[7] at Harwell, Durst[8] at Karlsruhe, Chigier[9] at Sheffield, Trolinger[10] at Arnold Research Organization, Samualson[11] at U.C. Davis, and Bray[12] at Southampton have all used the dual scatter, real fringe technique to study flames. Self[4] at Stanford has applied direct optical frequency determination with a single beam to measure boundary layers in MHD combustors.

For most combustion situations the flow velocity is low enough that the dual scatter, real fringe method is probably best. Counter electronics are preferred over the frequency tracker since low seeding concentrations are needed to avoid perturbing the combustion system.

Laser velocimetry is the most developed diagnostic technique discussed in this section. Nevertheless, a number of significant improvements can be made to overcome some of the current limitations that exist in applications to combustion problems. Some of these limitations, and suggested research to improve upon them are the following.

(1) *Refractive index fluctuations*. Fluctuations in the optical paths due to refractive index variation through a combustion environment may cause error and drop-out of signals in the dual scatter method or any method except the single backscatter method. The problem increases with path length and depends on the intensity and scale of refractive index fluctuations, and is difficult to quantify. This is probably not a problem in small flames but Whitelaw[7] had limited success with a large furnace at the International Flame Foundation. Single backscatter using Fabry-Perot, or more likely heterodyne, is not always convenient since it measures only the velocity component along the line of sight.

(2) *Particles*. Good scatterers are needed. Inert refractory oxides such as

MgO, Al_2O_3, TiO_2, and ZrO_2 are probably best for combustion studies. Reliable methods of dispersing powders into gas streams are also needed.

(3) *Background interference*. Most laboratory flames are reasonably clean and transparent in the visible, i.e., they don't radiate like black-bodies. Flame luminosity can probably be rejected sufficiently by spectral and spatial filtering. However, "real" flames, such as occur in practical furnaces and internal combustion engines, may be very luminous. For these cases, elimination of background interference will require more developmental research and some innovation. Background radiation from hot walls and from laser scatter from walls is a problem in boundary layer measurements but can be adequately eliminated if one has good spatial filtering.

(4) *Sampling bias*. Even in cold turbulent flows there may be a sampling bias. (See Tiederman.[13]) If the particle concentration is uniform then the flux of particles is higher when velocity is high than when it is low (in turbulent fluctuations) and the sample of the velocity probability function, $f(u)$, will be biased to high velocities. However, this may not be true in compressible flows because there will, in general, be a correlation between the instantaneous gas density, and hence particle concentration, and the instantaneous velocity.

Durst[8] has considered this effect in a turbulent flame. For a plane flame front, the gas density and particle concentration decreases and the velocity increases across the flame front in proportion to the temperature ratio. Thus the particle flux stays constant. Then if the flame front oscillates across the measurement volume, there would be no bias in the measured time-average velocity probability function at the point. However, the time-average can give an *apparent* large turbulence intensity which is really an average of $f(u)$ on either side of flame front. In practice, streamtube divergence in the vicinity of the flame front further complicates interpretation of data. Durst has considered this, but his treatment and inferences from his data are not entirely convincing.

(5) *Droplet and solid particle combustion*. There have been no applications of LDV to the study of droplet or particle combustion as yet, largely because of problems due to transparency in large systems. The absorption mean free path, $1/Na$ where N = concentration, a = area of particles, may only be ~1 meter in coal combustors. The fuel particles will often be too large (50–100 μ) to act as satisfactory flow tracers and are even larger than the fringe spacing. Whether one can use LDV on small refractory tracers in the presence of large fuel particles is not clear. Furthermore, we are also interested in the fuel particle motions themselves since the velocity of the particles relative to the gas strongly influences the supply of O_2, the removal of products, and hence the burning rate. We are also interested in the concentration, flux and size distribution of fuel particles. For large fuel particles, it may be better to use time-of-flight or flow visualization techniques.

(6) *I. C. engines*. Detailed analysis of the combustion process in I.C. engines will require time resolution on the order of 100 μsec. Such short sample times can result in unacceptable statistical fluctuations unless signal rates are high, or unless signals at a fixed crank angle are averaged over a number of combustion cycles. The latter procedure introduces cycle-to-cycle variations which must be suitably averaged as well. Trolinger[10] has attempted LDV measurements in a diesel engine and found that window fouling, even using heated quartz,

was troublesome. I.C. engine studies are probably the most difficult problem. With effort, one may be able to make some useful measurements, but the quality and detail will always be much less than in simpler flow situations.

B. Laser Raman and Rayleigh scattering

1. Description of Raman scattering diagnostics

Raman scattering has been evaluated in laboratory experiment and by theoretical analyses for applicability to combustion and fluid mechanics measurements. Indeed, this was the subject of a highly-focussed workshop on this topic held in 1973 and reported in a series of 29 papers published in the workshop proceedings.[14] The virtues of Raman scattering, described in more detail in the introduction to the proceedings, include those mentioned in general at the beginning of this section for all light-scattering diagnostics, as well as several important additional features including *specificity* (scattering signature indicative of molecular species *and* its level of excitation); *well-determined and independent response* (scattered intensity proportional to number density of scattering molecules and independent of density of other molecules); *accessibility of temperature information* (strong temperature dependence of signature in selected spectral regions, with temperature data independent of density); *capability to probe systems not in equilibrium* (internal mode excitations directly related to Raman signature); and *relative lack of interference* (signatures of different species often do not overlap, and even when some overlapping occurs, as with hydrocarbons, measures of composition are still obtainable from signals corresponding to a common chemical bond).

Thus, the motivation for studying laser Raman scattering for combustion is clearly the high specificity of the temperature, density, and composition information that is gained. In view of the small scattering cross sections for this process, it is equally clear that the technique can, in many instances, be difficult. However, as a trade-off for this experimental difficulty (ameliorated by modern developments in laser technology, photodetectors, signal processing techniques, etc.), we are offered a far easier interpretation of the resultant state property data than is obtained either with perturbing probes (because of complicated calibration and system interference problems) or with many other optical probes (because of the effect of system non-homogeneities on optical absorption, the effect of self-absorption and test-volume delineation problems on optical emission, and the effect of quenching on fluorescence). Considering the fact that we are ultimately concerned with probing hot, multi-component, chemically reacting gases, often in the presence of turbulence and non-homogeneities, the ease and uniqueness of data reduction is of paramount importance.

Many volumes on the basic theory of Raman scattering have been written, far too numerous to mention here. The two volumes edited by Anderson[15] are recent examples of excellent expositions of the science. From the point of view of diagnostics, the salient features of interest to use are contained in the following two brief phenomenological descriptions of rotational and vibrational scattering, upon which are based their relative utilities for the various types of temperature, density, and composition diagnostics described in the succeeding sections.

2. Rotational scattering

In describing scattering processes, we will be speaking about molecules in a general fashion, but dealing specifically (i.e., in terms of internal mode quantum numbers, symmetry, etc.) with simple diatomics, such as N_2, O_2 or CO. More complicated molecules can be dealt with when required, with more or less difficulty depending largely upon the molecular symmetry. Thus, CO_2 is relatively easy to handle because it is a linear molecule. However, water vapor is more difficult by far to treat than many molecules with greater numbers of component atoms (as, for example, methane), because the asymmetric top structure of water vapor presents a much more complicated symmetry and energy level structure than that for, say, methane.

The energy level of the conceptual diatomic molecule contains rotational levels spaced apart by energy differences growing with the energy level itself, i.e., the spacing between levels increases roughly proportional to the rotational quantum number J. Pure rotational Raman scattering transitions correspond to changes in the molecular rotational excitation whereby J changes by either +2 (called the Stokes lines) or by −2 (called the anti-Stokes lines). Because of the increasingly-spaced rotational energy level structure, these selection rules cause a series of Stokes and anti-Stokes lines to appear around the central (laser) incident photon energy. These lines are, to a first approximation, spaced equally far apart ($4B$, where B is the molecular rotational constant), except for the spacing between the incident energy and the first line (which is $6B$). Since the spacing between lines is proportional to the rotational constant, and since rotational constants for gases of similar molecular weight are themselves not far apart in magnitude, the rotational signature of a hot, multi-component gas is often a sum of various overlapped spectra.

Rotational Raman scattering intensities are given by relatively simple expressions dominated by the rotational population distribution function. The absolute magnitudes of rotational Raman scattering cross sections for some molecules are available,[16] but data for many molecules of interest are lacking.

Rayleigh scattering is the elastically scattered component which corresponds to the molecule in question remaining in the same rotational (and vibrational) state after the scattering event. It is a stronger process than rotational Raman scattering by several orders of magnitude for a given molecule, and also can be observed (although more weakly) for atoms. Rayleigh scattering intensities can be calculated through use of index of refraction data, providing useful calibration information for other scattering phenomena. Clearly, the Rayleigh signatures of various gases all overlap, so that it is not possible to determine that contribution due to a particular molecule if the molecular gas composition is not known.

3. Vibrational scattering

The vibrational energy levels of our conceptual simple diatomic molecule are spaced roughly equally apart, but with a separation that changes slightly with the magnitude of the vibrational quantum number v because of molecular vibrational anharmonicity. Thus, in addition to the Stokes ($\Delta v = +1$) and anti-Stokes ($\Delta v = -1$)

"spikes" that we might expect from the selection rules, we actually have a series of spikes on both the Stokes and anti-Stokes side of the exciting line. These correspond, respectively, to $v=0 \to v=1$, $v=1 \to v=2$, etc., and $v=1 \to v=0$, $v=2 \to v=1$, etc. The wavelength spacing between the series of spikes (called "hot" bands or upper state bands) is about 8 Å for diatomics such as N_2 when irradiated with light in the middle of the visible, and so they are relatively well-resolved by moderate dispersion apparatus.

When we consider also the vibration-rotation interaction possessed by all molecules, represented in the energy level structure by an interaction term proportional to $vJ(J+1)$, and the possible rotational transitions which are superimposed on the $\Delta v=1$ vibrational transition (given by the selection rules $\Delta J=0, \pm 2$), we find that the $\Delta J=0$ selection rule gives a "saw-tooth" envelope to the spike structure described above. [The $\Delta J=\pm 2$ rule gives weak (by roughly two orders of magnitude, compared to the $\Delta J=0$ peaks) wings diffusely surrounding the $\Delta J=0$ saw-tooth signature. Except for spectral interference effects, these wings are generally not of much concern for diagnostics, although they have been used in some experimental measurements.[17]]

Vibrational cross section data are available for a great many molecules, and continues to be produced at a rapid rate.[18] However, data for many "simple" species in flames, such as the radicals OH, NH, CH, etc., are totally lacking —providing an excellent opportunity for optical physicists to contribute to the fundamentals of combustion diagnostics.

4. Temperature measurements

In general, Raman temperature measurements are based upon one of the following schemes[19]:

Rotational Scattering Structure

a. Contour or envelope fit of band
b. Intensity ratio of spectral regions (including use of individual lines) of band (using monochromator, filters, or interferometer)
c. Shift of band peaks

Vibrational Scattering Structure

a. Contour or envelope fit of band
b. Intensity ratio of spectral regions of bands (using monochromator or filters)
c. Stokes/anti-Stokes intensity ratio (using monochromator or filters)
d. Shift of band peaks: Q-branch ($\Delta J=0$)
   O- and S-branches ($\Delta J=\pm 2$)
e. Width of specific Q-branch bands, such as ground state ($v=0 \to v=1$) band.

These methods are, for the most part, analogous to the type of temperature measurement methods utilized in infrared spectroscopy.

One clear attraction of the rotational scattering compared with vibrational scattering is the strength of the total rotational band in comparison with the vibrational Q-branch. However, pure rotational Raman scattering generally is not considered a good candidate for temperature diagnostics of a multi-component gas because the rotational scattering signatures of many of the species may overlap to a very substantial degree. In some isolated instances, specific rotational signatures may predominate in certain spectral regimes, but analysis of the rotational data for multi-component mixtures is usually difficult, and, if the composition of the mixture is a function of the desired measurement quantity (i.e., a function of temperature), deconvoluting the rotational structures can become highly complex. One possible new method for measuring temperatures from rotational structure is based upon an interferometric technique described in the next subpart.[22] Temperature measurements for a flame have not yet been attempted with this method.

Although the strength of vibrational Raman scattering is not as great as that of rotational scattering, its relative freedom from spectral interferences for different molecular species makes it more suitable for flame diagnostics. The "saw-tooth" appearance of the Raman vibrational spectra is characteristic of many diatomic molecules of interest in flame studies, e.g., N_2, O_2, NO and CO. A significant criterion for useful temperature measurement schemes for these molecules can be based upon this structure, through consideration of the separation between successive fundamental vibrational Q-branches. If the experimental spectral resolution of the monochrometer or interference filters is sufficiently good, compared to the peak separations, to produce spectrally resolved temperature-sensitive features, then an experimentally useful temperature measurement technique can be implemented for these molecules.

A promising practical temperature measurement scheme would incorporate interference filters to isolate and record individual peak intensities within the vibrational Q-branch of a particular problem. For such applications, however, the particular system would have to be tailored to the experiment. Gases of different molecular weight produce widely different spectral widths and as such produce different demands on spectral position and resolution.

Since the "saw-tooth" profile observed at the Raman Q-branch is influenced by the rotational population distribution as well as the vibrational population distribution, temperatures measured from these profiles can be influenced by the existence of thermal non-equilibrium. Conversely, since the Raman spectrum contains information about the entire energy population distribution, we can in principle measure the presence and degree of thermal non-equilibrium in a gas.

5. *Density and composition measurements*

Both rotational and vibrational Raman scattering can be used for density and species composition measurements, with some of the qualifications addressed in the application to temperature measurement in Part 4. Since temperature measurements involved ratios of spectral bandpasses, some system performance characteristics are normalized out. In density measurements, spectral intensities are recorded and must be compared to a standard, such as N_2 at one atmosphere.

Generally, vibrational Raman scattering is preferred for composition measurement because the individual spectral bands from various species are easier to spectrally isolate. For elevated temperatures, however, the rotational wings (0-branch and S-branch) of the vibrational band (Q-branch) from a major constituent may spread far enough to overlap the Q-branch for other species.[20] Although this has not appeared as a major problem in flame studies to date, future attempts to measure certain trace species at very high temperatures by Raman scattering may be difficult.

Composition measurement by rotational Raman scattering is difficult with filters or monochrometers because of the overlapping rotational spectra of several species. Analytical reduction of such complex spectra is possible, but when the compositions are changing, subsidiary measurements of either temperature or chemical composition may be required. However, the interferometric "comb" method of Barrett[21] is capable of isolating the rotational lines of individual species, even when the species are of relatively similar molecular weight. This interferometric method also has more light gathering capacity than monochrometers or normal filtering techniques, and as such should prove advantageous for composition measurements in low density flows. It has, however, not been used successfully yet for flame measurements or other hot, multi-component gas diagnostics. This method can also be used, with modification, to measure temperatures.[22]

Much effort has been spent on detecting low concentrations of gases. The development of multi-pass detection systems, such as the Hartley-Hill "light trap cell,"[23] and Barrett's "comb" detector are examples of innovative ideas where physicists have helped advance the capabilities of Raman detection. Applying these laboratory ideas to real engineering experiments has posed another series of problems to the spectroscopists, with exhaust emission work[24] typical of the effort required for this interface. Interference from fluorescence and back scattering has caused substantial difficulties here which will, no doubt, be solvable but are nonetheless frustrating.

6. *Applications to combustion*

Raman spectroscopy is a promising new tool for combustion research because it can determine the presence of different chemical components and their internal energy population distribution, all with spatial and temporal resolution.

Already, application of Raman scattering to the study of non-reacting turbulent flows has generated new insights. Single pulse Raman scattering experiments have been performed in a turbulent isothermal jet by Lederman and Bornstein[25] which showed large temporal fluctuations in concentration that would not be measured by averaging techniques. Raman scattering was also used by Hartley[26] in experiments directed at determining pressure effects on global turbulent mixing in multi-species mixtures. Boiarski[27] measured the temperatures and density behind a shock wave using Raman scattering.

Only a few experimental studies of Raman scattering from flames have been reported. Leonard[20] used a pulsed nitrogen laser to record Raman vibrational bands of N_2 and CO_2 at ~1000 °K in the post-flame region of a propane-air burner. Using pulsed ruby laser light to record N_2 vibrational bands, Lederman and Bornstein[25] measured temperatures from 400 °K to 2000 °K across a methane-air flame. Lapp *et al.*[28] used an argon-ion laser to record vibrational bands of N_2, O_2, H_2, and H_2O from H_2-O_2 flames. Setchell[29] measured CO profiles in a rich methane-air flame using an argon-ion laser and light-trapping techniques.

The detection sensitivity is a function of the specific experiment, and has been explored by both Lapp[28] and Setchell.[29]

Very accurate data from flames is possible using curve-fitting routines for the Q-branch spectral shape. This technique can be simplified to use of filter spectral discrimination. The sensitivity for this type of probe has been illustrated by computer simulation and, to some extent, demonstrated in laboratory experiments with preliminary success.[30]

Application of Raman scattering to more complex combustion problems is just beginning. No results of such studies have yet been reported, but the need for such engineering applications is becoming recognized. The U. K. Gas Council expressed the desirability of Raman experiments for combustion research in 1971,[31] and again in 1973[32] in reviews of turbulent diffusion flame analysis. They recognized that Raman scattering showed the greatest potential for measuring the turbulent properties needed by the theorists for closure to their analytical treatments. More recently, in a National Science Foundation Combustion Workshop,[33] the need for such measurements was recognized and accorded high priority.

The point here is that Raman scattering *can* provide access to new needed experimental combustion data; it *is* being used in global turbulence problems and in simple flames; and it *does* have the sensitivity to measure specific quantities of interest to the theorist.

The questions of turbulent combustion and of turbulence model closure schemes are addressed in Section 1 of this report with recommendations as to priority turbulence quantities to be measured. In part IV of this section, we recommend specific combinations of Raman spectroscopy and LDV experiments which might be capable of measuring these quantities.

An intermediate step in this effort is exemplified by the work at Caltech[34] whereby Rayleigh scattering is used in a He-N_2 turbulent mixing experiment. Since the Rayleigh scattering cross section for He is two orders of magnitude smaller than that for N_2, Rayleigh scattering may be capable of N_2 detection in such mixtures with excellent temporal resolution. Rayleigh scattering has also been used by Graham *et al.*[35] for measuring transient concentrations in turbulent flows.

Other important advances in applying Raman scattering to combustion problems may come from further development of more complicated Raman processes such as resonant Raman and coherent Raman scattering. Resonant Raman shows promise for enhancing the signal of one species over others, but is still in a very primitive state of development. Coherent Raman has been applied recently to combustion flows and is discussed further in Part III.A.4.

C. Laser fluorescence spectroscopy

1. Description of fluorescence diagnostics

Fluorescence has been used as a spectroscopic tool since the latter part of the nineteenth century. With the increasing availability of lasers in recent years, activities based on the use of fluorescence techniques for obtaining fundamental spectroscopic data have increased markedly. Like Raman scattering, fluorescence is an optical scattering phenomenon which can achieve both spatial and temporal resolution. Unlike Raman, however, fluorescence is an absorption and re-emission process requiring the incident laser wavelength to correspond to specific energy shifts for the molecule. The absorption–re-emission process also requires a finite residence time for the molecule which, except for very low density gases (less than one Torr pressure), results in pressure broadening and collisional quenching effects which can be virtually impossible to account for. Nevertheless, there are specific applications where the relatively large scattering cross section for fluorescence (see Table 1) offsets its inherent difficulties, and quantitative information may be retrievable.

The application of fluorescence techniques to combustion measurements has become practical only with the advent of tunable lasers. These devices can provide the high intensities and narrow line widths required for the selective excitation of fluorescence. Fluorescence from vibration-rotation levels in excited electronic states is (1) characteristic of the structure of the molecule and can be used to establish its identity, and (2) characteristic of the relative population of molecular energy levels and therefore can also be used to establish measures of its temperature.

A typical fluorescence spectrum for a simple diatomic molecule consists of a resonance line with a series of Stokes lines at longer wavelengths and anti-Stokes lines at shorter wavelengths. The number of anti-Stokes lines is indicative of the ground-state vibrational level from which excitation occurred. Most basic fluorescence studies measure the spacing and splitting of the Stokes and anti-Stokes lines in order to provide structural information and molecular constants, which can also be used to identify molecular species. Intensity alternations from line to line can also provide additional structural information. Fluorescence diagnostics are sometimes made using the off-resonance lines as well, in order to minimize interference effects.

Collisional quenching can limit the sensitivity of fluorescence measurements, since the measurements correspond to a series of discrete steps involving electric dipole absorption, redistribution of the ensuing excited state population by collisional and radiative transfer, and, finally, electric dipole (or, occasionally, somewhat violated "forbidden") emission. Thus, collisional effects occupy a key position in the train of events corresponding to a fluorescence process and, if collisional quenching is sufficiently severe, they can profoundly change the character of the re-emission.

The important point to be recognized here is that quenching is highly dependent upon the composition and pressure of the total gas mixture being probed, and not just dependent upon fluorescing species. However, in many cases (even at high pressures, where quenching can be significant), the fluorescence measurements are still orders of magnitude more sensitive than Raman scattering. On the other hand, Raman processes are (except in the approach to resonance) independent of the environment of the active scattering species. These processes, rather than depending upon discrete electric dipole steps, are governed by changes in the molecular polarizability.

2. Pressure effects on fluorescence

Generally, the Doppler and pressure broadening contributions to line shapes can be taken into account in their effect upon fluorescence. However, quenching effects even at pressure of only a torr can seriously reduce or eliminate fluorescence emissions from certain excited molecular states.

Limitations imposed by quenching can be illustrated using the Stern-Volmer relation,[36] where the quenching rate constant can be estimated from hard-sphere collision theory. Quenching cross sections for small molecules[36] generally range from 10^{-2} to 10^2 Å^2. For diatomic radicals, excited state lifetimes fall between 10^{-9} and 10^{-7} seconds. At a pressure of one atmosphere and 300°K, collisional quenching would reduce fluorescence output by nearly two orders of magnitude for a lifetime of 10^{-7} seconds and a quenching cross section of 1 Å^2.

3. Collisional quenching in combustion diagnostics

Quenching can lead to significant reductions in fluorescence output. However, the high laser intensities and relatively high fluorescence excitation cross sections (see Table 1) associated with many small molecules still make it a useful technique for combustion studies even at elevated pressures. The results reported for C_2 by Vear, Hendra, and Macfarlene[37] and for CH by Barnes *et al.*[38] from atmospheric flame studies support this premise. With present lasers and signal processing systems, measurements of radical concentrations well below 10^{12} molecules/cm^3 are possible in flames at atmospheric pressures.

Even with strong quenching, molecular fluorescence is generally several orders of magnitude stronger than vibrational Raman scattering. The use of fluorescence can be advantageous for the detection of very low-level combustion intermediates such as radicals, while the Raman measurements, which are not affected by collisional quenching, would be more useful for the higher concentration species.

Strong quenching can complicate or limit the quantitative interpretation of fluorescence. However, qualitative information on when and where intermediate species appear in a combustion zone can be very important, and can be derived from fluorescence measurements without having to perform a detailed analysis of quenching effects. Even at high pressures, emissions from some radicals are not completely quenched, and can be observed. Lavoie, Heywood and Keck[40] have observed OH, CH, and C_2 emission bands through a window in an internal combustion engine. Also, Rodig and Zalud[41] have monitored emissions from C_2, NH, CN, and NH_2 in the combustion chamber of an operating diesel engine.

Fluorescence quenching can vary for different excited states in the same molecule. For example, SO_2 fluorescence excited at 2288 Å is strongly quenched by H_2O vapor while fluorescence excited at 2138 Å is only slightly quenched. Fluorescence excited at 2138 Å has been used successfully to measure SO_2 in the atmosphere at levels down to a few parts per billion in the presence of water vapor.[42]

4. Applications to combustion

Laser fluorescence appears quite promising for measuring local species concentrations of small molecules in flame and combustion systems at low pressures where collisional quenching of fluorescence is minimal. To interpret fluorescence measurements quantitatively at pressures where quenching effects are significant, it is necessary to know the appropriate quenching cross sections. In general, very little data on quenching effects are available, and more attention needs to be directed toward the study of quenching processes involving combustion species. Virtually no information is available on the quenching of radicals. Data for the deactivation of radicals and stable flame constituents through collision with CO_2, CO, H_2O, N_2, O_2, H_2, and other species are important and should receive attention. Hollander, Lijnse, Jansen, and Franken[43] have successfully measured quenching cross sections for excited atomic states in complex flames by solving sets of simultaneous algebraic equations for different flame-gas compositions. This same approach should also prove useful for studying the quenching of molecular states in flames. This information is important not only from the viewpoint of interpreting laser fluorescence measurements, but would be of value for understanding energy transfer processes in flames.

Although very little work on fluorescence diagnostics has been done in combustion applications, the recent studies on C_2 and CH in atmospheric flames[37,38] indicate the utility of this technique for the study of small radicals in combustion processes. Fluorescence techniques can be adapted to other molecular species, such as OH, NH, NH_2, C_3, CN, and NO_2 on the basis of currently available tunable lasers and spectroscopic data. There are many other combustion species, such as CH_2, CH_3, HCO, C_2N, CN_2, and HO_2, which are of considerable interest, but cannot be studied in combustion processes until additional fundamental spectroscopic data are available.

Basic spectroscopic studies involving combustion species is an area where physicists can make important contributions to combustion science. Theoretical predictions of spectral characteristics for small radicals would also be useful as a guide for locating spectral features experimentally.

The effectiveness of the laser-fluorescence techniques is enhanced when they are used in combination with both optical (preferably, tunable-laser) absorption and Raman-scattering measurements. Information derived from the latter two techniques can facilitate interpretation of fluorescence measurements. Tunable-laser line–center absorption with derivative signal processing has been shown to have extremely high sensitivities for combustion intermediates and products.[44] Absorption measurements provide concentrations integrated over the absorption path. Use of fluorescence measurements along the absorption path can be used to aid in the deconvolution of the integrated absorption to obtain local concentrations in combustion systems. Temperatures and concentrations for the high-concentration species determined by Raman measurements can also be used to aid in the interpretation of fluorescence data for the low-level radicals.

Tunable laser developments will have strong influence on the range of molecules that can be monitored using laser fluorescence techniques. Presently, dye lasers can cover the spectral range from below 4000 Å (and, with frequency doubling techniques, from ca. 2500 Å) to about 1 μ, with optical parametric oscillators covering the ranges from 6000 Å to beyond 3 μ in the infrared. Further extension of the tunability range at both the infrared and ultraviolet ends of the spectrum would enhance considerably the capabilities of the fluorescence approach. Of particular interest is the spectral region down to 2000 Å, which would include the NO signature as well as those for several radicals of current interest. Vapor-phase frequency-mixing techniques are beginning to show promise at the shorter ultraviolet wavelengths, but further development of nonlinear materials for use with tunable systems in both the ultraviolet and infrared also appears warranted. High laser intensities and narrow line widths are also important for high detection sensitivities.

D. Particle scattering

Observation of scattered light often provides the means to measure certain properties of assemblages of liquid or solid particles in a flow. Intentionally seeded particles of specified size distribution can be added to a flow such that Doppler-shifted light can be used to infer particle, and resultingly, gas flow velocities. This field of Doppler velocimetry is discussed in detail in Part A above. However, as particularly occurs in reacting flows, droplets and particles may be present for which size distribution and concentration are not known. In these cases, it is often desirable to use particle scattering theory to deduce size distribution. The subject of small particle scattering has had constant and increasing interest since Mie theory became available and one should expect no abatement of this interest. Mie theory is the well-established theoretical tool to describe light-particle interaction and scattering for specified boundary conditions and its predictions and the manner in which it is used are well documented.

1. Description of basic methods

Many different types of optical measurements are presently employed in attempts to determine particle size in the range $0.01\ \mu < d \leq 100\ \mu$.[45] Most of these techniques can yield mean size, but it is very difficult to determine the size distribution.

Some properties that can be used to measure particle size are[46] (a) the ratio of the intensity of scattering at two angles, (b) the ratio of polarization at a given angle when the incident light is unpolarized, (c) the depolarization at a given angle when the incident light is polarized, (d) the angular position of an extremum in the intensity of scattering when the incident light is monochromatic, (e) the angular position of various colors when the incident light is white, (f) the total amount of scattering as measured by a transmission test, and (g) variants and combinations of the above. The survey paper by Heller[47] and its references are good sources of the methods derived from Mie theory that are available to the experimentalist.

The power of the Mie theory is unfortunately severely limited by two problems normally encountered in a practical experiment. These problems are the lack of homogeneous particle size and possible presence of multiply-scattered light.

The Mie theory describes the interaction of monochromatic electromagnetic radiation with a single particle. In simple experiments it is possible to interpret scattering of a number of particles providing they are all the same size, i.e., a monodispersion. Difficulty arises, however, because in practical systems, such as fine sprays or flame-formed soot, particles of various sizes are present simultaneously and many of the distinctive features of the scattering of monodispersions are no longer available.

The second consideration that often limits the use of Mie theory is the presence of an aggregation of particles of such a high concentration and/or large spatial extent that the incident light undergoes multiple scattering events before it is observed. The complexities in interpreting multiply-scattered light are sufficiently great that no successful attempts to interpret such measurements in terms of the properties of the scattering sites have been reported.

For optically dense polydispersions, Dobbins[48] has developed a method of interpreting spectral transmission tests in such a way as to circumvent both the limitation normally imposed by the presence of the polydispersion and the occurrence of high optical depth. He shows that the mean scattering coefficient of a polydispersion is weakly dependent on the shape of the size distribution function for given mean diameter. By a suitable choice of two wavelengths of light Dobbins shows that it is possible to deduce the mean particle size and particle distribution.

2. *Applications to combustion*

The formation of particulate phases in combustion processes and the generation of very fine fuel sprays provide numerous situations where light scattering is useful for size distribution measurements. Millikan[49] has measured carbon particle temperatures in a methane air flame based on observations of particle size and optical extinction. Erickson[50] and Kunugi[51] have attempted to measure the growth of soot in flames by measurements of angular distribution of scattering. Difficulty in interpreting these results was encountered because of uncertainty in the refractive index of soot and because of the presence of agglomerates.

Temkin and Dobbins[52] used the method of mean size determination by spectral transmission tests to measure the attenuation and dispersion of sound in an oleic acid-nitrogen aerosol in the size/frequency range where particles are most efficient in attenuating sound.

More recently Belden and Penney[45] developed a laser particle counter for determining water droplet sizes in low quality steam. Their system is capable of detecting particles in the range $0.5\ \mu < d < 5\ \mu$. Their instrument used forward scattering to reduce its dependence on refractive index and particle shape. An interesting problem observed in their experiment was the possible alteration of particles by the laser beam. This possibility could be serious with particles such as water or liquid fuels which can vaporize. The interaction of high intensity laser light with liquid phase droplets is a problem which deserves attention.

III. OTHER NON-PERTURBING AND PURPOSELY PERTURBING PROBES

The currently popular non-perturbing diagnostic techniques which we have considered above were discussed with the objective of obtaining point information about the thermodynamic state properties, temperature and concentration, and about velocity. One should not limit his considerations to these techniques alone because they do not represent a panacea for the combustion community. Indeed, sensitivity limitations of Raman scattering prohibit detection of trace, or even in some cases, small, concentrations of particular gases of interest. Similarly, lifetimes for particular species may be sufficiently short to preclude direct application of some of these techniques. Fluorescence, as described above, has many limitations due to the pressure effects of quenching and broadening. Both Raman scattering and fluorescence are limited by laser power or wavelength choice. Therefore, though these techniques may answer many questions of combustion science and technology, there are still many radicals they cannot detect, many environments with which they are not compatible, and many spatially limited data requirements they cannot retrieve without considerable innovation. In this section, therefore, we introduce some even newer ideas, and some old ideas with new applications.

A. Image-forming techniques

A number of image-forming techniques have been under development which have the capability of supplying qualitative flow field information as well as quantitative, spatially correlated information for turbulent mixtures. These include holography for multi-dimensional photography of droplets and particles, radiography for diagnostics where optical windows cannot be employed, and Ramanography and coherent Raman scattering for viewing single species distributions in flow.

1. Holography

Holography has become an established technique for the recording and study of three-dimensional distributions of particles and, more recently, droplets. The production of a hologram requires that light reflected from or scattered by the object field be mixed with a mutually coherent reference beam. The interference patterns resulting from the sum of these two waves constitute the hologram. When the hologram is re-illuminated with the reference wave, the recorded interference fringes diffract it into the form of an image identical to the original object. Since holograms can be made with pulsed lasers, the technique provides a unique method to freeze dynamic events in three dimensions.

The most interesting application of holography to combustion research lies in the study of droplets in sprays. Trolinger[10] obtained holographic records of spray combustion in a diesel engine and identified the major shortcomings of such a technique. After a few cycles of the engine, the windows were burned and coated to the extent that light passing through them was badly deteriorated and did not constitute a satisfactory reference wave, even when the field was nearly transparent. Also, when using in-line holography, the density and temperature gradients in the fuel/air mixture modulated the phase of the wave making it a poor reference. Using off-axis holography, Trolinger was able to attain resolution better than 10 μ during engine operation without employing tedious techniques. Resolution of a few microns seems feasible.

A more sophisticated technique, "cine-holography," was also reported[10] whereby holographic images are recorded at a rate of 1000/sec. This technique was used to record a holographic reconstruction of an exploding alcohol droplet. Applications to diesel fuel droplets and emulsion droplets could provide useful information.

2. *Radiography*

Radiography, particularly neutron radiography, is a technique which has been used for many years to locate and image certain tracers inside complex hardware with no optical access. The isotope of helium, He^3, for example, has a neutron capture cross section many orders of magnitude larger than most gases, and, even in reasonable concentrations larger than many metals. For example, He^3 can be viewed inside steel tubing and vessels with excellent resolution. Very recently, the technique has been applied to the study of gasdynamic phenomena by using a pulsed neutron source (e.g., reactor). Johnston[53] has dynamically imaged the moving interface between He^3 and He^4 in a simple gas blowdown problem. He is currently pursuing the development of "neutron cinematography" for obtaining such information for very complex gas flow problems. Requirements for He^3 concentration levels are sufficiently low that it can be used as a tracer in gas mixtures and still yield resolvable information. The application of this technique to the study of gas motion in a dual chamber stratified-charge engine would be very interesting. The location of the interface between the prechamber gases and main chamber gases during combustion could provide important qualitative and quantitative information to modelers.

3. *Ramanography*

Ramanography is a technique for two-dimensional recording of Raman scattering from a particular gas in a flow field. Hartley[54] has demonstrated this technique by passing a "sheet" of light from a pulsed, frequency-doubled, ruby laser through a turbulently mixing flow field, imaging the plane of laser light, selectively filtering the desired Raman scattered frequency, amplifying the image with high gain image intensifiers, and recording that image. The resulting image is a two dimensional map of species concentration in the flow field. In Hartley's experiments, hydrogen jets could be viewed in a nitrogen flow field, with turbulent eddies identifiable and concentration variations readily distinguishable. This technique is still in its infancy, and at present is instrumentation-limited. The experiments above were performed with a 5-joule ruby laser, frequency doubled, a 10^6-gain image intensifier, and hydrogen partial pressures greater than 100 psi. In order to detect gas concentrations at significantly lower partial pressures, an even higher power laser is required. The image tube was virtually photon-limited so higher gains there are of no value.

The application of Ramanography to the study of simple turbulent flow fields could readily supply turbulent space correlations for concentration fluctuations. Such data are certainly of more than academic interest (see the modeling section) because they can supply correlations for closure of turbulence models.

4. *Coherent Raman scattering*

Taran[55] has developed a flow visualization technique, capable of distinguishing individual gaseous species, by coherent Raman anti-Stokes scattering. The method is based on a four-wave mixing process: two collinear light beams of frequency ω_1 and ω_2 generate a collinear (anti-Stokes) wave at frequency $2\omega_1 - \omega_2$ when traversing a gas containing a Raman active molecular species with vibrational frequency $\omega_v = \omega_1 - \omega_2$. The intensity of the new wave is proportional to the square of the number density of resonant molecules. This scattering is much more intense than spontaneous Raman scattering. Taran was able to visualize H_2 in a flame with detection sensitivity limits down to 100 parts per million.

The limitations of the system are two-fold. First, although the method applies to all molecular gases besides H_2, their use would require tunable laser sources at high power. Taran is currently developing such a system for studying the combustion products of kerosine droplets. Second, the flow field image must be recorded in-line with the laser light (unlike Ramanography, which though much weaker, can be recorded 90° from the laser light). The resulting image is therefore a two dimensional image integrated to some extent along the line of sight of beam excitation, over the range where the laser beam is focused. Removal of the laser frequency from the collected image must also be accomplished. Despite these limitations, coherent Raman anti-Stokes scattering can be useful for a large number of problems because of the strength of the signal. In fact, ultimate detection sensitivity limits as low as 10^{-1} to 10^{-4} ppm should be achieved.

These image forming techniques are in their infancy and are just beginning to offer new insight into complex flow phenomena. Such techniques are exemplary of the innovative nature of physics input into instrumentation and diagnostics developments. Their continued application to even more complex combustion phenomena should be encouraged.

B. Purposely-perturbing and tracer techniques

1. Optically and acoustically excited

There may be a number of problems in combustion science which can be readily studied by means of purposely perturbing a flow and reading its response to the perturbation. Optical excitation, for example, may be used for studying the excitation or de-excitation of rate-limiting species, for example, OH*. Therefore, by changing the chemical kinetics in a controlled way, information on radical chemistry may be deduced. Even though the overall density of a reaction zone may be large, the chemistry may be controlled by much smaller concentrations of excited species (e.g., radicals) such that it may require relatively little optical energy to modify the reaction. When this is the case, detection sensitivity would be quite good. Pulsed optical excitation may also be used to obtain relaxation data for select chemical species.

Acoustic excitation, on the other hand, may be more applicable to the study of turbulent combustion. It may be possible to relate the acoustic field in a turbulent exhaust jet (e.g., gas turbine) to the combustion-generated noise field. Summerfield[56] suggests that jet noise has two principal components distinguishable by their source of generation. One component, "Lighthill noise," is associated with jet turbulence, interaction of turbine blades and flow, etc. This component is essentially removable. The second component is combustion-generated noise, etc., turbulence which propagates through the system and appears as noise in the exhaust. The random noise spectrum is an integral of the localized sound fields over the combustor. Therefore, by selective acoustic excitation one may be able to relate the local acoustic fields and the measured integrated acoustic field.

2. *Radioactive and isotopic tracers*

For some problems, particularly catalytic combustors and fluidized bed combustors, optical access for laser diagnostics is not possible. For such problems, the addition of trace quantities of radioactive tracers may prove to be the only means of *in situ* diagnostics. These techniques are not new, but their application to new, energy efficient, combustion devices has not been attempted.

Selected isotopic substitutions, on the other hand, may be used for the study of more complex chemistry. Deuterium-hydrogen exchanges have been used in the past to learn more about specific chemical reactions. More recent specific interests in fuel-bound nitrogen, and its role in NO formation, may provide impetus for replacing nitrogen-in-air with isotopic nitrogen for laboratory research. The source of the NO in the reaction zone would then be distinguishable. Such isotopic tracer techniques may be combined, in some cases, with Raman scattering which is capable of distinguishing $N^{14}O$ from $N^{15}O$.

IV. SUMMARY AND PROPOSED ILLUSTRATIVE EXPERIMENTS

During the course of this APS Summer Study, we have explored combustion diagnostics both in a general fashion and with a strong emphasis on those non-perturbing optical techniques which we feel are most suited for current exploitation by combustion scientists. (Note: Diagnostics applied specifically to emulsions appears in the section on emulsions.) Within this framework, we have concentrated to the greatest extent upon optical light-scattering techniques, viz., laser Doppler velocimetry, laser Raman scattering, laser fluorescence, and particulate scattering, for two major reasons. Firstly, these techniques address themselves to the direct measurement of the fundamental quantities in combustion science—fluid velocity, temperature, density, composition, and particulate properties. Secondly, these techniques have undergone an advanced stage of development lately, from the point of view of laboratory demonstrations in some cases and with actual commercial devices in others. These developments have originated in many instances with workers outside the field of combustion science—some physicists, some electrical engineers, some fluid mechanicians *et al.*,—and so the meshing of the desired measurement variables (from the combustion modelers) and the measurement capabilities (from the diagnostic workers) has been the major effort of our portion of the APS Study.

The other techniques mentioned in this report (holography, radiography, etc.), were not discussed in depth due to the decision to focus upon the light-scattering techniques just mentioned, as well as the feeling that they have already been well-reviewed. Similar considerations led to only brief reference for optical absorption and emission techniques as well as such conventionally applied techniques as hot wire anemometry, gas sampling and thermocouple probing. However, we re-emphasize here our belief that the best-designed combustion experiments will utilize a variety of simultaneously imposed diagnostic schemes. In this fashion, not only will we gain additional information about the combustion processes, but we will also be able to calibrate conventional, slightly perturbing probes by means of simultaneously measured data obtained with (presumably) more expensive non-perturbing probes.

A. Examples

Based on the above discussion, various optical techniques such as laser Doppler velocimetry, and laser Raman spectroscopy or combinations of the two should be employed to determine the following aspects pertaining to chemically reacting flows:

1. The effect of combustion on turbulence should be determined. (See Modeling Section, Part IIB.) Eventually one must be able to measure all pertinent one-point average quantities involving velocities and chemical species concentrations such as $\overline{u_i^2}$, $\overline{y_\alpha u_i}$, $\overline{y_\alpha y_\beta}$, $\overline{y_\alpha^2 y_\beta^{1/2}} \exp - E/T$, in order to completely determine the validity of any turbulent combustion theory. It may prove more practical to measure directly the various probability density functions instead of the moments (see Modeling Section—statistical turbulence). Once these probability density functions such as $f(t, x, u_i)du_i$, $F_\alpha(y_\alpha)dy_\alpha$, $F_{\alpha\beta}(y_\alpha, y_\beta)dy_\alpha dy_\beta$, $F_{\alpha i}(y_\alpha, u_i)dy_\alpha du_i$, etc., are determined, all moments can be readily computed without any further measurement. In the preceding sentence, f is a probability density function of the fluid elements, whereas F_α, $F_{\alpha\beta}$, $F_{\alpha i}$, etc., are the probability density functions or joint-density functions of the chemical species concentrations, y_α, and the velocity component, u_i.

2. We can make experimental advances to check various fundamental aspects of combustion theories without having to produce actual turbulent streams with complicated high-temperature combustion. For instance, one of the basic aspects of several of the existing theories is the question of the overall scale of turbulence in the situations where there exists more than one characteristic scale inherent to the flow field. A simple experiment is presently being carried out at the University of Illinois to answer this question. In this experiment, two parallel turbulent streams are produced with equal mean velocities but different turbulence scales and turbulence energies. The mixing region of these two streams is being probed for the variation of mean length scale and mean turbulence energy.

In addition, we can study experimentally the behavior of the probability density functions, f, F_α, $F_{\alpha\beta}$, $F_{\alpha i}$, etc., in a cold simple turbulent flow field of chemically inert gases. In fact, we can first simply consider all scalars to be analogous and assume the temperature represents a single chemical species. The various probability density functions involving the temperature and velocity fluctuations can then be examined. Such a measurement should contribute much toward testing the fundamentals of various theories designed for eventual combustion analysis.

3. A low-temperature chemically reacting flow should be probed. Measurements of the probability density functions for these low-temperature reacting flow fields should be a second step in understanding certain basic properties of reacting turbulent streams, and in testing the fundamental aspects of the proposed theories.

The above mentioned experimental measurements should be useful for understanding the properties of chemically reacting flow fields and also for testing *all* chemically reacting flow theories whether they are phenomenological or statistical-type theories.

In order to gel our thinking on the application of laser scattering techniques to combustion science, and to provide strong interaction with modeling theorists, we have explored some examples of proposed illustrative experiments. It is useful to elaborate upon one such experiment designed to infer turbulent combustion properties from measurements on a simple laboratory burner.

The proposed burner is simply a 10-cm-diameter pipe with a 1-cm-spacing wire grid placed close to one end. A premixed moist CO-air flame is sustained by means of a small (several mm diameter) disk placed at the center of the end of the tube as an igniter and flame holder. The combustion and product gas zone downstream of this disk can then be studied with good optical access.

The major goal of this experiment would be to ascertain the presently unknown relative magnitudes of a number of the correlation terms in the conservation equations, in order to better understand the relative importance of the various competing and interactive physical (i.e., fluid mechanic) and chemical processes. Ideally, the simultaneous measurement is desired of the turbulent fluctuation of the velocity and the various state properties, e.g., u_i', ρ', T', and y_α', which are, respectively, the fluctuation values of the velocity, density, temperature, and mass fraction of species α.

Although the Reynold's stress terms, $\overline{u_i'u_j'}$, can be measured with two-component LDV systems, no single experiment has been devised to measure the turbulent transport of mass $\overline{u_i'y_\alpha'}$ or energy $\overline{u_i'T'}$. Seeding requirements for LDV seem to conflict with background scattering problems associated with other laser diagnostics. However, some feasible ways of performing simultaneous LDV and Raman measurements of the turbulent transport terms can be suggested.

Consider a sampling volume for the proposed burner to be approximately $\frac{1}{8}$ mm on a side. Then, using appropriately chosen LDV seeding densities we can calculate exposure time response for a given flow velocity range. For typical turbulent laboratory flames, the minimum LDV acquisition time may be on the order of 20 μsec. Now, superimpose upon this LDV acquisition time a double-pulse 50 nsec Raman acquisition as shown in Figure 1. The Raman pulses immediately precede and immediately follow the LDV pulse. The seeding rate is chosen (at ca. 10^5 particles/cm^3) such that the probability of an LDV particle being in the sample volume during LDV acquisition time is sufficiently large, but the probability of the particle being in the sample volume during Raman acquisition time is relatively small. The time between Raman pulses is sufficiently short (20 μsec) such that even the smaller eddies (down to $\frac{1}{8}$ mm) have longer characteristic time scales. Thus information on concentration or temperature from the

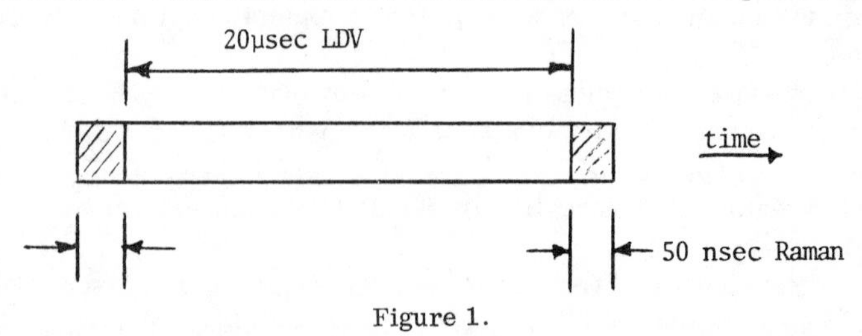

Figure 1.

two Raman pulses should be identical. If one Raman signal is significantly larger than the other, it probably had an LDV scattering particle in the volume and should be discarded.

Another method to accomplish the goal of essentially simultaneous LDV and RS measurements with low probability of interference between them (i.e., assuming modest seeding rates) is to trigger the pulsed laser used for Raman data acquisition by the LDV signal processor in such a fashion that a Raman trigger pulse is only given after an LDV data point is successfully obtained. Experiments such as these should be feasible with currently available equipment (with considerable sophistication and development) and should be capable of direct measurements of turbulent transport properties.

B. Recommendations

As a result of our review, we recommend the following diagnostic developments for combustion research, in order of priority:

1. Raman scattering from flame gases
   theoretical treatment of temperature effects
   experimental studies of flame constituents
   improved sensitivity, particularly pulsed
   direct application to practical systems (turbines, simple engines)
2. Combined Raman/LDV measurements
   momentum transport measurements in non-reacting isothermal turbulent flows
   energy transport measurements in non-reacting non-isothermal turbulent flows
   momentum and energy transport in combustion environments—particularly to investigate flame generated turbulence
3. Fluorescence from flame gases
   improved sensitivity, broader application
   solution of quenching effects with possible application to local pressure fluctuation measurements
4. Flow visualization of turbulent environments
   Ramanography of two-dimensional shear layer
   broader application of coherent Raman scattering with dye lasers
   application of holography to droplet sprays, particularly where emulsified fuels are used
   application of radiography to dual chamber stratified charge engine flow

V. REFERENCES

[1]See for example, Penner, S. S., *Quantitative Molecular Spectroscopy and Gas Emissivities* (Addison-Wesley Publishing Co., Inc., Reading, Massachusetts, 1959).

[2]Trolinger, J. D., *Laser Applications in Flow Diagnostics*, Agardograph No. 186, March 1974.

[3]van de Hulst, H. C., *Light Scattering by Small Particles* (Wiley, New York, 1957).

[4]Self, S. A., in *Proceedings of Second International Workshop on Laser Velocimentry*, edited by H. D. Thompson and W. H. Stevenson, Purdue Univ. (1974).

[5]Mossey, P., General Electric Co., private communication.

[6]Durst, F., Melling, A., and Whitelaw, J. H., Combustion and Flame 18, 197 (1972); also Baker, R. J., Bourke, P. J., and Whitelaw, J. H., *Fourteenth Symposium (International) on Combustion* (Combustion Institute, 1973), p. 699.
[7]Baker, R. J., Hutchinson, P., and Whitelaw, J. H., ASME paper 73-HT-34; also, Baker, R. J., Hutchinson, P., and Whitelaw, J. H., Harwell report No. AERE-R-7492.
[8]Durst, F., and Kleine, R., "Velocity Measurements in Turbulent Flows by Means of Laser Doppler Anemometers," in *Proceedings of 6th German Flame Conference* ("Deutcher Flammentag"), Essen, Germany, September 1973.
[9]Chigier, N. A., Sheffield University, private communication.
[10]Trolinger, J. D., Bentley, H. T., Lennert, A. E., and Sowls, R. E., "Application of Electro-Optical Techniques in Diesel Engine Research," SAE Paper No. 740125, 1974.
[11]Samualson, U. C. Davis, private communication.
[12]Bray, U. of Southampton, private communication.
[13]McLaughlin, D. K., and Tiederman, W. G., Phys. Fluids 16, 2082 (1973).
[14]*Laser Raman Gas Diagnostics*, edited by M. Lapp and C. M. Penney (Plenum Press, New York, 1974).
[15]*The Raman Effect*, Volumes 1 and 2, edited by A. Anderson (Marcel Dekker, Inc., New York, 1971).
[16]For example, see: Penney, C. M., St. Peters, R. L., and Lapp, M., J. Opt. Soc. Am. 64, 712 (1974) and Hyatt, H. A., Cherlow, J. M., Fenner, W. R., and Porto, S. P. S., J. Opt. Soc. Am. 63, 1604 (1973).
[17]Hillard, M. E., Jr., Emory, M. L., and Bandy, A. R., "An Experimental Study in the Application of the Raman Scattering Technique as a Remote Sensor of Gas Temperature and Number Density in Hypersonic CF_4 Flow," presented at the AIAA 7th Aerodynamic Testing Conference, Palo Alto, September 13–15, 1972.
[18]For example, see: Murphy, W. F., Holzer, W., and Bernstein, H. J., Appl. Spectrosc. 23, 211 (1969); Fouche, D. G., and Chang, R. K., Appl. Phys. Lett. 18, 579 (1971); Penney, C. M., Goldman, L. M., and Lapp, M., Nature Physical Science 235, 110 (1972), and Hoell, J. M., Jr., Allario, F., Jarrett, O., Jr., and Seals, R. K., Jr., J. Chem. Phys. 58, 2896 (1973).
[19]Lapp, M., in Ref. 14, p. 107.
[20]For example, see: Leonard, D. A., Ref. 14, p. 45.
[21]For example, see: Barrett, J. J., Ref. 14, p. 63.
[22]See Harvey, A. B., Ref. 14, p. 147.
[23]Hartley, D. L., and Hill, R. A., Applied Optics 13, 186 (1974).
[24]Bresowar, G. E., Lt., and Leonard, D. A., "Measurement of Gas Turbine Exhaust Pollutants by Raman Spectroscopy," AIAA Paper No. 73-1276 (1973).
[25]Lederman, S. and Bornstein, J., "Temperature and Concentration Measurements on an Axisymmetric Jet and Flame," Project SQUID (ONR) Tech. Report PIB-32-PU (1973); Lederman, S. and Bornstein, J., "Species Concentration and Temperature Measurements in Flow Fields," PIB-31-PU (1973).
[26]Hartley, D. L., "Application of Laser Raman Scattering to the Study of Turbulence," AIAA Journal 12, 816 (1974).
[27]Bolarski, A. A., "Temperature and Density Measurements Behind an Incident Shock Wave Utilizating Laser Raman Spectroscopy" (to be published).
[28]Lapp, M., Ref. 14, p. 130. See also Lapp, M., Goldman, L. M., and Penney, C. M., Science 175, 1112 (1972).
[29]Setchell, R. E., "Analysis of Flame Emissions by Laser Raman Spectroscopy," Western State Section, The Combustion Institute, May 1974. Also: Sandia Report SLL 74-5244.
[30]Lapp, M., General Electric Co., private communication.
[31]Jessen, P. G., "On the Use of Laser-Raman Spectroscopy in Combustion Research," Gas Council, London, England, Report LRS TN N207 (1971).
[32]Melvin, A., "The Present State of Theories of Turbulent Diffusion Flames," Gas Council, London, England, Report LRS T 513 (1974).
[33]Glassman, I. and Sirignano, W. A., "Summary Report of the Workshop on Energy-Related

Basic Combustion Research," sponsored by the NSF, Princeton Univ. Dept. of Aerospace and Mechanical Sciences Report No. 1177, 1974; also, Glassman, I., in "Research Objectives in the Field of Aircraft Gas Turbines," Dynalysis of Princeton, Princeton, N.J., December 1969.

[34]Roshko, A., and Storm, E., private communication. For description of mixing apparatus and some experimental discussion, see Storm, E., Ref. 14, p. 369.

[35]Graham, S. C., Grant, A. J., and Jones, J. M., "Transient Molecular Concentration Measurements in Turbulent Flows Using Rayleigh Light Scattering" (submitted to AIAA Journal).

[36]Steinfeld, J. I., "Quenching of Fluorescence in Small Molecules," Accounts of Chemical Research 3, 313 (1970).

[37]Vear, C. J., Hendra, P. J., and Macfarlane, J. J., "Laser Raman and Resonance Fluorescence Spectra of Flames," J. Chem. Soc. Comm., 381 (1972).

[38]Barnes, R. H., Moeller, C. E., Kircher, J. F., and Verber, C. M., "Dye-Laser Excited CH Flame Fluorescence," Appl. Opt. 12, 2531 (1973).

[39]Penney, C. M., Morey, W. W., St. Peters, R. L., Silverstein, S. D., Lapp, M., and White, D. R., "Study of Resonance Light Scattering for Remote Optical Probing," NASA-CR-132363 (1973).

[40]Lavoie, G. A., Heywood, J. B., and Keck, J. C., "Experimental and Theoretical Study of Nitric Oxide Formation in Internal Combustion Engines," Comb. Sci. Tech. 1, 313 (1970).

[41]Rodig, J., and Zalud, F., "Some Contributions to Experimental Combustion Research," Proc. Instr. Mech. Engrs., p. 203 (1969–1970).

[42]Schwarz, F. P., Okabe, H., and Whittaker, J. K., "Fluorescence Detection of Sulfur Dioxide in Air at the Parts per Billion Level," Anal. Chem. 46, 1024 (1974).

[43]Hollander, T. J., Lijnse, P. L., Jansen, B. J., and Franken, L. P. P., "Quenching of Excited Strontium Atomic State Measured in H_2- and CO-Flames," J. Quant. Spectrosc. Rad. Trans. 13, 669 (1973).

[44]Sulzman, K. G. P., Lowder, J. E. L., and Penner, S. S., "Estimates of Possible Detection Limits for Combustion Intermediates and Products with Line-Center Absorption and Derivative Spectroscopy Using Tunable Lasers," Comb. and Flame 20, 177 (1973).

[45]Belden, L. H., and Penney, C. M., "Optical Measurement of Particle Size Distribution and Concentration," General Electric Report No. 72 CRD066 (1972).

[46]Dobbins, R. A., "Applications of Light Scattering in Research and Technology," AIAA Paper 67-35, January 1967.

[47]Heller, W., "Theoretical and Experimental Investigation of Light Scattering of Colloidal Spheres," in *Electromagnetic Scattering*, Proceedings, edited by M. Kerker (Macmillan Co., New York, 1963), pp. 101–120.

[48]Dobbins, R. A., and Jizmagian, G. S., "Particle Size Measurement Based on Use of Mean Scattering Cross Sections," J. Opt. Soc. Am. 56, 1351 (1966).

[49]Millikan, R. C., "Sizes, Optical Properties and Temperature of Soot Particles," in *Temperature, Its Measurement and Control in Science and Industry, III*, 2 (Reinhold, New York, 1962), p. 497.

[50]Erickson, W. D., Williams, G. C., and Hottel, H. C., "Light Scattering Measurements on Soot in a Benzene-Air Flame," Combustion and Flame 8, 127 (1964).

[51]Kunugi, M., and Jinno, H., "Determination of Size and Concentration of Soot Particles in Diffusion Flames by a Light Scattering Technique," Eleventh Symposium (International) on Combustion (Combustion Institute, 1966).

[52]Temkin, S. and Dobbins, R. A., "Measurements of Attenuation and Dispersion of Sound by an Aerosol," J. Acoust. Soc. America 40, 1016 (1966).

[53]Johnston, S. C., Sandia Laboratories, unpublished results.

[54]Hartley, D. L., "Raman Scattering Experiments and Ramanography," Ref. 14, p. 311.

[55]Regnier, P. R., Moya, F., and Taran, J. P. E., "Gas Concentration Measurements by Coherent Raman Anti-Stokes Scattering," AIAA Journal 12, 826 (1974).

[56]Summerfield, M. A., Princeton, private communication.

[57]Osgerby, I. T., "Literature Review of Turbine Combustion Modeling and Emissions," AIAA Journal 12, 743 (1974).

3. EMULSIFIED FUELS

I. INTRODUCTION

Emulsions are dispersions of one liquid phase within another liquid phase. They exhibit physical properties which are often quite different from either of the constituent phases. In the case of emulsified fuels, various combinations of water-in-oil (W/O), oil-in-water (O/W), or oil, water and methanol emulsions exhibit properties which may be effectively utilized for fire safety, for improved pipeline transmission, and for more efficient and less polluting combustion. The use of emulsified fuels in combustion systems may offer improvements in emission and performance characteristics without requiring major revision in existing technological design. This is important in view of the inertia represented by industrial retooling. The broad scope of emulsified fuels and the innovative nature of its application should pose interesting problems for physics research.

The sections that follow will introduce the fundamental properties of emulsions, particularly stability and rheology; discuss the current methods of emulsification; present the status of emulsified fuel combustion research with discussions of possible benefits in practical systems, and fundamental experimental and analytical approaches to understanding combustion characteristics of emulsions—particularly the postulated micro-explosion phenomena whereby the internal water phase vaporizes and shatters the external continuous fuel phase; discuss other societal and technical advantages of using emulsions — particularly for pipeline transportation and fire safety; and finally summarize the relevant physics research areas that can impact effective utilization of emulsions.

II. PROPERTIES OF EMULSIONS AND METHODS OF EMULSIFICATION

The best overall characteristics of an emulsified fuel spray (drop size distribution, concentration and composition of dispersed phase, etc.) will not only depend on optimum combustion requirements determined by laboratory experiments, but also, to a large degree, on the practical problems associated with the production, handling and storage of the emulsion. In these questions the characteristics of the emulsion in terms of stability, rheological properties and methods of emulsification, will be of considerable interest. It is the purpose of this section to present a brief discussion of these aspects with emphasis on those areas which need more research.

A. Stability

Emulsions possess a great deal more than the minimum surface energy between the two constituting phases, and are therefore inherently thermodynamically unstable. In general, if no particular precautions are taken, as soon as two droplets come in contact they coalesce to form a single droplet, and this process proceeds until the two phases revert to two separate liquid masses ("breaking" of the emulsion). The contact between the droplets can be produced by collision in a flow field or by gravity or electrical forces. For some practical systems, emulsion stability may be required for only a few seconds; however, if the emulsion is to be stored and transported before use, then stability requirements become more demanding. The generation of stable emulsions is an empirical art for which surfactant addition and method of manufacture are the major tools.

Surfactants decrease the interfacial tension and hence the interfacial free energy of the system. Usually interfacial tensions between hydrocarbons and water are in the neighborhood of 30 dynes/cm^2. Addition of surfactants can easily reduce this by one order of magnitude or more. A surfactant forms a layer on the dispersed droplet surface which prevents coalescence of the dispersed phase. Because surfactants differ in their ability to lower interfacial tension, to form molecular layers at interfaces, and to form thicker layers or skins at surfaces, the selection of appropriate ones for a particular system or use can be critical. There exists a large body of information on surfactants and for details the reader is referred to References 1–4.

Perhaps of greater interest to a physicist is the problem of dispersed drop-drop interaction within the emulsion and the dynamics of coalescence, both at a microscopic level (i.e., interaction of two droplets) and at a macroscopic one (i.e., the evolution in time of the dispersed phase droplet size distribution function). In spite of some results obtained in the analysis of these phenomena,[5–9] it appears that more work is needed to develop a model to predict stability times of emulsions. The collision of two droplets has been studied,[6] but only in the case of relatively large droplet diameters (0.3 cm). Fluid mechanic calculations of the deformation of the approaching spheres and of the flow in the film separating them are very crude, and the factors that determine coalescence or rebound upon contact are poorly understood. Other features, such as the comparatively high instability of emulsions containing droplets of different sizes, (high coalescence rate between a large and a small droplet), and the effect of viscosity (which appears to be a stabilizing agent) have not been sufficiently clarified. The stability and coalescence of emulsion fuel droplets in the range 1–10 μm is of interest for both stationary and flow conditions.[10] The recombination mechanism is also of interest in the design of emulsification methods (see below).

B. Rheology

The rheology of emulsions and other dispersed systems has long interested physicists. In 1906, Einstein derived a relationship for the viscosity, μ, of a suspension of small, spherical particles in terms of the viscosity of the suspending fluid, μ_0, and the volume fraction of the particles, Φ:

$$\mu = \mu_0(1 + \tfrac{5}{2}\Phi).$$

Because it was based on the assumption of no interaction between particles, this equation applies up to Φ-values of only a few percent. For more concentrated dispersions of practical interest, various workers have tried to modify this equation and quite a number of relations have been derived, often by empirical or semi-empirical means. Two recent review articles are listed as References 11–12, and additional recent work is described in References 13–16. The key point is that a moderate or high oil-concentration emulsion exhibits non-Newtonian behavior.

It appears that our understanding of the rheological properties of a moderate oil-concentration emulsion is adequate as far as the use of emulsified fuels is concerned. If high oil-concentration emulsions (of the order of 95 percent) are to be used for oil transportation or fire safety purposes, there is a need for additional research in this area. Topics include the derivation of rheological constitutive equations, computation of flow fields and power dissipation, both under laminar and turbulent conditions.

C. Methods of emulsification

An important additional hardware requirement for an emulsion combustion system from the standpoint of cost, maintenance, and control over emulsion characteristics is the emulsifying device. Any potential gains in efficiency of combustion equipment and attendant energy savings can be negated if the cost or energy requirements for emulsification is too large. The principal component which contributes to the control of drop size distribution for a given combination of continuous phase, dispersed phase and surfactant is the emulsifying device. As was discussed previously, a desirable droplet size distribution for uniform dispersion in the fuel spray is one of narrow width with a mean diameter sufficiently smaller than the atomized fuel drops to aid in preventing breakdown of the emulsion.

There are three areas of focus in choosing an emulsifying apparatus:

(i) Is the emulsion to be created and stored or created at the point of use?

(ii) What drop size distribution is required?

(iii) How much energy is required for emulsification?

The minimum amount of emulsification energy is the increase in the total surface energy of the dispersed phase, but in practice a much larger amount of energy is required. It is conceivable that a better understanding of the physics of the emulsification process will result in a reduction of the energy requirements.

In this section the various techniques for producing an emulsion are reviewed with emphasis on those mechanisms which appear to require physics research.

1. Mechanical mixing [1,2,17,18]

In the case of mechanical mixing the fluids to be emulsified are introduced into a chamber provided with a suitable stirrer which induces a high level of turbulence.

Emulsification is brought about by the intense shear that results, and is mainly controlled by the nature of the turbulence in the flow. Improvements in design in terms of efficiency and drop size control may be derived from a better understanding of turbulence and of the mechanism of drop breakup in a turbulent field. Both would probably benefit by experimental study coupled with a detailed modeling effort. Droplet recombination and coalescence should also be investigated, especially for higher concentrations of the internal phase.

In situ emulsification on a smaller scale appears to be possible through the use of conventional colloid mills or homogenizers. These have already reached a high degree of development and it is doubtful whether efforts to obtain further improvements are warranted. However, a better understanding of power dissipation in these devices would be of interest.

2. *Ultrasonic emulsification*

A basic problem with all the mechanical emulsifiers considered is that emulsification is achieved by brute force, in the sense that work is performed on the entire volume of liquid rather than being localized on the interface to be disrupted. As a consequence, the amount of energy required for emulsification is a factor of 10^3–10^5 greater than the surface energy created.[1] If these power requirements are to be substantially lowered, it seems that a more subtle approach to emulsification is required. Ultrasonic emulsification is a promising alternative.

In spite of progress in the field of ultrasonic emulsification in the past two decades,[10] the basic mechanism or mechanisms governing the process are rather poorly understood. This situation detracts from the possibilities inherent in the method and in many instances leads to non-optimum working conditions and the acquisition of unfavorable results. This is unfortunate because in principle (and sometimes in practice, at least on a laboratory scale) ultrasonic emulsification has some unique advantages in terms of the power required and the uniformity of dispersed drop size distribution.[20,21] Two basic mechanisms are responsible for this effect. One is the familiar Rayleigh-Taylor type of instability that can occur at the interface of two fluids of different density under the intense acceleration produced by a sound field. The other is associated with cavitation that may take place in one or both liquids at sufficiently high sound intensity. The relative importance of these two effects has not been clearly established and most of our knowledge is based on uncertain and indirect evidence. For instance, one would like to know what is (if any) the detailed mechanism of action of cavitation. Once this understanding has been achieved, conditions can be optimized in terms of pressure, temperature, etc., and it may also be possible to obtain higher efficiencies of emulsification, for example, by seeding the liquids with bubbles of the appropriate size. Such developments may allow greater insight into the mechanism of cavitation itself, which is already a traditional area of physical research. Other important topics are the investigation of recombination due to acoustic streaming, the effect of shape of the sound field, and the dependence on its intensity and frequency. Finally, it is of interest to mention that a large-scale ultrasonic emulsifier (100 barrels/min.) for high internal phase oil-in-water emulsions shows a potential for a favorable economic position for ultrasonic emulsification.[22]

3. Spray techniques[23]

Another method which is potentially more efficient than conventional mechanical emulsifiers is that of the injection of one liquid into the other in a form of a high pressure jet. Here turbulence is generated in a more localized volume and the power requirements therefore can be lowered. Problems of interest to physicists are of a similar type as those described in the combustion section on the formation of sprays.

The efficient production of an emulsion with stringent requirements on drop size distribution may require a combination of the various methods discussed above. The use of a suitable surfactant may be necessary. In addition, the preparation of the high internal phase emulsions (of the order of 95 to 98 percent) that appear necessary for oil transportation and fire safety poses some unique problems. It may be necessary to produce emulsions with lower internal phase content and to subsequently boost the internal phase fraction. A fast and efficient method to separate the excess continuous phase without breaking the emulsion would then prove extremely useful.

D. Methods of characterizing emulsions

A brief review of some of the experimental techniques used in characterizing emulsions and their limitations is relevant since a systematic study of stability, flow properties and emulsification techniques cannot be conducted without efficient diagnostic procedures.

Many of the experimental problems that one faces in emulsion studies are similar to those described later for sprays. However, since an emulsion can exist at rest, there is considerable room for methods of analysis purely of the drop size distribution with lesser concern for its spatial and velocity dependences. On the other hand, in the experimental study of the flow of emulsions, the deformation of the drops and their distribution in the flow field become important.

The most widely used method for establishing drop sizes is direct visual or photographic observation with the aid of a microscope. As in the case of sprays, the procedure is quite lengthy and time consuming. An automatized method which employs the change in liquid conductivity when a drop passes in a small capillary (Coulter counter)[1] has been perfected, but cannot be used when the continuous phase has low electrical conductivity (for instance, water-in-oil emulsions). A similar technique makes use of the difference in the light transmission through a capillary in which the emulsion is made to flow, when a drop traverses the light path.[24] Both these techniques require a previous dilution of the emulsion and have some obvious shortcomings such as the difficulty in assessing sampling errors, the loss of droplet clusters which may have strong influence on the emulsion properties, and the effort required to reconstruct the size distribution function from the data. Alternative approaches might focus on properties directly sensitive to the distribution function. Promising areas of research in this respect are the optical and electrical properties of emulsions,[25] sound absorption characteristics[26, 27] and correlation with dynamical behavior (e.g., effect of drop size on thermal instabilities[28]). Quite different approaches may also be useful. An example is a method of "particle chromatography" recently reported, based on the behavior of particles of different sizes at an advancing freezing front.[29]

For emulsions under flow conditions the diagnostic techniques are not nearly as developed as for the case of emulsions at rest. In general, photography and high speed cinematography can be used only for low density emulsions. An interesting way to avoid this difficulty is presented in Reference 30. An increased effort in this area appears necessary in fostering a better understanding of stability and flow of emulsions.

III. COMBUSTION OF EMULSIFIED FUELS

This section discusses the present understanding of and postulations regarding the combustion properties of emulsions, suggests potential benefits which may result from emulsified fuel use, and defines areas in which further research is needed to determine the extent to which these benefits can be realized.

A. Fundamental combustion properties

1. Multicomponent fuels

In all practical uses of liquid fuels, except the carbureted spark ignition (SI) engine, there has been little or no attempt to entirely pre-vaporize and premix the fuel with air before injection into the combustion chamber. Indeed, even in the carbureted SI engine there is evidence[31] that the pre-mixed fuel/air charge contains a significant number (and size distribution) of unvaporized fuel droplets. In other systems, (for example, the fuel injected SI engine, diesels, furnaces, gas turbines), a spray of fuel is introduced and mixed with air during the combustion process. Thus all uses of liquid fuels involve combustion of fuel droplets and partially mixed gases resulting from droplet vaporization.

In addition to the fact that the combustible mixture is generally heterogeneous in nature, liquid fuels are invariably a mixture of several hydrocarbons. (For example, gasoline has a boiling point range from about 150 °F to nearly 400 °F). This adds further complication to the spray (droplet) vaporization and combustion processes. Fundamental research on single droplets of multicomponent fuels[32] has shown that the droplet composition changes by simple "batch" distillation during vaporization. Significant internal circulation in the droplet tends to limit internal temperature and composition gradients. Hence, the more volatile components vaporize first, concentrating the higher boiling point compounds in the remaining liquid drop. Thus, depending on the chemical and physical properties of the fuel and the surrounding environment, there is a minimum droplet diameter above which combustion will be established in the gas phase before the droplet has entirely vaporized.

Let us consider the combustion of a homogeneous, low density, stationary dispersion of multi-component fuel droplets. Each droplet burns in a configuration known as a diffusion flame. A flame sheath exists around each droplet where fuel vapor is consumed (as it vaporizes and diffuses outward from the droplet surface) by oxygen diffusing inward toward the droplet. Fuel and oxygen meet at the flame sheath in near stoichiometric proportions resulting in the maximum flame temperatures which could be achieved using stoichiometric premixed gases. Indeed, these high flame temperatures are associated with each fuel droplet even for very lean overall fuel/air mixtures. This fact is important to NO pollutant formation, which is chemically well described in the lean combustion of nitrogen-free fuels by the Zeldovich Mechanism[33]:

$$N_2 + O \rightarrow N + NO\,, \tag{1}$$

$$N + O_2 \rightarrow NO + O\,. \tag{2}$$

This mechanism is strongly temperature dependent because of the high activation energy of Reaction (1). Indeed, if reaction temperatures can be kept below about 1800 K (by burning locally fuel rich or very lean) little if no NO will result from the presence of nitrogen in the air. Stoichiometric flame temperatures of all hydrocarbon/air mixtures are well above this temperature.

Thus droplet (diffusion controlled) combustion inevitably leads to NO formation. However, Bracco[34] has shown that even reducing droplet size (by increasing number density at fixed total mass) will reduce NO formation.

Diffusion controlled combustion also leads to formation of soot and carbon particulates. The fuel vapor diffusing away from the liquid surface is subjected to high temperatures in the absence of oxygen. This leads to gas phase dissociation and pyrolysis of the material to form unsaturated hydrocarbons including acetylinic compounds and finally to soot.

As described earlier, there is also a concentration of higher boiling point compounds in the liquid phase during droplet combustion. It is believed that through liquid phase reactions this inevitably results in formation of larger carbon particles (1–200 μ) called cenospheres. The important point to be made is that once soot is formed, it is very difficult to remove by combustion processes; large residence times at high temperatures in an oxygen rich environment are needed.

2. *Emulsified fuels*

An interesting feature which distinguishes the combustion of an emulsified fuel droplet from that of a pure fuel is a phenomenon referred to in the Russian literature as "micro-explosions." The initial work of Ivanov and Nefedov[35] suggested the occurrence of this phenomenon, and to date there has been no other corroborating fundamental evidence.

These workers studied the vaporization, ignition and combustion of water-mazut (mazut is a grade of heavy oil), water-kerosine and water-benzene emulsions by suspending single droplets in hot inert or oxidizing environments. Observations were made using rapid cinematography and temperature-time measurements (obtained by suspending the droplet on a thermocouple). Results were compared to observation of combustion of pure fuel droplets in the size range of 0.8–3 mm.

Photomicrographs of the emulsified mazut showed droplets of water (10–50 μ) uniformly distributed in the fuel. Cinematography illustrated that the emulsified mazut droplet undergoes "micro-explosive" enhanced vaporation when thrust into a hot oxidizing environment. This process precedes ignition of the droplet, continues after ignition and apparently leads to complete dispersion of the primary droplet during the combustion sequence. Indeed droplet ignition times were suggested to be significantly reduced by the addition of the water in emulsion form. The droplets are "broken up into small fuel particles, gaining more speed in relation to heating up, evaporation and igniting." (The Russian authors coined this phenomenon as "micro-explosions.") Ivanov and Nefedov suggested this process will produce "a better mixing with oxidant and a more complete combustion." Similar experiments were conducted with kerosine—30-percent-water emulsions, and the combustion time was observed to be reduced to one half that of an equivalent pure kerosine droplet.

This work does not provide any illumination as to possible effects of high pressure, high density sprays, flow conditions, emulsion quality, material and size distribution of dispersed phase on the micro-explosion phenomenon. However, there is the possibility that emulsified fuel usage may have significant impact on processes dominated by heterogeneous combustion, through improvement of the rate of vaporization and micro-mixing.

3. Dispersed phase materials other than water

There is no reason to suggest that dispersed phases other than water cannot be used in emulsions to produce the micro-explosion phenomenon. Indeed, any low boiling point compound which produces sufficient expansion during evaporation to shatter the primary droplet, which is immiscible in the primary fuel, and which does not produce noxious products during combustion could be used. An example of such a compound is methanol (Methanol could be an abundantly produced material and has even been suggested as the basis for an entire fuel economy[36]). While methanol is only somewhat soluble in fuels such as gasoline, small additions of water render the methanol immiscible. Methanol is a fuel and thus does not significantly detract from the total energy which can be stored per unit volume. The heat of vaporization of methanol (262.8 cal/gm) is less than half that of water (540 cal/gm) and thus local cooling effects produced during vaporization are reduced. The critical pressure of methanol (99 atm) is well above values experienced in most combustion systems, and thus vaporization will produce expansive disruption of the primary droplet in which it is dispersed.

Because methanol's boiling point (65 °C) is less than water, the micro-explosion phenomenon will initiate earlier in the primary droplet heat-up period. This may be important in combustion systems which may suffer from significant ignition delay due to vaporization (such as the diesel). Furthermore, it is interesting to note that since the surface tension of methanol (22.61 dynes/cm) is appreciably less than that of water (73.05 dynes/cm) and closer to that of the hydrocarbon continuous phase, one would expect the possible range of dispersed droplet sizes to be extended to smaller diameters. This may be important to the micro-explosion phenomenon and to the stability of the emulsion itself. However, it is easy to lower the surface tension of water by the addition of known surfactants while it is not known if such surfactants are effective for methanol. A more practical reason for using methanol or methanol/water in fuel emulsions is that such systems would even be useful in freezing climatic conditions. The addition of methanol to water-containing emulsions may also limit corrosive behavior.

B. Potential benefits from combustion of emulsified fuels

It appears that system studies of water in fuel emulsions have been conducted under less than controlled experimental conditions, with no serious attempt to make more than minor design revisions to accommodate their use. Neither fundamental nor practical studies of the micro-explosion phenomenon using other than pure water as the dispersed phase have been reported in the literature. Thus the improvements in performance and emission characteristics from the use of emulsions must at present be considered as potential. The extent to which this potential may be realized will only be evaluated through further fundamental and specific system studies. Present available system results and some discussion of the potential for benefits in specific applications will be discussed in the following subsections.

1. *Continuous combustion systems*

a. Furnace applications. The recent development by Cottell of an ultrasonic cavitation technique to produce water-in-fuel emulsions just prior to combustion[37] and its application to furnace combustion[38] has spurred new interest in the use of emulsion combustion in heat generation applications. However, earlier work presented by Scherer and Tranie at the 14th International Symposium on Combustion,[39] and published Russian research of Komissarov *et al.*[40] previously established some aspects of emulsified fuel combustion in furnaces. Scherer and Tranie combusted 20 percent water in No. 2 and No. 5 fuel oil emulsions in domestic oil-fired boilers with power output between $1.5\times 10^4-2\times 10^6$ Btu per hour. Emulsions were made just prior to injection into the furnace, and adjustments of excess air, type of atomizer, atomization pressure, etc., were made with each boiler. Cenospheres and soot were appreciably reduced when water/fuel emulsions were burned. Soot reduction was systematic with the same excess air and was not less than 5 : 1. Excess air in these furnaces could be reduced by more than factors of two to obtain the same smoke number from the exhaust. (These authors suggested the enhanced vaporization effect by micro-explosions, but did not reference the work of Ivanov and Nefedov.) Similar conclusions were obtained by the Russian study. It was also stated that the same furnace efficiency could be obtained at significantly reduced air and fuel flow. This result has been more recently inferred from studies of a Cottell emulsified combustion furnace system at Adelphi University.[41]

Evaluation of emulsified fuel combustion of numbers 2 and 6 fuel oils in residential furnaces is presently underway at the Environmental Protection Agency, Research Triangle Park, under the direction of R. E. Hall. Similar research on applications to residual fuels is in progress at the University of Southampton by Professor I. Smith.

Possible benefits from the use of emulsified fuels in furnaces are more likely to occur in industrial and residential sizes (< 250 MBtu/hr). These furnaces depend primarily on convective heat transfer. The reduction of soot formation realized from enhanced vaporization and mixing (from micro-explosions) limits the deposition of carbon on the furnace convective heat transfer surfaces. Indeed, water may also remove carbon deposits through the carbon-steam reaction. This is important to maintaining convective heat transfer at the optimum design condition.

Soot emissions are presently controlled by adding large amounts of excess air. If excess air can be reduced while maintaining convective heat transfer, stack losses will be decreased and some improvement of overall furnace efficiency may be realized. In fact, smaller furnace sizes do not usually employ economizer sections (additional heat exchangers) to reduce stack losses, primarily because of the earlier mentioned deposition problems. Indeed, if carbon deposition can be eliminated in residential size furnaces, economizer sections could be effectively used to further reduce stack losses. Possible improvements of overall furnace efficiency cannot be identified without more detailed system studies.

b. Gas turbines. While aircraft gas turbine emissions are at present only a small fraction of the worldwide total, localized levels near airports (primarily due to excessive engine idling[42]) and emissions (primarily NO_x) at cruise conditions[43] in the stratosphere are of great concern. Water injection in gas turbines at cruise power has been shown to be a feasible method of reducing NO_x[44]; total hydrocarbons (THC) and CO emissions under full power are inconsequential. However, at idle conditions THC and CO emissions are considerable, apparently from poor mixing and droplet vaporization. The possibility of reducing THC and CO by enhanced mixing with emulsified fuels should be explored. Some recent evidence that enhanced droplet vaporization may reduce these emissions can be drawn from the drastic effect humidity has on the measured levels of pollutants at idle.[45] An early Russian study indicated that increased humidity leads to more rapid vaporization of hydrocarbon fuel droplets.[46] There may also be a reduction in smoke production in gas turbines by burning emulsions, much as is evidenced in furnaces. However, the question of improved engine performance remains moot. There have been no reported studies on the combustion of water-in-fuel emulsions in gas turbines.

2. *Intermittent combustion systems*

a. Internal combustion spark ignition (SI) engines. The benefits of water addition to internal combustion engines have been of passing interest for some years. As early as 1947, the addition of finely atomized water (or ethyl and methyl alcohol) to spark ignition engines was recognized as a method of eliminating hard knock (detonation). The scheme is attractive since water injection results in an *increase* in power output and a *decrease* in engine coolant requirements.[47] It was suggested that water is an effective internal coolant because of its high heat of vaporization. Addition of water in the liquid state significantly lowers chamber temperature and slightly increases chamber pressure. These arguments can be substantiated by performing simple adiabatic flame temperature calculations at constant volume for stoichiometric mixtures of iso-octane/air with varying quantities of liquid water. Lower cylinder temperatures are directly associated with reduction in necessary external cooling requirements of the engine while the vaporized water produces extra mass in the cylinder and hence higher pressure at fixed temperature. Reduced cooling requirements directly reflect decreased heat losses, which are a significant fraction (~25 percent) of the difference between ideal and practical efficiencies of the SI engine.[48]

Furthermore, the lowering of peak cylinder temperatures is a very positive way of reducing NO_x formation due to the strong temperature dependence of the Zeldovich mechanism. This effect of water injection in SI engines has been adequately established.[49] It is thus feasible to use leaner fuel/air ratios with water injection which can also reduce the CO engine emissions. Of more practical importance, valve burning and engine overheat associated with lean combustion can also be avoided.

The knock suppressant character of water addition is also of great importance, particularly with the introduction of low octane fuels and elimination of lead as an anti-knock additive. Plausible suggestions as to why water acts as an effective knock suppressant are the following.

(a) The chemical induction period to auto-ignition is increased by the temperature reduction at top dead center. Thermal ignition is thereby suppressed.

(b) Water is an effective third body for hydrogen atom recombination and may afford a fast chain breaking step.[50] The concentration of hydrogen atoms is important to the chain branching mechanism leading to auto-ignition.

In the case of knock suppression with methyl or ethyl alcohol, possible effects of (a) must be much reduced since the heats of vaporization are less than half that of water, and the auto-ignition delay is exponentially dependent on temperature. Thus one might speculate (b) to be the more plausible mechanism for alcohol addition. The effective octane number increase achieved with water or methanol addition may permit use of higher compression ratios on engines designed for non-leaded fuel. Alternately, a substantial increase in the "gasoline" yield per barrel of crude might be possible if lower octane fuels could be used in existing engines with the addition of water or alcohol.

If water addition in the form of a secondary atomization to intake manifold or direct injection into the SI cylinder can accomplish the above combustion improvements, why develop a water/gasoline emulsion as an alternative fuel? There are two practical reasons:

i. It is believed that direct water injection on intake stroke leads to build-up of water in the oil crank case of the (unmodified) engine. This apparently occurs from water condensation on the cylinder walls and leakage by the cylinder rings during the compression stroke. This might not occur when the water is carried in the emulsified fuel.

ii. Direct water injection will require an additional storage tank and pumping system. These additions are not necessary if the water is carried in a stable emulsified fuel.

The use of water or alcohol fuel emulsions may result in augmented droplet vaporization (the micro-explosion phenomena) as described earlier. While little effect on charge distribution would result if the engine were carbureted, the improved vaporization would reduce droplet combustion and permit a more lean overall operating condition. Indeed, improved droplet vaporization from the use of emulsions coupled with fuel injection to reduce unequal charge distribution might be a most favorable alternative.

The full potential of application of emulsified fuel technology cannot be deduced from the limited results presented in the literature.[51–54] However, both Ewbank[51] and Cottell[52] claim at least a 10 percent increase in fuel economy on automobile engines. It is possible that emulsion combustion may even permit one to meet pollution standards without catalysts.[55]

b. Internal combustion compression ignition (CI) engines. In the light of the increased concern for fuel economy, interest in the automotive diesel is also increasing due to the fact that the thermal efficiency and operating characteristics of the Diesel cycle are inherently better than the Otto cycle. However, major problems with smoke, odor, and NO_x emissions remain unsolved. It is in the potential solution to these problems that emulsified fuels may prove beneficial. In order to discuss the potential role of emulsified fuels in diesel combustion, we must first discuss the character of heterogeneous combustion in diesels—a character which could be influenced by the use of emulsified fuels. Emulsified fuel usage in diesel engines has been reported in the literature by only a limited number of investigators.[56–58] The reported results indicate no advantage of using emulsions other than a slight reduction in smoke. Current research is underway at Princeton University by F. Dryer, I. Glassman, and Dr. W. Naegeli to investigate diesel applications.

In a sense, diesel engines are ultra-lean stratified-charge engines with very heterogeneous combustion (i.e., wide variations in local air/fuel ratio from the overall air/fuel ratio). Diesels commonly operate very much leaner than stoichiometric, and for homogeneous combustion of such mixtures, one would expect negligible NO_x, even at diesel compression ratios (18 to 1).

The heterogeneous character of diesel combustion is a result of injecting a high pressure spray into a high pressure (typically 350—850 psia) oxidizing environment.[59] Fuel droplet sizes produced during injection range from 50 to 70 μ. Because of hydraulic effects in the fuel injection system, tailoring of droplet size distribution is difficult. In fact, during the last part of the injection process, even larger droplets may be formed due to the relatively small pressure differential acting on the fuel. The concentration of components in a fuel mixture may differ in various locations of the spray. Indeed, Lamb[60] found that for multi-component fuel sprays, concentration of the higher boiling point components occurs in remaining droplets as the spray evaporates. Evaporation of the spray first occurs at its periphery, since there the temperature field in the chamber is least depressed by the injection process and droplet residence times are longest. Multiple ignitions occur at various locations in the periphery where the conditions (fuel component/air mixture, temperature) are most suitable for auto ignition. Flame propagation from these ignition sources is responsible for consuming the remaining heterogeneous field. It has been suggested[61] that vaporization of the spray is very rapid. Therefore, it would be relatively unimportant in controlling the rate of energy released during the main combustion process. Thus, mixing of fuel-rich gas pockets with the lean surroundings, together with relevant chemical kinetic data, would be the key to ultimate engine behavior. However, a second theory[62] suggests that while mixing is of significance, the vaporization of the spray itself is of major consequence to the rate of energy release.

In any case, one is dealing with diffusion type combustion of either fuel rich gaseous pockets[63] or evaporating droplets. It is well established[64] that diffusion type combustion results in flame fronts with temperatures approaching the stoichiometric adiabatic value regardless of the overall mixture ratio. Together with the initial high temperatures produced during compression, diffusion type combustion would result in flame temperatures approaching *3000* K. When one realizes that NO production attributed to the Zeldovich mechanism becomes significant above 1800 K, it is not surprising that combustion systems dominated by diffusion (mixing) processes produce large quantities of NO_x.[65,66] Smoke production is also related to heterogeneous effects.

There has been considerable study of diesel combustion system parameters—indirect versus direct fuel injection (discussed above), turbo-charging, injection timing, combustion chamber configuration, etc., and their relation to diesel exhaust emissions.[67] However, it must be re-emphasized that NO_x is the emission one must be most concerned about in meeting the original 1976 model-year light-vehicle standards. Lower NO_x emissions have been best achieved by retarding the injection timing and the direct injection design responds more favorably than other designs to this process. NO_x reduction is due principally to shorter residence times and lower maximum temperatures reached in the combustion chamber. However, extreme retardation appears to be necessary and this produces significant increases in hydrocarbon/smoke emissions with considerable penalty in fuel economy. Exhaust gas recirculation (EGR) has also shown some promise; however, because of a smoke problem, this system can be effectively applied only in the low-load range of operation. Corrosion of inlet system components may also be restrictive. Exhaust treatment (catalyst approaches) is not possible in the case of the diesel, since this requires high CO/O_2 ratios in the exhaust gas. Another approach to the reduction of NO_x, smoke, and odors in the diesel is through the addition of water and other compounds to reduce the heterogeneity in diesel combustion, thereby reducing local temperatures.

Water addition could again be accomplished through direct injection, intake charge saturation or the use of water-in-fuel emulsions. Provided material compatibility and emulsion stability can be satisfied the latter process would most certainly be least complicated.

Indeed, if the micro-explosive character of emulsified fuel burning can be exploited, the use of emulsified fuels may improve the performance and emission characteristics of diesel engines. This follows if droplet vaporization controls the energy release process, since the number density of droplets will increase with a decrease in the mass average size. It has been suggested that for droplet type combustion, many small droplets produce less NO_x than the equivalent mass of larger droplets.[34] Thus in addition to the internal coolant properties of water and more distributed thermal properties achieved in the cylinder, either of the proposed mechanisms controlling energy release predict a reduction in NO_x emissions.

These same qualities of the modified combustion are important even if the fuel pocket mixing phenomenon[61] controls energy release. Micro-explosions should produce smaller droplets and thus more distributed, small fuel pockets. In either event, soot formation may be reduced, as has been observed in oil-fueled furnace studies.[39] Again the reason is the decrease in the overall heterogeneity of the system. It is well known that soot formation is more prevalent in heterogeneous diffusion flames than in homogeneous combustion systems.

As mentioned earlier, the conclusions of previous diesel experiments suggest that no benefits were obtained when emulsions were used other than a slight reduction in smoke. In fact the diesel ignition delay was increased so much that any NO_x reductions were completely negated by the necessary advanced timing.

One should not be unduly discouraged by these rather limited results. The emulsions employed were not clearly described. However, the statement that the water-fuel emulsion had to be continually stirred to prevent separation suggests that the dispersed water droplets were of considerable size in comparison to those described by the Russians.[35]

The use of the Cottell ultrasonic technique, which emulsifies just prior to injection, should circumvent this difficulty. A small ultrasonic device is certainly feasible even in a practical automotive system. Most certainly the "micro-explosion" phenomenon must be a function of both size and number density of the dispersed water droplets. Indeed, the emulsion used by Valdmanis and Wulfhorst[57] may have been so unstable as to partially break down during the injection process. This would lead to local cooling in the vicinity of the fuel spray and a corresponding reduction in the optimum micro-explosion intensity of a specific fuel-water emulsion mixture. Thus the potential reduction of the physical ignition delay, τ_p, associated with droplet vaporization may not have been achieved, and in fact, an increase in chemical delay, τ_c, (from local cooling of the mixed fuel/air vapor regions) should have occurred. In the event that the emulsion was properly dispersed, the physical delay should have been decreased and the chemical delay increased because of the local lower temperatures created from water vaporization. In this case τ_c would be such that $\tau_c > \tau_p$ and would be the major part of τ_i. τ_c may be modified in many ways, e.g., by changing the cetane number of the fuel or by increasing the compression ratio. Neither of these possibilities has been explored. Indeed, the negative conclusions of earlier work hinge entirely on the fact that the diesel ignition delay becomes so large as to offset any positive effects.

Indeed, if the relative importance of τ_c is prohibitively increased by the cooling effects of the water added in the emulsion, it must result from the reasons given earlier for knock suppression by water in an SI engine. The effect of the high heat of vaporization may be significantly reduced while maintaining the micro-explosion characteristics of an emulsified fuel by using an insoluble organic such as methanol to produce the emulsion. In addition to improvements of the micro-explosion phenomenon from the physical properties of methanol, there may also be an effect on τ_c resulting from a change in the effective cetane number of the fuel.

Because of the dominance of heterogeneous combustion effects in diesel combustion, and the potential benefits which might be realized from the use of emulsion type fuels, the diesel application deserves careful consideration.

C. Additional research on combustion of emulsified fuels

Research to date has not established the potential for energy savings and emission control which emulsified fuel combustion may present. No fundamental information on the characteristics of emulsified fuel combustion and the parameters which influence those characteristics exists. The emulsified fuels (with the exception of those produced ultrasonically) which have thus far been combusted appear to be of very poor quality, having to be continuously stirred to prevent breakdown.[57] No evidence exists that even the ultrasonically created emulsions survive breakdown under the large dynamic shear forces produced during spray atomization. In addition, all research with the exception of reference 35 has been of the cut-and-try approach with only small modifications to the conventional combustion device. No attempts have been made at optimizing either device or fuel for improved performance.

Emulsified fuel combustion research remains as virgin territory for the physicist with unanswered technical problems of fundamental nature as well as unconfirmed performance potential in system applications. Some suggested research areas in fundamental studies and in systems applications are presented below.

1. Fundamental studies

Additional fundamental research on water-fuel emulsions and their combustion characteristics is needed. The stability of the emulsion under storage, pumping, and spray atomization is an important area which cannot be completely separated from the combustion characteristics. This area was addressed earlier. The fundamental research which is necessary to determine the characteristics of emulsified fuel combustion includes the following.

a. *Droplet combustion studies*. Single droplet combustion research would appear useful to study the fundamental parameters and, in particular, to assess the validity of the micro-explosion model derived on the basis of the results of Ivanov and Nefedov.[35] One may evaluate comparatively the combustion process of pure and emulsified fuel droplets as well as the effects of the size distribution and material of the dispersed phase. Emulsion characteristics should be better defined than in the earlier work. The studies should be extended to include pressures and temperatures more aligned with practical combustion systems (I.C. engines, etc.). While suspended droplet studies[68] should yield qualitative information of some interest, the technique is restricted to droplet sizes (0.3 mm to 0.8 mm) which are not of particular relevance to practical systems.

In a fuel spray combustion process, droplets formed are typically less than 100μ in size and interaction occurs among the droplets during ignition, vaporization, and combustion. Neither the size nor possible interaction effects can be studied in suspended droplet experiments. However, a technique described in a recent review article[68] shows promise of permitting single droplet examination in the size range of $10–200\mu$. This approach can also be used to simulate dilute sprays where interaction between droplets can be observed. In a dense spray, interactions are more severe, but most probably cannot be examined with any experimental control. This method[69] of producing streams of droplets of well defined size and number density would appear suitable for quantitative examination of droplet behavior of homogeneous and emulsified fuels.

The droplet generator is a modification of that originally described by Lindblad and Schneider.[70] A uniformly sized, equally dispersed stream of droplets is generated by a constant pressure liquid jet forced through a capillary which is mechanically vibrated. Accurately sized droplets from 10 μ to 200 μ are produced at rates from 1 per ten seconds to 200,000 per second. A recent application of this technique to the study of free droplet homogeneous fuel combustion[71] has been successful.

b. *Spray combustion studies.* With the exception of carbureted spark ignition engines all practical devices discussed earlier use sprays as a means of delivering fuel to the combustion chamber. In view of the limited amount of information obtained from single droplet studies, there is general agreement that complementary studies of sprays are necessary. Again comparison of single fuel and emulsified fuel spray combustion is important. Comparative combustion experiments on homogeneous and emulsified fuels would reveal effects of, e.g., flame geometry, radiation, temperature distribution, ignition characteristics, and chemical species distribution.

There is a large body of basic data available on single fuel sprays, mainly stemming from interest on the part of the liquid propellant rocket designers (see, for example, references 72 and 73). The behavior of emulsified fuel sprays and single fuel sprays should be similar with some effects due to the non-Newtonian nature of the emulsion. However, the possible occurrence of partial breakdown of the emulsion under high dynamic shear rates must be considered. The distribution of both primary droplets and the dispersed phase should be determined.

Present diagnostics available for characterizing spray phenomena, particularly droplet size and spatial distribution, are somewhat limited. Sampling techniques which disturb the streamlines in the probe vicinity (called perturbing techniques in the previous section) are widely used for particle size measurements.[72,74,75] Typically they involve the collection of a sample of the spray on a glass slide or in a liquid with subsequent visual counting. The shortcomings of these techniques are obvious (they discriminate against the smaller particles which tend to be entrained by the gas, larger droplets may shatter upon impact with a solid or liquid surface, the flow pattern is strongly affected, the drops collected on a slide will not be spherical so that a correction to the measured diameter is required, coalescence may occur, etc.), but they are comparatively easy to apply and have been widely used in the past. Although not usually considered a perturbing technique, the frozen wax technique can find a place under this heading due to the uncertain effects that the incipient change of phase and possible non-Newtonian characteristics may have on the shedding of the smaller droplets. These difficulties may be reduced by making the wax freeze after the spray is fully formed as in the frozen drop method, but the information gathered in this way solely concerns size distribution. One of the most attractive features that these methods offer is that the particle size can be analyzed by sieving the solidified droplets, but no efficient sieves exist for particle diameters below 50–75 μ. Other perturbing techniques have been devised which may allow semi-automated data collection of drop size distribution, but room for improvement exists. Significant theoretical development is needed to estimate with greater confidence the effects that the perturbation has on the accuracy of the data obtained with these techniques. Methods of analyzing the drop behavior without the necessity of perturbative suspension should be examined.

Photography has of course been extensively used in spray studies.[72,73,76] Its limitations reside in the poor resolution of the dense parts of the spray, the necessary magnification to resolve the particles and the consequent loss in field depth and covered area. More recently, holography[77,78] has been successfully used to obtain more detailed information on the three dimensional structure of larger spray volumes. A general difficulty with holography is the effort required to analyze the data. These methods enable one to reconstruct the droplet size distribution function from the observed values of each particle. A more efficient approach would be to try to measure the distribution function directly. Some progress could be made in this direction by the use of optical scattering and absorption measurements backed by a sufficiently sophisticated theory.[79-82] At present, the main obstacles to the use of this approach are the occurrence of multiple scattering in the dense portions of the spray and its spatial inhomogeneity, but it appears possible to overcome these difficulties by a combined theoretical and experimental effort. A recent approach[83] makes use of different scattering angles and wavelengths used simultaneously. Holographic techniques have recently been applied to the study of spray combustion.[84] Clearly, the problems encountered in data reduction for the non-combustion cases are compounded for reacting flows.

Methods to determine the velocity distribution of the droplets are needed. So far photographic techniques have been used, and more recently holography.[77,85] The idea in both methods is to record two successive images separated by a known time interval. The holographic method is superior because it allows a better determination of the three components of the droplet velocity. However, the effort required to quantify the data from the recorded images is enormous. It appears that radically different techniques are necessary in this area for fast and relatively accurate measurements. An adaptation of the techniques familiar to atmospheric physicists may prove useful.

2. *System studies*

Although fundamental studies are necessary to improve our understanding of emulsified fuel combustion, the real impact of the use of emulsified fuels can only be evaluated through careful system studies on practical combustion devices. Because of the uniqueness of each combustion system, variations in optimizing parameters may even occur from application to application. What is of paramount importance is that each applied research study be responsive to the fundamental research program described earlier and not be limited by the lack of versatility of the test instrumentation. For example, in the earlier referenced diesel studies on emulsions[57] no attempt was made to identify the effective cetane number of the emulsified fuel, and the test engine compression ratio was not modified from that used for pure fuel. The only parameters varied were quality of the emulsion and injection timing. It is not surprising that such a constrained study produced ambiguous results. In all the systems that are tested the experimenter must be able to modify system parameters to permit the important effects and benefits of emulsified fuels to be optimized. Care must be taken in deciding how results with emulsified fuels will be compared with pure fuel measurements in these system studies.

In the suggested system studies one should recognize the possibility that emulsion combustion may produce the same results as water injection, with no further advantages. Therefore any emulsion test in a practical combustion device should be compared to water injection tests.

Specific system tests in furnaces, turbines, and internal combustion engines need not be detailed here. The potential benefits are discussed in detail in Part IIIB. Their verification should be central to any system test using emulsified fuels.

IV. OTHER ADVANTAGES OF EMULSIONS

A. Pipeline transmission of crude oil

Major energy sources are often far distant from major energy uses. Middle East oil and its users in Europe, the Far East and North America is a well-known example of this. Although supertankers are the most economical method of bulk transportation of such low unit value liquids for most inter-continental routes, pipelines must be used overland. Alaskan oil transmission is an example where both possibilities exist, but supertankers are extremely risky because of ice conditions in the Arctic Ocean. The resulting need to move Alaskan oil over land has caused considerable controversy because of the risky technical and environmental implications. Technological advances are needed which will minimize the environmental hazard, reduce pipeline construction requirements, or relax pumping requirements. Emulsions or dispersions of several sorts show promise of worthwhile applications to the Alaskan pipeline.

Overland transportation of crude oil in the Arctic is very difficult and expensive because of the strong dependence of viscosity on temperature, the geographical isolation from industrial sources and a geologic condition known as permafrost. The line now under construction in Alaska is designed to carry hot crude oil (130° to 150 °F) in a pipe which will be constructed on expensive supports for almost half its length. The pipeline will be thermally isolated from the earth's surface and will hence prevent thawing of the permafrost which would in many places lead to severe environmental damage and also possible failure of the pipe. The natural gas produced with the crude oil from this same Prudhoe Bay field will eventually have to be transported in another pipeline, either paralleling the oil line or going through Canada. The decision on both the route and the design of this latter line are still under study and thus there will be a considerable delay in bringing the gas into use.

Many alternative methods of transportation for this oil have been proposed. It is not our purpose to review all of these but only to describe two which utilize emulsions or dispersions. The first was initially described in 1970[86-90] while the second, which is an outgrowth of the first, has only been described to a very limited audience so far.

The first requires the preparation of crude oil-in-brine emulsions and the pumping of these at temperatures below 0°C through a fully buried pipeline. Under this condition heat cannot be transferred to the permafrost to melt it. The amount of salt in the brine (typically 10 percent NaCl) determines the temperature at which freezing of the continuous phase (brine) would start. At the temperature of operation (−2 to −7 °C), the dispersed oil droplets would probably be solidified, creating a situation similar to well-known slurry pipelines. An advantage of any cold pipeline over a hot one is that the liquid petroleum gas, LPG, components of the natural gas can be safely redissolved in the cold oil and thus transported to market. While there are several engineering advantages and disadvantages of this kind of cold pipeline relative to a hot one, there also seem to be distinct environmental advantages for a cold one. Not the least of these is the greater seismic stability of a buried line compared to one supported above the ground.

The second method of transportation requires the preparation of a crude oil dispersion in methanol and then pumping it at temperatures below 0°C through a fully buried pipeline. The methanol can be manufactured from methane by well-known processes. Thus this method allows both the crude oil and natural gas components of petroleum to be transported simultaneously in the same carrier. These dispersions must be stabilized by surfactants in the same way that other emulsions must be stabilized. Both cold pipeline systems require significant amounts of re-cooling of the material being transported to overcome frictional heating. The crude oil-in-methanol dispersion requires less re-cooling than the crude oil-in-brine emulsions because of its more favorably rheological properties.

B. Fire safety

The increasing utilization of energy and the continuing shift of supply from domestic to far distant sources will require moving larger quantities of fuels over larger distances in the future. Presently, some 25 percent of the total energy (and greater than 50 percent of the petroleum) used nationally is also consumed in devices requiring a mobile energy source (aircraft, automobile, etc.) and this type of usage is predicted to grow significantly. The increased use and transport of liquid fuels (both petroleum products and liquified gases) suggests a rise in the probability of accident and damage, both in terms of fire and spillage impact on the surroundings.

A number of approaches to the fire safety problem have been considered and explored under government sponsorship over the past ten years. Most of these studies have been directed toward the military aircraft fire safety problem. However, the significance to domestic applications is equally important.

The fire hazard which mobile fuels represent in the event of transportation accidents result from

(1) the possibility of rupture of the fuel tank;
(2) the availability of uncontrolled ignition sources;
(3) the possible formation of highly flammable mists or aerosols of fuel when sufficient impact energy is available;
(4) the increase in size of the contaminated area by liquid flow from the ruptured containment vessel.

Much can be done to reduce the number of possible ignition sources but rapid release of flammable vapors from the spillage is particularly difficult to control. While fuel containment appears as a most suitable solution to the fire hazard problem, this approach has a prohibitive weight penalty in the case of aircraft. Weight is also detracting from the standpoint of inertial effects (during acceleration and deceleration) in the case of ground transportation.

The use of gelled fuels as a crash-safe fuel is of more recent interest and is still receiving limited attention. This approach would reduce flame propagation and aerosol formation. However, gelled fuels have several major disadvantages:

(1) the rheology of gelled fuels is very temperature sensitive;
(2) considerable pressure drop is experienced during pumping;
(3) *JP*-4 fuel phase tends to separate out during pumping;
(4) gelled fuels are impossible to reconstitute once they are broken;

(5) flash/fire temperatures and explosive character are not altered favorably.

From a technical and practical standpoint, emulsification of the fuel in a continuous phase of a non-flammable material (such as water) has several major advantages over gelled fuels:

(1) the rheology is rather temperature insensitive;
(2) the removal from containment vessels with low surface energy is facilitated because such emulsions do not wet these surfaces;
(3) stability during pumping and fuel transfer is improved;
(4) the emulsions can be reconstituted with ease.

Unlike the conventional gelled fuel, high internal phase emulsions also have the following important advantages:

(5) the flash point of the fuel can be significantly increased by emulsification; i.e.,

T_{flash} *JP*-4 aircraft fuel = 77 °F,

T_{flash} 97% *JP*-4 in 3% = 155° F,

surfactant and continuous phase

(6) spillage of the emulsified fuel can be simply dispersed from the vicinity of the crash by further dilution of the continuous phase with additional water;
(7) high internal phase emulsions produce the lean explosion limit vapor concentrations of pure substances only after much longer exposure times.

Considerable research and development efforts on high internal phase emulsions for use in aircraft have already been conducted.[91-94] However, these efforts have been directed primarily toward meeting specific military needs (for uses associated with gas turbine powered systems). Objectives of these studies were to develop an emulsion of *JP*-4 as a dispersed phase in an inert low percentage continuous phase. Specifications for the emulsion were very restrictive, demanding high stability, good flame retardation, minimum performance degradation, minimum corrosive character and survivability under certain combat conditions.

While these studies produced a satisfactory screening of the many available surfactants and resulted in development of several suitable emulsions which at least marginally met the above objectives, several deficient research areas were noted (to which the physical scientist can easily devote his talents):

(1) High internal phase emulsions generally exhibit enhanced corrosive character. While some material screening has been accomplished further work is necessary. No research on chemical inhibition of the corrosiveness has been noted.
(2) Available rheology data are not entirely sufficient to permit designers to predict adequately performance of high internal phase emulsions.
(3) Further analysis of the breakdown and stability of high internal phase emulsions is necessary to prevent breakdown during the high shear rates imposed by pumping.
(4) Some question remains as to emulsion stability during atomization. There may be undesirable modification of the pollutant emissions generated on combustion of such sprays.

Another safety application of oil-in-water emulsions takes advantage of their dispersive character. Emulsions of crude or fuel oil-in-water as tank cargo may disperse in the ocean in case of accident or loss at sea. Such dispersions are more readily attacked and decomposed by micro-organisms than are oil slicks.

The use of high internal phase emulsions as an approach to fire safety in automotive transportation has apparently not been considered previously. Tank rupture, aerosol formation from impact and the very low flash point of gasoline produce severe fire hazard upon automotive collision. Indeed, death by fire has become a major fraction of the total number of transportation fatalities each year. The properties of a high internal phase emulsion of gasoline (if it can be made) would be important for the following reasons:

(1) increased flash and fire points of the emulsion over pure fuel may change the flame spread rate over spills drastically (from gas phase controlled spread to liquid phase controlled spread)[95];

(2) reduction of aerosol formation upon impact would significantly reduce the possibilities of ignition;

(3) reduction rate of evaporation of the emulsion would aid in delaying possible build up of explosive concentrations of gasoline vapors in enclosed areas (i.e., in the automobile trunk);

(4) the spilled fuel from a ruptured tank could be easily dispersed by simple water jets.

Thus in addition to the research areas considered important to the aircraft fire safety research, screening of possible surfactants and compatible materials for other applications remains to be done.

A consideration of the possible technical and societal advantages of the use of emulsified fuels has suggested to this study the interesting possibility of combining considerations of fire safety and combustion improvement. To do so would require the development of an inversion device that would take oil-in-water emulsions from tank storage, convert into water-in-oil emulsion before or during atomization and injection into the combustion system. It is in just such technical areas where the innovative application of fundamental physics could result in successful practical developments.

V. CLOSING REMARKS

The focus on emulsified fuels as part of this APS Summer Study is not meant to imply that the introduction of emulsified fuels into practical combustion systems is the most challenging physics problem or the most promising novel combustion technique for efficient fuel utilization. Rather, it is an attempt to present an objective review of the current state-of-the-art in a topic receiving considerable publicity because of its novelty and because of its innovative, yet simple, character. Some of the projected benefits of emulsified fuel use are based on speculations which may or may not prove to be true. Nevertheless, a sufficiently convincing technical base does not exist from which to project quantitatively the impact of emulsified fuel use, and such a base should be established.

Interest in emulsified fuel as a means of water addition, fire safety, and for fuel transmission alone justify much of the research described in this section. Additional benefits which may be derived from modification of combustion characteristics through the use of emulsified fuels are postulated on the existence of the micro-explosion phenomena and its potential effects on the fuel vaporization process. With the exception of the Ivanov and Nefedov research, occurrence of this phenomenon is presently supported by indirect observation. Whether or not the micro-explosion phenomenon indeed occurs to the extent necessary to beneficially modify combustion properties will only be firmly established with additional well controlled fundamental experiments.

VI. REFERENCES

[1]Becher, P., *Emulsions: Theory and Practice*, 2nd edition (Reinhold Publishing Co., 1965).

[2]Sherman, P., *Emulsion Science* (Academic Press, New York, 1968).

[3]Schwartz, A. M. and Perry, J. W., *Surface Active Agents; Their Chemistry and Technology* (Interscience Publishers, New York, 1949).

[4]Schonfeldt, N., *Surface Active Ethylene Oxide Adducts* (Pergamon Press, 1969).

[5]Hill, R. A. W. and Knight, J. T., "A Kinetic Theory of Droplet Coalescence with Application to Emulsion Stability," Trans. Faraday Soc. 61, 170 (1965).

[6]Scheele, G. F. and Leng, D. E., "An Experimental Study of Factors which Promote Coalescence of Two Colliding Drops Suspended in Water," Chem. Eng. Sci. 26, 1867 and 1881 (1971).

[7]Shilah, K., Sideman, S., and Resnick, W., "Coalescence and Breakup in Dilute Polydispersions," Can. J. Chem. Eng. 51, 542 (1973).

[8]Babukha, G. L., *et al*., "Experimental Study of Stability of Colliding Particles," Heat Transfer Sov. Res. 5, 11 (1973).

[9]Maraschino, M. J. and Treybol, R. E., "The Coalescence of Drops in Liquid-Liquid Fluidized Beds," AlChE J. 17, 1174 (1971).

[10]Hughmark, G. A., "Drop Breakup in Turbulent Pipe Flow," AlChE J. 17, 1000 (1971).

[11]Brenner, H., "Rheology of Two-Phase Systems," Ann. Rev. Fluid Mech. 2, 137 (1970).

[12]Cox, F. G. and Mason, S. G., "Suspended Particles in Fluid Flow Through Tubes," Ann. Rev. Fluid Mech. 3, 231 (1971).

[13]Sherman, P., "Rheology of Interfaces and Emulsion Stability," J. Colloid & Interface Sci. 45, 427 (1973).

[14]Franke, N. A. and Acrivos, A., "The Constitutive Equation for a Dilute Suspension," J. Fluid Mech. 44, 65 (1970).

[15]Yaron, I. and Gal-Or, B., "On Viscous Flow and Effective Viscosity of Concentrated Suspensions and Emulsions. Effect of Particle Concentration and Surfactant Impurities," Rhel. Acta 11, 241 (1972).

[16]Mannheimer, R. J., "Anomalous Rheological Characteristics of High-Internal-Phase-Ratio Emulsions," J. Colloid & Interface Sci. 40, 370 (1972).

[17]Mlynek, Y. and Resnick, W., "Drop Sizes in an Agitated Liquid-Liquid System," AlChE J. 18, 122 (1972).

[18]Izara, J. A., "Prediction of Drop Volumes in Liquid-Liquid Systems," AlChE J. 18, 634 (1972).

[19]Neduzhii, S. A., "Investigation of Emulsification Brought on by Sonic and Ultrasonic Oscillations," Sov. Phys. Acoustics 7, 221 (1962).

[20]Hislof, T. W., "New Ultrasonic Emulsifying Devices," Ultrasonics 8, 88 (1970).

[21]Greguss, P., "A New Sonic Emulsifying System: A Preliminary Report," Ultrasonics 10, 276 (1972).

[22]Ultrasonic Emulsification of Oil Tanker Cargo, Report prepared by Sonics International

Inc. (7101 Carpenter Freeway, Dallas, Texas 75247) for the Federal Water Pollution Control Administration, Department of the Interior (1970).

[23]Perrout, M. and Lowtay, R., "Drop Size in a Liquid-Liquid Dispersion: Formation in Jet Breakup," Chem. Eng. J. 3, 286 (1972).

[24]McFadyn, P. and Smith, A. L., "An Automatic Flow Ultramicroscope for Submicron Particle Counting and Size Analysis," J. Colloid & Interface Sci. 45, 573 (1973).

[25]Thompson, D. S., "Inelastic Light Scattering From Log Normal Distribution of Spherical Particles in Liquid Suspensions," J. Phys. Chem. 75, 783 (1971).

[26]Allegra, J. R. and Hawley, S. A., "Attenuation of Sound in Suspensions and Emulsions: Theory and Experiments," J. Acoust. Soc. Am. 51, 1545 (1972).

[27]Plotrovska, A., "Propagation of Ultrasonic Waves in Suspensions and Emulsions: 1. Investigation of Emulsions and Pigment Suspensions by an Acoustic Method," Ultrasonics 9, 14 (1971). "2. Relation Between Ultrasonic Properties and Certain Characteristics of the Medium," *ibid.* p. 235.

[28]Scalon, J. W. and Segel, L. A., "Some Effects of Suspended Particles on the Onset of Benard Convection," Phys. Fluids 16, 1573 (1973).

[29]Kuo, V. H. S. and Wilcos, W. R., "Particle Chromatography," Separation Sci. 8, 375 (1973).

[30]Allak, A. M. A. and Jeffreys, G. V., "Studies on Coalescence and Phase in Thick Dispersion Beds," AlChE J. 20, 564 (1974).

[31]Hills, F. J., personal communication, Mobil Research & Development Corporation, Paulsboro, N.J., 1974.

[32]Wood, B. J., Wise, H., and Inami, S. H., "Heterogeneous Combustion of Multicomponent Fuels," NASA TN D206 (1959).

[33]Zeldovich, Ya. B., Sadovniskov, P.Ya., and Frank-Kamenetski, D. A., "Oxidation of Nitrogen in Combustion," Academy of Sciences USSR, Moscow-Leningrad, 1947.

[34]Bracco, F. V., "A Model of Diesel Combustion and NO Formation," Central States Spring Meeting, The Combustion Institute, Ann Arbor, Michigan, March 23, 1971.

[35]Ivanov, V. M. and Nefedov, P. I., "Experimental Investigation of the Combustion Process of Natural and Emulsified Liquid Fuels," Trudy Instituta Goryachikh Iokopayemykh, Vol. 19 (1962) (in Russian) NASA TTF-258 (1965).

[36]Reed, T. B. and Lerner, R. M., "Methanol—A Versatile Fuel for Immediate Use," Science 182, p. 1229 (1973).

[37]Löf, G. O. G., "Oil and Water do Mix! and It Saves Fuel," Science News 104, 33B (1973).

[38]Cottell, E., "Major Fuel Oil Savings/Less Pollution Proved with the Cottell Ultrasonic Combustion System," Technical Bulletin No. 16, Tymponic Corporation, Plainview, New York.

[39]Scherer, G. and Tranie, L. A., "Pollutant Formation and Destruction in Flames, and in Combustion Systems, Paper No. B3, *Fourteenth International Symposium on Combustion*, Penn State University, August 20–25, 1972 (not published in Symposium Proceedings).

[40]Komissarov, L., Ivanov, V. M., Smetannikov, B. N., and Levaneskii, V. S., "Prospects for Use of Water-Soaked Fuels as Emulsions in Heat and Electric Power Plants," Nov Methody Szhiganiya Topl. Vop. Teor. Goreniya 103–11 (1972) (in Russian).

[41]Dooher, J. H., personal communication, Adelphi University, July 1974.

[42]Sawyer, R. F., "Atmospheric Pollution by Aircraft Engines and Fuels, A Survey," AGARD-AR-40, NATO, March 1972.

[43]Grobeman, J. and Ingebo, R. D., "Jet Engine Exhaust Emissions of High Altitude Aircraft Projected to 1990," NASA TMX-3007, March 1974.

[44]Nelson, A. W., "Detailed Exhaust Emissions Measurements of Three Different Turbofan Engine Designs," Paper No. 36, 41st Meeting of AGARD Propulsion and Energetics Panel on Atmospheric Pollution by Aircraft Engines, April 1973.

[45]Osgerby, I. T., "Literature Review of Turbine Combustor Modeling & Emulsions," AIAA J. 12, 743 (1974).

[46]Ledoseyev, V. A., Polishchuk, D. I., and Ispareniye, D. I., "Evaporation of Some Organic

Fluids," Prudy Odesskogo Gosudarstvennogo Instituta. Seriya fizicheskikh nauk, Nos. 5, 7 (1958) (in Russian).

[47]Obert, E., "Detonation and Internal Coolants," SAE Quarterly Transactions, p. 52 (1948).

[48]Cleveland, A. E. and Bishop, I. N., SAE Paper No. 60150A.

[49]Nicholls, J. E., El-Messiri, I. A., and Newall, H. K., "Inlet Manifold Water Injection for Control of Nitrogen Oxides —Theory and Experiment," SAE Transactions 78, 167 (1969).

[50]Eberius, H., Hoyermann, K., and Wagner, H. S., "The Third Body Effect of H_2O in the recombination of H Atoms," Ber. Bunsenges Phip Chem 73, p. 962 (1969) (in German).

[51]Norbye, J. P., "Ultrasonic Fuel Reactor —Gives Cleaner Exhaust, More Miles to the Gallon," Popular Science Monthly, November 1972, p. 22.

[52]Westgate, R., "Mix Gas with Water for More Miles Per Gallon," Popular Science, July 1974, p. 108.

[53]Slavolyubov, S. S., "Increasing Economy and Lowering Toxicity of Carbureted Engines by Preliminary Emulsification of the Fuel," Tr. Perm. Gas. Sel'skokhoz Inst. 76, p. 9 (1971) (in Russian).

[54]Lippeatte, L., personal communication, Tymponics Corp., Plainview, N.Y., March 13, 1974.

[55]Ewbank, W., personal communication, July 16, 1974.

[56]Abthoff, J. and Luther, A., ATZ 71, p. 124 (1969).

[57]Valdmanis, E. and Wulfhorst, D. E., "The Effects of Emulsified Fuels and Water Induction on Diesel Combustion," SAE Vehicle Emissions Part III, Progress in Technology 14, 570 (1970).

[58]Zapevalov, P. O. and Robustov, V. V., "Emulsification of Liquids by Injection for Burning Artificially Water-Cut Fuel," Nauk Tr., Omsk. Sel-Khoz 84, 132 (1971) (in Russian).

[59]Schmidt, A. F., *The Internal Combustion Engine* (Chapman and Hall, London, 1965), p. 78.

[60]Lamb, G. G., "Vaporization and Combustion of Multicomponent Fuels Droplets," NACA Report 1300, p. 31, 1959.

[61]Lyn, W. T., "The Spectrum of Diesel Combustion Research," Diesel Combustion Symposium, Proceedings, Institution of Mechanical Engineers, London, Vol. 184, Pt.3J, p. 1 (1970).

[62]Obert, E. F., *Internal Combustion Engines* (3rd ed.) (International Textbooks Co., Scranton, Pa., 1968).

[63]Spalding, D. B., "Theory of Particle Combustion at High Pressures," ARS J. 29, p. 828 (1959). Modified by D. E. Rosner, AIAA J. 5, p. 163 (1967).

[64]Hottel, H. C. and Hawthorne, W. R., "Diffusion in Laminar Fuel Jets," Third International Symposium on Combustion (Williams & Wilkins, Baltimore, 1940), p. 254.

[65]Starkman, E. S., (ed.), *Combustion Generated Air Pollution* (Plenum Press, New York, 1971).

[66]Glassman, I., "Current Understanding of NO_x Formation in Mobile & Stationary Combustion Sources," Proceedings of Symposium on Current Status of the NO_x Problem, Middle Atlantic Consortium on Air Pollution (1973).

[67]Pischinger, R. and Cartellieri, W., "Combustion System Parameters and Their Effect Upon Diesel Engines Exhaust Emissions, Paper No. 72056, 8AE Transactions 18, p. 2251 (1972).

[68]Williams, A., "Combustion of Droplets of Liquid Fuels: A Review," Combustion and Flame 21, 1 (1973).

[69]Hieftje, G. M. and Malmstadt, H. V., "A Unique System for Studying Flame Spectrometric Processes," Anal. Chem. 40, 1860 (1967).

[70]Lindblad, N. R. and Schneider, J. M., "Method of Producing and Measuring Charged Single Droplets," Rev. Sci. Instrum. 38, 325 (1967).

[71]Keston, A., personal communication, United Aircraft Research Laboratories, May 23, 1974.

[72]Putnam, A. A., *et al.*, "Injection and Combustion on Liquid Fuels," Battelle Memorial Institute Report, March 1957, WADC Technical Report 56-344, ASTIA Document No. AD118142.

[73]Harrje, D. T., ed., "Liquid Propellant Rocket Combustion Instability," NASA SP-194, 1972.

[74]Rimberg, P. and Keafer, D., "Accuracy of Measuring Aerosol Concentration with Particle Counters," J. Colloid Interface Sci. 33, 628 (1970).

[75]Steen, W. M. and Chattajee, A., "Technique for Measuring the Size and Number of Droplets," J. Phys. E 3, 1020 (1970).

[76]Clark, C. J. and Dombrowski, N., "On the Photographic Analysis of Sprays," J. Aerosol Sci. 4, 27 (1973).

[77]Fourney, M. E., Mathin, J. H., and Waggoner, A. P., "Aerosol Size and Velocity Distribution via Holography," Rev. Sci. Instr. 40, 205 (1969).

[78]Royer, H., Albe, R., and Sutterlin, P., "Visualization of Water Droplets by Holography," Om. Comm. 4, 75 (1971).

[79]Pankuch, D. S., "Information Content of Extinction and Scattered-Light Measurements for the Determination of the Size Distribution of Scattering Particles," Appl. Opt. 11, 2844 (1972).

[80]Neiburger, M., Levin, Z., and Rodriguez, L., "Method of Measuring Drop Size Distributions in Laboratory Clouds," J. Appl. Meteoreol. 11, 550 (1972).

[81]Mani, J. V. S., "Drop Size Determination in Liquid Sprays," Indian J. Technol. 9, 233 (1971).

[82]Petrov, G. D., Sokolov, R. N., and Vasil'ev, V. A., "Particle Size Distribution in Various Zones of a Spray Jet," J. Eng. Phys. 18, 77 (1970).

[83]Harris, F. S., "Aerosol Size Distribution from Light Scattering Measurement," Conference on Cloud Physics, Fort Gillins, Colo., 1971.

[84]Belz, P. A. and Daugherty, N. S., Jr., "In-Line Holography of Reacting Liquid Sprays, Proceedings of the Symposium on Engineering Application of Holography," Los Angeles, February 1972, p. 293.

[85]Paritt, K. W., *et al.*, "Holography or Fast-Moving Cloud Droplets," J. Phys. E 3, 971 (1970).

[86]Marsden, S. S. and Rose, S. C., "Cold Emulsion Line Proposed in Arctic," *Oil and Gas J.* (Oct. 11, 1971) 69, no. 41, pp. 100–106.

[87]Marsden, S. S. and Rose, S. C., "Pipelining Tars and Crude Oils Containing Dissolved Natural Gas at Sub-freezing Temperatures in Order to Avoid Environmental Damage," U.S. Patent No. 3, 670, 752 (June 20, 1972).

[88]Marsden, S. S., "A New Kind of Crude Oil Pipeline for the Arctic," presented at Joint Meeting, MMIJ-AIME, Tokyo, Japan, May 24–27, 1972; Petroleum Review 27, No. 322 (October 1973), 380.

[89]Marsden, S. S., "Arctic Pipeline Transportation of Petroleum as a Cold Dispersion in Brine," *Logistics and Transportation Review*, 8, No. 4, pp. 83–90 (1972).

[90]Marsden, S. S., "Chemical Aspects of a Cold Dispersion Pipeline," presented at SPE Symposium on Oilfield Chemistry, Denver, Colorado, May 1973, SPE 4359.

[91]Nixon, J., Beerbower, A., Philippoff, W., Lorenz, P. A., and Wallace, T. J., "Investigation and Analysis of Aircraft Fuel Emulsions," USAAVLABS Tech. Rep. 67-62, Nov. 1967.

[92]Urban, C. M., Bowden, N. N., and Gray, J. T., "Emulsified Fuels Characteristics and Requirements," USAAVLABS Tech Rep. 69-24, March 1969.

[93]Harris, J. C. and Steinmetz, E. A., "Investigation and Analysis of Aircraft Fuel Emissions," USAAVLABS Tech Rep. 67-70, Dec. 1970.

[94]Peacock, A. T., Hazelton, R. F., Gresko, L. S., and Christenson, L. D., "A Study of the Compatability of a Four Engine Commercial Jet Transport Aircraft Fuel System with Gelled and Emulsified Fuels," Report No. NA-70-11, DS-70-1, April 1970.

[95]Glassman, I. Sirignano, A., and Summerfield, M., "Physics of Flames," Princeton Report 952, Princeton University, Princeton, N.J. (1970).

EFFICIENT USE OF ENERGY

PART III

ENERGY CONSERVATION AND WINDOW SYSTEMS

Edited by

S. M. Berman

S. D. Silverstein

PART III

PREFACE AND ACKNOWLEDGMENTS

The goals of this study are to relate fundamental physics research and development to the practical problem of improving the energy transfer conditions of windows. Some of the group were in attendance for the entire period while others were present for only a short time, contributing some special expertise to the needs of the group. Both university and industrial scientists participated in the hope that timely attention to a proper understanding of window systems could enhance the national energy conservation program.

All of us would like to acknowledge the support of the National Science Foundation through the office of Dr. Paul Craig, of the Federal Energy Administration through the office of Dr. Kurt Riegel, of the Electric Power Research Institute through the office of Dr. Craig Smith, and especially that of Professor W. W. Havens of the American Physical Society which made this study possible.

One of the editors (S. D. Silverstein) would like to thank the General Electric Company for the generous allocation of his time which was necessary to complete this final report.

PART III
ENERGY CONSERVATION AND WINDOW SYSTEMS

INTRODUCTION

This report on window systems does the following:

Assesses the role of the architectural window as an important factor in reducing energy consumption for residential and commercial climate control.

Evaluates the cost effectiveness of many existing and modified window systems. Some of the suggestions, if implemented, would cost-effectively reduce energy consumption while maintaining high aesthetic standards.

Stresses, and quantitatively illustrates, that during the space heating season the solar energy transmission through windows is a significant positive source of fuel-free energy. The window acts like a low-temperature solar collector, and if it is properly designed the winter heat solar gain (weather averaged) will exceed the thermal losses for east, west, and south faced windows over most of the continental U. S.

Constantly draws the analogy between the architectural window and the low-temperature solar heat collector. All heat transfer analyses, cost-effectiveness analyses, discussions of new materials, and innovations have applications to various aspects of the solar heat collector problem.

Assesses the physics and technology of selective coating materials which act primarily as infrared reflectors to reduce thermal radiation transport.

Reviews properties of presently existing selective surface materials.

Explores methods for deposition of selective surfaces on plastics. Such laminates could serve as inexpensive retrofit materials for reducing thermal radiative transport through existing windows.

Reviews the fundamental aspects of the optical properties of such materials and performs a search which identifies classes of new promising semiconductor materials.

Background and perspective

Space heating in the U. S. accounts for ~21% of the total national energy consumption (NEC), while air conditioning accounts for ~1.5% of the NEC.

Consider a typical U. S. single-family residence characterized by

Single story, basement, attic, and two external doors.
Ceiling area ~1300 ft^2, wall area (subtracting off windows) ~100 ft^2, window area 150 ft^2, door area 20 ft^2, volume not counting basement and attic 10, 200 ft^2.

If all surfaces are uninsulated, the windows are single pane, and the air exchange rate for the home is one exchange per hour, the losses per 10^3 degree-days F and the percentages of the total loss for each component are the following:

	MBtu/10^3 degree-days	% Total
Ceiling	8.7	33
Walls	7.7	29
Windows	4.1	16
Doors	0.6	2
Air exchange heating	5.3	20
Total	26.4	

In the 40°–42° N latitude zone, ~14.6 MBtu of fuel-free energy can be obtained from weather-averaged solar influx through the 150 ft^2 of windows for the total heating season. At an oil furnace efficiency of 0.67, each MBtu requires 11.5 gallons of fuel oil. At $0.40 a gallon, the heating costs are $4.60/MBtu. The total heating cost of this uninsulated home in a 5500-degree-day area would be $600.

By introducing insulation for each of these exposure surfaces we can save approximately the following quantities of thermal energy per ft^2 per 10^3 degree-days F:

Ceiling: 4 in. insulation saves 5.4 KBtu/10^3 degree-days ft^2

Walls: 2 in. insulation saves 5.3 KBtu/10^3 degree-days ft^2

Windows: single pane to double pane saves 13.7 KBtu/10^3 degree-days ft^2

For calibration: New York City area has ~4.9, Dallas 2.4, Minneapolis 8.4, and Seattle 4.1 thousand degree-days.

The relative loss percentages of the exposed surfaces change with insulation. For the home considered above with insulated ceiling, walls, and single-pane windows the estimated thermal energy required drops to ~12.6 MBtu/10^3 degree-days and the breakup of the losses is:

Ceiling ~12%, walls ~17%, windows ~30%, doors ~3%, air exchange 38%.

The conclusions which can be drawn here are:

The first item to insulate in single-family homes is the ceiling—the savings here represent the largest return on capital investment.

Efforts should be made to reduce air exchange by weatherstripping doors and windows, and keeping the fireplace flue closed.

Window insulation can be significant in certain climates—cost-effectiveness analyses of the type shown in Section 2 A should be performed.

Of the 21% of the NEC used for space heating the division between residential and commercial is

~12.6% residential,
~ 8.4% commercial.

In commercial buildings, windows typically constitute a higher percentage of exposed surface area, and the roof a lower percentage. Average losses for windows over U. S. are not meaningful because of large variance in architectural design. Losses per unit window area as discussed in Sec. 1 should be applied.

A reasonable estimate of the percentage of the NEC corresponding to space heating loss through architectural windows is ~5%.

The thermal losses through windows are offset to a certain degree by solar gain—the amount depending upon latitude, climate, and window orientation. The solar heat gain makes some windows better in the net energy balance than the best insulated walls. Section 1 B reviews this in detail.

Of the 5% the NEC lost through windows,

Less than one-half of this loss or <2.5% of the NEC is cost-effectively recoverable by going to better insulated window systems and summer shading.

Approximately one-third to one-half of the thermal heat losses through windows are compensated by solar gains during the heating season.

Summary

The study group on windows was charged with the objective of making meaningful suggestions on ways to improve the energy performance of window systems. The suggestions contained within this report apply to the full spectrum of technology from research and development through manufacturing, marketing, and consumer utilization.

Rather than present a broad shopping list of nonsubstantive suggestions and possible research directions the study group decided to give recommendations backed by quantitative predictions of potential energy savings and cost effectiveness. The advantages of using the results generated from this type of study follow:

It focuses research and development (R&D) on the promising situation only, saving time, effort, and precious R&D monies.

Quantitative analysis of general parameters allows researchers to evaluate the potential impact of new ideas or innovations *ab initio*.

It allows architects, construction engineers, or consumers to make objective decisions of presently available window systems for optimum energy savings and cost effectiveness in a given locality.

To implement the strategy, the study group decided that substantial efforts of the group should be devoted to original technical research for the purposes of determining quantitatively the relative roles of the various intrinsic (structural and materials) and extrinsic (environmental) window parameters on the energy transfer through windows.

The significant energy transfer mechanisms through an architectural window are:

Thermal heat transfer losses—coupled conduction-convection, thermal radiation.

Solar influx—using solar energy transport through architectural windows as a fuel-free energy source in heating season.

Infiltration—leakage due to air gaps.

In Section 1 A the thermal heat transfer is discussed and the results of detailed window calculations for variable window geometries, construction materials, and weather factors are given. The variables considered individually and in combination are:

Variable-emissivity surfaces to reduce thermal radiation transport—these emissivities cover the full range available to doped semiconductor or metallic films.

Effects of multiple glazing (storm windows, double glazing, etc.).

The effects of gap thicknesses in multiple glazing.

Gaps of multiple-glazed units filled with heavier-molecular-weight gases to reduce convection—here it is shown that this option is effective only when coupled with low-emissivity surfaces.

Effects of wind velocity, which drives the forced convective losses at the exterior window surface—suggestions are made regarding inexpensive retrofit wind-reducing screens which can be put on the perimeters of windows.

In Sec. 1 B, analyses of the potential solar heat gain through architectural windows are presented which include considerations of:

Various seasons—winter, where solar heat gain represents a fuel-free energy source, and summer, where solar heat gain is energy consuming when air conditioning is used.

Various geographical localities representative of latitude and weather—New York City, Seattle, Dallas, Sault Ste. Marie.

Various window directional orientations.

The most significant results of Sec. 1 B are:

Solar influx through architectural windows represents a very beneficial fuel-free energy source for auxiliary space heating.

Capital costs to make windows better in conserving energy than insulated walls are not high.

These results have an important bearing on prospective energy-conservation legislative codes which restrict glass areas of buildings, e.g.:

California Administrative Code, Title XXV, Chapter I, Article 5, "Energy Insulation Standards"—single-glazed glass area is restricted to 16% of the gross heat floor area with no distinction of window orientation.

An example of the cost-effectiveness analysis for a specific grouping of window systems is illustrated in Figure 1. This analysis is given for the New York City area climate. Some points to help understand this chart are:

Fuel costs are taken at \$0.40/gallon fuel oil, with a furnace efficiency of 67%. This implies a heating cost of \$4.60/MBtu.

For each window system shown there is a bar with symbols HT, N, E-W, S. HT represents the direct heat transfer lost. N, E-W, S represent the net energy transfer for each of the cardinal directions. The difference between the various directions and HT represents the weather-averaged maximum potential solar heat gain. Intermediate window orientations can be obtained readily by interpolation within the bars.

To determine the cost effectiveness of one system relative to another, one first compares the relative fuel costs per ft² and then uses the interest-depreciation scheduling formula given in Appendix B. In evaluations for systems which should have a long life, use is made of the example of the interest-depreciation formula quoted in Appendix B: 20-year life at 10% interest and 6% fuel inflation rate implies that the initial fuel savings per year must be greater than 6.4% of the initial incremental capital expense for the improvement to be cost effective.

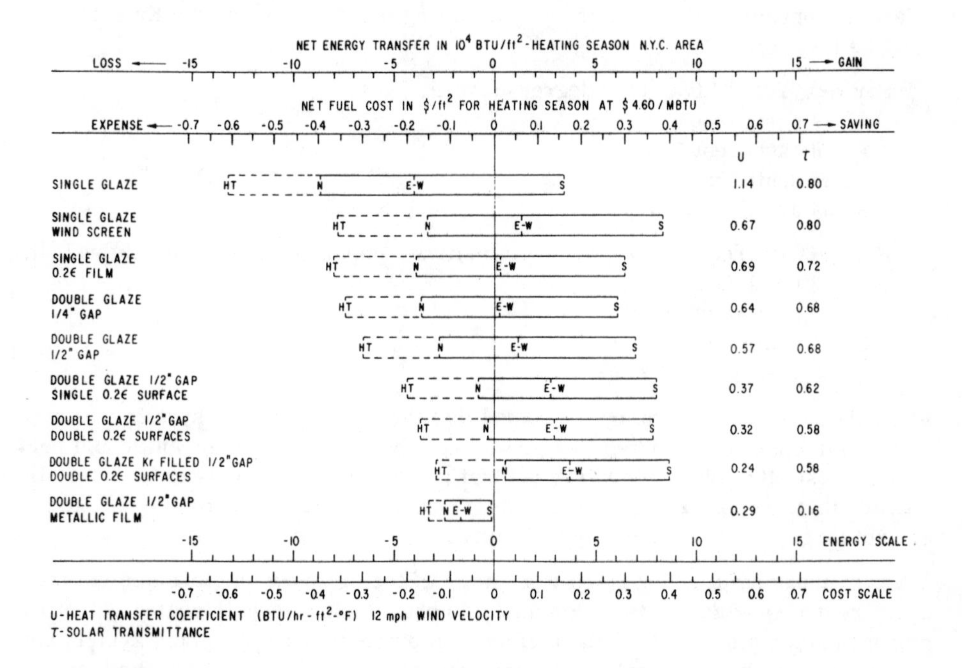

Figure 1. Cost-effectiveness analysis for a specific group of window systems.

Example: Comparison of double-glaze unit with and without single low-emissivity surface. To be cost effective, the incremental selling price per ft^2 for the system with the film must be less than (each dollar here corresponds to a fuel savings of \$0.064 per ft^2):

For New York City @ 4900 degree-days F,
\$1.40 north face
\$1.25 east-west face
\$0.78 south face
\$1.14 average all faces

For 8500 degree-days F, with solar influx comparable to latitude 45°–47° N,
\$2.58 north face
\$2.19 east-west face
\$1.87 south face
\$2.21 average all faces

Due to the additional solar exclusion, the double-low-emissivity-film window will not save enough additional energy to be justified. It is doubtful that manufacturing techniques could be cost-reduced to justify a retail selling price of \$1.14/ft^2; however, \$2.21 for the colder climate is within reason. From this evaluation, the conclusion is made that there is a potential, though probably not substantial, market for low-emissivity-film-coated windows in the U. S.

As a second comparison consider the Kr-filled double-low-emissivity-film window compared to the uncoated air-filled double-glazed unit. To be a cost-effective improvement, the incremental selling price per ft^2 for the Kr system must be less than:

For New York City @ 4900 degree-days F,
\$2.34 north face
\$1.88 east-west face
\$1.25 south face
\$1.84 average all faces

For 8500 degree-days F, with solar influx comparable to latitude 45°–47°N,
\$4.22 north face
\$3.59 east-west face
\$2.97 south face
\$3.59 average all faces

From these figures, it can be concluded that the double-film Kr system will not be cost effective for climates like or warmer than New York City, but could perhaps be cost effective for the very coldest parts of the U. S. It is to be noted, of course, that due to low population in the coldest climates the potential markets are not large; hence development incentives are reduced.

A very meaningful comparison to make is of the ordinary single-glaze vs. the ordinary double-glaze units. This analysis should apply to well-sealed retrofitted storm windows also. Taking the half-inch-gap uncoated double-glaze system and comparing it to the standard single-pane system, it can be seen that the incremental selling price per ft^2 must be less than:

For New York City,
$4.38 north face
$3.75 east-west face
$3.13 south face
$3.75 average all faces

As current storm windows retail for $1.50–$2.50 per ft^2 installed, they represent good investments in most climates. A $2.00 per ft^2 storm window which is well sealed should be a cost-effective improvement for a climate with greater than 2600 degree-days F (at a fuel rate of $4.60/MBtu). For storms with high infiltration rates which cause substantial cooling of the interior pane, a U value of $\simeq$0.65 would be more appropriate. For New York City, the incremental fuel savings would be reduced $0.043/$ft^2$ in going from a U value of 0.57 to 0.65. The average selling price for all faces would then have to be less than $2.97 to be cost effective in the New York City climate. A $2.00/$ft^2$ with a U value of 0.65 will be cost effective in areas colder than 3000 degree-days F.

From these examples it can be seen that there is a potential, though probably not substantial, market for low-emissivity films on windows with or without Kr filling providing that:

Manufacturers can produce high-quality uniform films at *very low* costs, a measure of which is given above.

Marketing is restricted primarily to northern regions.

In Section 2, selective surface materials for application to architectural and/or solar heat collector windows are discussed in detail.

Selective infrared reflecting coating materials have broad application to many energy-storage systems where thermal radiative losses are significant:

In architectural windows, very low cost and high film-thickness uniformity are required to achieve significant consumer utilization.

In solar heat collector windows high transmission of the solar irradiance is mandatory—while iridescence is of less concern. Low cost is also important.

Thermal energy storage tanks—here the low emissivity properties and low costs are of primary concern.

In Section 2A, a review of the presently available selective surface materials is given. Included here are:

Discussions of the properties of, and deposition techniques for, the doped wide-band-gap semiconductors such as $SnO_2:Sb$, $SnO_2:F$, $In_2O_3:Sn$. Materials such as these are in common use today to increase the efficiency of the low-pressure sodium lamps by infrared reflection. They are also widely used in the electronics industry for transparent contacts in photoactivated devices. The infrared reflectivity and high electrical conductivity both arise from the same mechanism of high mobility-free carriers introduced by the impurity doping.

Discussion is given of various metallic films which are primarily marketed with the objective of reducing summer solar heat gain where air conditioning is required.

Discussion is given of the plastic-metallic film laminates which are presently marketed. Here it is stressed that the hoped for low emissivity properties of such laminates are most often negated by the infrared absorptions of the plastic covering material.

Section 2B deals with the problems of bonding both semiconductor and metallic films to transparent plastic materials. A research suggestion to accomplish better adhesion which is elaborated upon is the use of difunctional organic molecules which have a reactive group at one end that can bond with polymer surfaces, and at the other end a metal atom to which the selective surface is bonded.

Plastic laminates have considerable potential:

For architectural windows—they could be easily cut and fitted (retrofitted) to existing windows. Specific polymers are discussed which have low infrared absorption at the thermal radiative wavelengths of interest.

For solar collector windows where weight loading due to glass is often of concern—here polymers must be chosen which would not be subjected to ultraviolet degradation.

Section 2C deals with the basic physics of selective infrared reflecting wide-band-gap semiconductors.

The mechanisms for the infrared reflectivity are discussed to identify the significant electronic parameters which lead to the desirable selective characteristics.

An extensive search of most known semiconductor materials has been made with the objective of finding new candidates exhibiting the desirable selective surfaces.

Thirteen possible candidate materials have been identified for which further research is recommended. These candidate materials are:

CaB_6	SiN	Sb_2P_3
CuCl	S	PbO
CuI	ZnO	Bi_2O_3
MnO	ZnS	
NiO	TiO_2	

1. WINDOW SYSTEMS AND CLIMATE CONTROL PERFORMANCE

A. Heat transfer through existing and modified architectural windows*

1. Introduction

As a consequence of the recent rapid rise of the costs of energy, there has been a much greater public awareness of the necessity of efficient energy utilization. Heating of buildings in the winter in northern climates is a necessity of life, while cooling them in the summer, although not essential, is certainly pleasant—and desirable if costs are not prohibitive. As scientists, we should re-examine the long-accepted technologies in architectural design and building materials in an attempt to find new ideas which would be economically cost effective and aesthetically acceptable, and would also lead to more efficient use of energy in residential and commercial structures.

It is very important in the research and development of new products to focus on programs which have a reasonable chance of economic success and consumer acceptance. There are a number of architectural window innovations which are suggested in the window report. Undoubtedly, many readers will have additional thoughts of their own. A detailed understanding of the mechanisms of heat transfer through architectural windows is quite important, as such knowledge allows us to quantitatively estimate the change in heat transfer which would occur upon changing window design parameters. Then, the improvement in heat transfer can be related directly to energy cost saving figures in dollar/ft^2. The cost savings per unit area from the reduction of energy consumption imposes a practical economic upper limit on the allowable capital costs for implementing an innovation. Accordingly, such cost analyses can serve as good rejection filters for contemplated window innovations which are intrinsically too expensive to be cost effective. In addition, the results given here can be helpful to consumers who are seeking to make objective, economic, energy-conserving decisions among available existing window systems.

In this section, we will give the results of some rather detailed window modeling calculations which we have made. We will make no attempt here to document the calculational details, but rather refer the interested readers to a separate publication[1] where these details are contained in appendices.

In the winter heating cycle, there are two components of thermal energy transport through an architectural window. The first component, the one which we have performed detailed calculations of, is the coupled convective-conductive-thermal radiative heat transfer from the interior of the building (hot reservoir)

---

*This section by S. D. Silverstein, General Electric Co.

to the exterior (cold reservoir). This component is direct heat loss from the system, and this loss of energy must be replenished by an interior source to maintain constant interior temperatures. There are, of course, substantial direct heat losses through the roof and walls of buildings, which except for ultramodern office buildings usually exceed the loss through the windows (see the discussion in the introduction "Background and perspective" for relative losses in typical buildings). Our results will always be expressed per unit window area so as to apply to any building architecture. The second energy transport through windows is the input of solar energy during daylight hours. This solar input is equivalent to an additional non-fuel-consuming internal energy source. At first thought, one may guess that the solar energy input through windows during the winter would be small—this is not the case, as has been shown in Section 1 B of this window report.

In the "Comparisons and conclusions" section of this part of the report, we use both the thermal losses and solar gains to make cost-effective analyses of different window types for the New York City area climate.

2. *Window heat transfer results*

Before we discuss the results of the calculations, there are a few qualifying points which we would like to emphasize.

One does not have a typical single-pane or double-glazed system. The storm window is an example of a double-glazed system with a wide air gap. The small variations in typical window systems will evidently lead to commensurate deviations in the heat transfer results calculated from specific idealized window models. Moreover, in the calculations we have made extensive use of semiempirical convective formulae which again have intrinsic inaccuracies. Accordingly, it would be unrealistic for us to conclude that our calculated models could give absolute heat transfer coefficients to accuracies better than ~15 percent. Actually, we are most interested in the relative values of the heat transfer for various systems rather than the absolute values. Here, the calculated relative values should prove more accurate and with them we can determine the relative improvements of one system over another. We found that the calculations were considerably simplified if mathematical expansions were used which limited the accuracy to ~10 percent. This limitation then allowed many simplifying analytical approximations.

It is important to recognize that infiltration (leakage due to gaps), where it exists in substantial amounts, will naturally play an important part in the heat transfer.[2] However, infiltration represents a much too subjective parameter to insert into our calculations. Accordingly, we assume that the windows considered are well weatherstripped and neglect such effects.

The calculations do not specifically take into account the heat loss by direct conduction through the window frames. For a wood frame window this is negligible, and the relevant area is just the glass area. For a painted metal frame window, with thermal isolation in the case of double glazing, the contribution per unit area of the metal frame will be essentially the same as the contributions per unit area of the glass. The physical reason for this is that the conductive heat transfer coefficient of the glass is so much larger than the relevant radiative and convective coefficients that temperature gradients in the glass can be neglected.[3] This, of course, is also true for the metal. For an unpainted metal frame the surface emissivity will be low. Therefore, for metal frame windows a reasonable estimate of the frame loss can be obtained by considering the frame area as additional glass area which is either uncoated for a painted frame or coated with a low-emissivity surface for an unpainted metal frame.

The convective heat transfer coefficient of the interior air-glass surface can vary considerably due to air currents in the room if the window surface is not shielded by an effective wind screen in the form of venetian blinds, curtains, or a window shade. The situation is obviously magnified if a convective heating radiator is located directly in front of a window. This latter situation is a common architectural design for many reasons, such as history, structural convenience, the tendency not to block windows and hence heating ducts by furnishings, and finally a comfort factor for mixing the cold air at the window. It is difficult for us to quantify the additional energy losses due to such interior forced convection. We believe the losses incurred here can be significant and *further studies on alternatives to the location of heating ducts are suggested.* In our calculations we assumed either that the interior air currents are small enough or that the windows are sufficiently shielded that free convection at the vertical air-glass surface is the appropriate condition to describe the convective heat transfer coefficient at the inner glass surface.

a. Single-pane system. The single-pane system with variable surface emissivities is the simplest system to consider. A schematic of this is illustrated in Figure 1.1. One solves the heat transfer problem by using the continuity of the net thermal current in three regions: the room to the inner glass surface, which has comparable contributions of convective and radiative components; conduction through the glass; and finally from the outer glass surface to the exterior. The latter term is dominated by the forced convection associated with the wind on the outer glass surface.

The net heat transfer coefficient U is the parameter conventionally used to describe the insulating properties of various building materials. The reason for

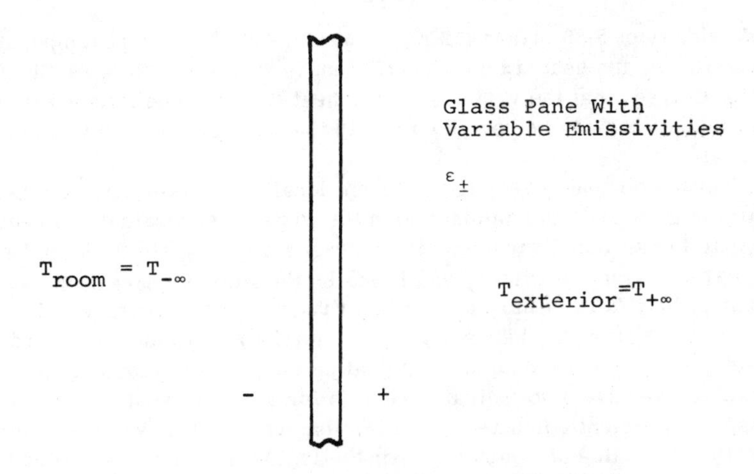

Figure 1.1. Schematic of single-pane window system.

this is that U is usually a slowly varying function of temperature over the range of normal ambient temperatures. Hence, it is not sensitive to climatic variations and accordingly represents a good parameter for comparison of different insulating systems. If Q_{loss} represents the thermal power lost through the window per unit area due to heat transfer (W/m², Btu/hr ft²), the heat transfer coefficient U is defined by

$$U = \frac{Q_{loss}}{(T_{room} - T_{ext})} \quad (\mathrm{W/m^2\,{}^\circ K;\ Btu/hr\,ft^2\,{}^\circ F}).$$

In Table 1.1, we show the results for the single-pane system as a function of the emissivities[4] of the interior (EM) and exterior surfaces (EP) of the glass pane for variable wind velocities. The external temperature here is taken at −2.22 °C (28 °F). Other temperature profiles are given in the more detailed report.

Due to the roughened exterior siding of a building, turbulent flow conditions must be used for computing the convective heat transfer at the higher wind velocities. For turbulent air flow, the forced convective heat transfer coefficient $h_{g+\infty}$ can be written as

$$h_{g+\infty} = 0.032 \frac{k}{L} R_L^{0.8}.$$

Here k is the thermal conductivity, L is the window width, and R_L is the Reynolds number,

$$R_L = V_\infty \rho L/\mu .$$

The parameters appearing in the Reynolds number are the wind velocity V_∞, the air density ρ, and the viscosity μ. By using the standard temperature dependences of the gas parameters, we can relate the turbulent-forced convective heat transfer coefficient at different wind velocities and absolute temperatures by the relation

$$h_{g,+\infty}(V_2, T_2) = h_{g,+\infty}(V_1, T_1) \left(\frac{V_2}{V_1}\right)^{0.8} \left(\frac{T_1}{T_2}\right)^{0.7}.$$

At a wind velocity of 5.36 m/sec (12 mph), at 0 °C (32 °F) ambient temperature for a 1-m window, the heat transfer coefficient is 27.3 W/m² °K (4.81 Btu/hr ft² °F). We note, of course, that the variation of the heat transfer coefficient with temperature is small and can be neglected within the intrinsic accuracy of our heat transfer analyses.

Wind velocities obviously vary for different localities. Typically, the percentage of calm is greater in the summer than the winter. Accordingly, a reasonable estimate for seasonal wind velocity averages would be about 15–20% higher than the yearly average in winter, and lower by the same amount in the summer. Cities such as New York, Chicago, and San Francisco have average yearly wind velocities close to 10 mph[5]; hence winter and summer estimates of 12 and 8 mph, respectively, would be reasonable in evaluating the heat transfer coefficients.

In the tables, we have also indicated results for a 2-mph wind velocity with a heat transfer coefficient computed in the laminar region (3.7 W/m² °K, 0.65 Btu/hr ft² °F). This value of 2 mph is a hypothetical value for a wind velocity which can conceivably be realized, on the average, by the use of either retro-

Table 1.1. Single-pane-window parameters for winter heating. EM and EP are the emissivities of the interior and exterior pane surfaces; T_G is the glass temperature, given in both Centigrade and Fahrenheit units; U is the heat transfer coefficient (SIU—W/m^2; EU—Btu/hr ft^2). The hemispherical emissivity of plain glass is 0.88, and that of infrared-reflecting films of SnO_2:Sb is ≃0.2. The window height is 1 meter. Room temperature =21.11 °C, 70.00 °F. Exterior temperature = −2.22 °C, 28.00 °F.

		T_G		U		Q_{LOSS}	
EM	EP	C	F	SIU	EU	SIU	EU
			WV=2.0 mph—laminar flow				
0.10	0.20	7.5	45.4	1.91	0.34	44.5	14.1
0.20	0.20	8.2	46.8	2.06	0.36	48.0	15.2
0.50	0.20	10.0	50.1	2.42	0.43	56.4	17.9
0.88	0.20	11.7	53.1	2.75	0.49	64.2	20.4
0.10	0.88	5.2	41.3	2.43	0.43	56.6	18.0
0.20	0.88	5.8	42.5	2.65	0.47	61.8	19.6
0.50	0.88	7.6	45.6	3.22	0.57	75.1	23.8
0.88	0.88	9.3	48.7	3.78	0.67	88.1	28.0
			WV=6.0 mph—turbulent flow				
0.10	0.20	2.3	36.1	3.07	0.54	71.7	22.7
0.20	0.20	2.8	37.0	3.41	0.60	79.6	25.3
0.50	0.20	4.1	39.5	4.33	0.76	101.1	32.1
0.88	0.20	5.6	42.1	5.32	0.94	124.1	39.4
0.10	0.88	1.7	35.1	3.20	0.56	74.7	23.7
0.20	0.88	2.2	35.9	3.57	0.63	83.2	26.4
0.50	0.88	3.4	38.1	4.57	0.81	106.7	33.8
0.88	0.88	4.8	40.6	5.67	1.00	132.3	42.0
			WV=9.0 mph—turbulent flow				
0.10	0.20	1.3	34.4	3.29	0.58	76.8	24.4
0.20	0.20	1.7	35.1	3.67	0.65	85.7	27.2
0.50	0.20	2.9	37.2	4.74	0.84	110.6	35.1
0.88	0.20	4.2	39.5	5.92	1.04	138.1	43.8
0.10	0.88	1.0	33.7	3.37	0.59	78.7	25.0
0.20	0.88	1.3	34.4	3.77	0.67	88.1	27.9
0.50	0.88	2.4	36.3	4.90	0.86	114.2	36.2
0.88	0.88	3.6	38.5	6.16	1.09	143.6	45.6
			WV=12.0 mph—turbulent flow				
0.10	0.20	0.7	33.3	3.42	0.60	79.9	25.3
0.20	0.20	1.1	34.0	3.84	0.68	89.5	28.4
0.50	0.20	2.1	35.8	5.00	0.88	116.6	37.0
0.88	0.20	3.2	37.8	6.31	1.11	147.2	46.7
0.10	0.88	0.5	32.9	3.48	0.61	81.2	25.8
0.20	0.88	0.8	33.5	3.91	0.69	91.2	28.9
0.50	0.88	1.7	35.1	5.11	0.90	119.2	37.8
0.88	0.88	2.8	37.1	6.48	1.14	151.3	48.0
			WV=15.0 mph—turbulent flow				
0.10	0.20	0.3	32.6	3.51	0.62	82.0	26.0
0.20	0.20	0.7	33.2	3.95	0.70	92.1	29.2
0.50	0.20	1.5	34.8	5.18	0.91	120.8	38.3
0.88	0.20	2.6	36.6	6.59	1.16	153.7	48.8
0.10	0.88	0.1	32.3	3.56	0.63	83.0	26.3
0.20	0.88	0.4	32.8	4.00	0.71	93.4	29.6
0.50	0.88	1.3	34.3	5.26	0.93	122.8	39.0
0.88	0.88	2.2	36.0	6.72	1.19	156.9	49.8

fitted porous wind-reducing screens fitted to the perimeter of existing windows or a structure similar to an external venetian blind.

We have not included the small corrections for the solar irradiance which will be absorbed by the glass pane, hence raising its temperature and slightly affecting the U coefficients for situations involving daytime solar exposure. The solar irradiance which passes through the window does not enter the U value computed here, but is an additional amount of energy which is supplied to the interior. In winter, this additional solar energy is beneficial because it serves as a cost-free energy source; however, in summer it is detrimental to air conditioning. This apparent dichotomy can be solved readily in residential buildings by selective summer use of white solar reflecting window shades or curtain linings. An interesting device which can be made simply would be a reversible window shade[6] or venetian blind. One side would be white and the other black. In winter the black side would face the exterior to optimize the absorption of solar energy. In summer the white side should face the exterior to reflect the solar input.

There are a number of interesting effects which we note in these results.

1. Introducing a low-emissivity surface to reduce thermal radiation transfer requires the selective surface to be applied on the inside of the glass rather than the outside. This is due to the fact that forced convection dominates the exterior heat transfer, forcing the temperature of the glass to be much closer to the exterior temperature than the interior temperature. As the radiation transfer is proportional to $(T_1^4 - T_2^4)$, the radiation transfer on the outside[7-10] will be considerably smaller than that on the inside; hence further reduction of this small quantity through a low-emissivity surface will have a negligible effect. Hence, a low-emissivity surface on the exterior of a single-pane system will probably never be justifiable on a cost basis.

2. Although not illustrated in Table 1.1 due to the high exterior temperature, at lower exterior temperatures the temperature of the glass drops substantially below the frost point for moderate winds. This situation is further aggravated by the introduction of low-emissivity interior surfaces, which further lowers the glass temperature. Hence, for all normal wintertime relative humidities, ice will tend to form on the interior surface at a higher exterior temperature, thus raising the emissivity of the surface to that of ice (0.97). If the temperature is cold enough that the ice formation is stable, the reduction in heat transfer due to the selective surface will be negated.

3. It is interesting to note that the values of the heat transfer coefficients for plain glass at 2-mph velocities are lower than that which can be achieved by the introduction of selective surfaces with emissivities of 0.2. Moreover, we would not have the problems in ice formation here. Our retrofitted wind screen results are based on a speculation of 2-mph laminar flow. The speculation is not optimistically unreasonable, and the results serve to indicate *that some engineering development along these lines should be done.*

b. Storm window and double-glazed systems. The basic geometry for these styles is illustrated in Figure 1.2. Here, the double glass panes are separated by an internal enclosed air gap of thickness δ.

For this system, one must include the radiative and convective heat transfer across the enclosed air gap. An important parameter in the radiative transport across the enclosed gap is a function we have defined as C_{12}, which is a symme-

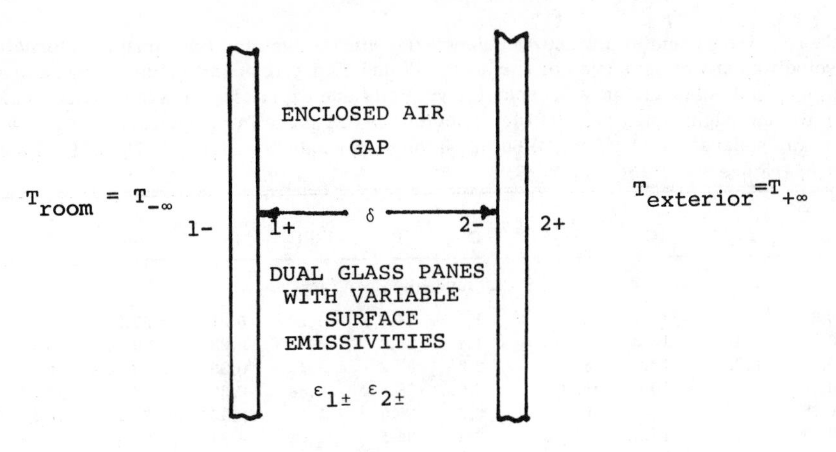

Figure 1.2. Schematic for double-glazed window.

tric function of the emissivities of the two internal pane surfaces:

$$C_{12} \equiv (1/\epsilon_{1+} + 1/\epsilon_{2-} - 1)^{-1}.$$

In Table 1.2 we give a matrix of the values of C_{12} as a function of the emissivities ϵ_{1+} (E1P) and ϵ_{2-} (E2M). If both surfaces are plain glass, $\epsilon_{1+} = \epsilon_{2-} = 0.88$, $C_{12} = 0.79$. In Tables 1.3 and 1.4, we give results for the various thermodynamic properties of air-filled double-glazed window systems as a function of variable gap widths, emissivities of surfaces, and wind velocities. The simplest results to focus on for comparisons are the U values, which as we discussed before are relatively insensitive to temperature. With regard to the window gap separation, we note that a 0.25-in. (0.63-cm) gap will have substantially higher net heat transfer value than a 0.5-in. (1.27-cm) gap. This degree of variation of the U values for these two separations depends upon the emissivities of the surfaces and varies between 12 and 52 percent. The improvement made upon further widening the gap more than 0.5 in. (1.27 cm) is small. The incremental cost for a half-inch as opposed to a quarter-inch gap in a double-glazed system should be negligible in a new installation; hence the wider-air-gap system is an obvious choice to a buyer.

Table 1.2. Matrix of C_{12} as a function of emissivities ϵ_{1+} ϵ_{2-}.

E1P, E2M	0.1	0.2	0.3	0.4	0.5	0.6	0.7	0.8	0.9	1.0
0.1	0.05	0.07	0.08	0.09	0.09	0.09	0.10	0.10	0.10	0.10
0.2	0.07	0.11	0.14	0.15	0.17	0.18	0.18	0.19	0.20	0.20
0.3	0.08	0.14	0.18	0.21	0.23	0.25	0.27	0.28	0.29	0.30
0.4	0.09	0.15	0.21	0.25	0.29	0.32	0.34	0.36	0.38	0.40
0.5	0.09	0.17	0.23	0.29	0.33	0.37	0.41	0.44	0.47	0.50
0.6	0.09	0.18	0.25	0.32	0.37	0.43	0.48	0.52	0.56	0.60
0.7	0.10	0.18	0.27	0.34	0.41	0.48	0.54	0.60	0.65	0.70
0.8	0.10	0.19	0.28	0.36	0.44	0.52	0.60	0.67	0.73	0.80
0.9	0.10	0.20	0.29	0.38	0.47	0.56	0.65	0.73	0.82	0.90
1.0	0.10	0.20	0.30	0.40	0.50	0.60	0.70	0.80	0.90	1.00

Table 1.3. Heat transfer through double-glazed window systems for variable external wind velocities and emissivities of surfaces. T_1 and T_2 are the equilibrium temperatures of the inner and outer surfaces, respectively. Units are expressed in both international units (SIU) and engineering units (EU). Calculations are made for 1-meter-high windows. Window gap separation = 1.27 cm, 0.50 in. Room temperature = 21.11 °C, 70.00 °F. Exterior temperature = −2.22 °C, 28.00 °F.

		T_1		T_2		U		Q_{LOSS}	
E1M	C_{12}	C	F	C	F	SIU	EU	SIU	EU
WV = 2.0 mph—laminar flow									
0.20	0.05	13.7	56.6	1.5	34.7	1.20	0.21	27.92	8.86
0.20	0.10	13.3	55.9	1.8	35.2	1.28	0.23	29.80	9.45
0.20	0.20	12.5	54.5	2.2	35.9	1.41	0.25	32.98	10.46
0.20	0.79	10.0	50.0	3.6	38.5	1.88	0.33	43.85	13.91
0.88	0.05	16.7	62.0	2.2	36.0	1.42	0.25	33.17	10.52
0.88	0.10	16.3	61.3	2.5	36.5	1.53	0.27	35.60	11.30
0.88	0.20	15.7	60.3	3.1	37.6	1.72	0.30	40.10	12.72
0.88	0.79	13.7	56.7	5.4	41.6	2.43	0.43	56.78	18.01
WV = 6.0 mph—turbulent flow									
0.20	0.05	12.9	55.3	−0.6	31.0	1.33	0.23	30.96	9.82
0.20	0.10	12.4	54.5	−0.4	31.2	1.43	0.25	33.31	10.57
0.20	0.20	11.5	52.8	−0.2	31.6	1.61	0.28	37.53	11.91
0.20	0.79	8.3	46.9	0.6	33.0	2.25	0.40	52.54	16.67
0.88	0.05	16.2	61.1	−0.2	31.6	1.61	0.28	37.53	11.91
0.88	0.10	15.7	60.3	−0.0	31.9	1.75	0.31	40.81	12.95
0.88	0.20	15.0	59.0	0.3	32.5	2.01	0.35	46.91	14.88
0.88	0.79	12.1	53.7	1.6	34.8	3.04	0.54	70.84	22.47
WV = 9.0 mph—turbulent flow									
0.20	0.05	13.0	55.4	−0.9	30.3	1.37	0.24	31.85	10.10
0.20	0.10	12.4	54.4	−0.8	30.5	1.47	0.26	34.30	10.88
0.20	0.20	11.5	52.6	−0.6	30.8	1.65	0.29	38.59	12.24
0.20	0.79	7.9	46.2	−0.0	32.0	2.31	0.41	53.90	17.10
0.88	0.05	16.2	61.2	−0.6	30.8	1.65	0.29	38.59	12.24
0.88	0.10	15.6	60.1	−0.5	31.1	1.79	0.31	41.65	13.21
0.88	0.20	15.0	58.9	−0.2	31.6	2.07	0.37	48.39	15.35
0.88	0.79	11.7	53.0	0.8	33.4	3.15	0.56	73.50	23.32
WV = 12.0 mph—turbulent flow									
0.20	0.05	12.9	55.2	−1.1	29.9	1.38	0.24	32.10	10.18
0.20	0.10	12.2	53.9	−1.1	30.1	1.47	0.26	34.34	10.89
0.20	0.20	11.3	52.3	−0.9	30.3	1.66	0.29	38.82	12.31
0.20	0.79	7.8	46.0	−0.4	31.3	2.37	0.42	55.24	17.52
0.88	0.05	16.0	60.9	−0.9	30.3	1.66	0.29	38.82	12.31
0.88	0.10	15.7	60.2	−0.8	30.6	1.82	0.32	42.55	13.50
0.88	0.20	14.9	58.9	−0.6	31.0	2.11	0.37	49.27	15.63
0.88	0.79	11.5	52.7	0.3	32.5	3.23	0.57	75.39	23.92
WV = 15.0 mph—turbulent flow									
0.20	0.05	12.8	55.1	−1.3	29.7	1.39	0.24	32.33	10.26
0.20	0.10	12.3	54.1	−1.2	29.8	1.50	0.26	34.95	11.09
0.20	0.20	11.2	52.2	−1.1	30.0	1.69	0.30	39.32	12.47
0.20	0.79	7.6	45.7	−0.6	30.9	2.40	0.42	55.92	17.74
0.88	0.05	16.1	61.0	−1.1	30.0	1.69	0.30	39.32	12.47
0.88	0.10	15.6	60.0	−1.0	30.2	1.84	0.32	42.82	13.58
0.88	0.20	14.9	58.8	−0.8	30.6	2.13	0.38	49.81	15.80
0.88	0.79	11.4	52.5	−0.0	32.0	3.30	0.58	76.89	24.40

Table 1.4. Heat transfer through double-glazed window systems for variable air-gap spacing and emissivities of surfaces. T_1 and T_2 are the equilibrium temperatures of the inner and outer surfaces, respectively. Calculations are for 1-meter-high windows. Room temperature = 21.11 °C, 70.00 °F. Exterior temperature = −2.22 °C, 28.00 °F. WV = 12.0 mph turbulent flow.

		T_1		T_2		U		Q_{LOSS}	
E1M	C_{12}	C	F	C	F	SIU	EU	SIU	EU
Window gap separation = 0.32 cm, 0.12 in.									
0.88	0.05	10.8	51.4	0.5	32.9	3.52	0.62	82.11	26.05
0.88	0.10	10.6	51.1	0.6	33.0	3.58	0.63	83.60	26.52
0.88	0.20	10.3	50.5	0.7	33.2	3.68	0.65	85.84	27.23
0.88	0.50	9.5	49.1	0.9	33.6	3.97	0.70	92.56	29.37
0.88	0.79	8.9	47.9	1.1	33.9	4.19	0.74	97.79	31.02
Window gap separation = 0.63 cm, 0.25 in.									
0.88	0.05	13.7	56.7	−0.3	31.5	2.50	0.44	58.22	18.47
0.88	0.10	13.4	56.1	−0.2	31.6	2.59	0.46	60.46	19.18
0.88	0.20	12.8	55.1	−0.0	31.9	2.78	0.49	64.94	20.60
0.88	0.50	11.4	52.6	0.3	32.6	3.26	0.58	76.14	24.16
0.88	0.79	10.4	50.7	0.6	33.1	3.62	0.64	84.35	26.76
Window gap separation = 1.27 cm, 0.50 in.									
0.88	0.05	16.0	60.9	−0.9	30.3	1.66	0.29	38.82	12.31
0.88	0.10	15.7	60.2	−0.8	30.6	1.82	0.32	42.55	13.50
0.88	0.20	14.9	58.9	−0.6	31.0	2.11	0.37	49.27	15.63
0.88	0.50	12.8	55.1	−0.1	31.9	2.75	0.49	64.19	20.37
0.88	0.79	11.5	52.7	0.3	32.5	3.23	0.57	75.39	23.92
Window gap separation = 2.54 cm, 1.00 in.									
0.88	0.05	16.1	61.0	−0.9	30.3	1.63	0.29	38.07	12.08
0.88	0.10	15.7	60.2	−0.8	30.5	1.76	0.31	41.05	13.03
0.88	0.20	15.0	58.9	−0.6	30.8	2.02	0.36	47.03	14.92
0.88	0.50	13.2	55.7	−0.2	31.7	2.62	0.46	61.21	19.42
0.88	0.79	11.9	53.4	0.2	32.3	3.07	0.54	71.66	22.73
Window gap separation = 5.08 cm, 2.00 in.									
0.88	0.05	15.9	60.6	−0.9	30.3	1.66	0.29	38.82	12.31
0.88	0.10	15.7	60.2	−0.8	30.6	1.82	0.32	42.55	13.50
0.88	0.20	15.0	58.9	−0.6	30.9	2.08	0.37	48.52	15.39
0.88	0.50	13.1	55.6	−0.1	31.8	2.69	0.47	62.70	19.89
0.88	0.79	11.8	53.3	0.2	32.4	3.14	0.55	73.15	23.21
Window gap separation = 7.62 cm, 3.00 in.									
0.88	0.05	15.9	60.6	−0.9	30.3	1.66	0.29	38.82	12.31
0.88	0.10	15.7	60.2	−0.8	30.6	1.82	0.32	42.55	13.50
0.88	0.20	15.0	58.9	−0.6	30.9	2.08	0.37	48.52	15.39
0.88	0.50	13.1	55.6	−0.1	31.8	2.69	0.47	62.70	19.89
0.88	0.79	11.8	53.3	0.2	32.4	3.14	0.55	73.15	23.21

It is beyond the scope of our discussion here to review the extensive experimental and theoretical work which has been done to obtain average conductive-convective heat transfer coefficients across vertical enclosed gas-filled parallel surfaces. The old semiempirical formulae of Jakob,[11] which unfortunately are the formulae quoted in most standard heat transfer texts,[12-14] have been substantially improved upon in more recent works.[15-18] We have used a combination of the relations developed by Eckert and Carlson[18] and Tabor[17] to describe the heat transfer coefficients across the gaps for the various conduction-convection regimes.

In the calculations, we retained the non-linear temperature dependence of the heat transfer coefficients. Numerical solutions were obtained from the resulting transcendental equations.

Before we review the results given here, a few comments are in order regarding the announcement[19] of "new" double-glazed windows. The window system described consists of dual heavily doped SnO_2 or In_2O_3 films on the interior surface of a sealed double-glazed unit. The window gap is 1.2 cm with the spacing filled with a heavy gas such as Kr.[19] The purpose of the Kr is to reduce the convective heat transfer across the enclosed gap. Dual SnO_2:Sb films correspond to a C_{12} value of 0.11. In the computations for the Kr-filled windows, we used formulae similar to those used in the air-filled windows which were suitably modified to take into account the transport parameters for Kr gas.[20] In Table 1.5, we give the results of our computation for a system filled with 1 atmosphere of Kr gas at STP. In these examples, we have considered 12-mph winds only, and have varied the gap thickness and emissivities of the interior surfaces. We note that the heat transfer across the interior is in the conduction regime for $\frac{1}{4}$-in. gaps, in the transition region between conduction and boundary layer convection for $\frac{1}{2}$-in. gaps, and in the boundary layer regime for $\frac{3}{4}$-in. gaps (see Eckert and Carlson[18] for further discussion). The U values will rise rapidly for gap thicknesses smaller than $\frac{1}{4}$ in. The choice of gap separation in this case is highly non-trivial solely on the basis of costs. In small quantities, 100 liters STP, Kr currently costs ~\$4.50 per liter. For very large purchases, 10^4 or more liters, the price drops down to the order of \$0.60 per liter.[21] The 1.27-cm-gap configuration using Kr corresponds to 1.17 liters of gas at STP per ft^2 of window area, which is a gas cost alone of \$0.71 per ft^2. Kr is a rare gas, and if demands on available supplies become too high the price will naturally show some instability. The results given here can be directly compared to those given in Tables 1.3 and 1.4. For a 12-mph wind, we see that the improvement using Kr as a replacement filler gas for air is about 14% for plain dual glass windows and 25% when the panes in both cases have a dual coating of SnO_2:Sb. Kr gas is rather expensive, and in the interest of increasing consumer acceptance via cost reduction, other lower-priced gases *should be investigated by researchers*. A further point to note here is that the change in gas has the greatest effect when coupled with low-emissivity surfaces. This is due to the fact that when C_{12} is small the heat transfer across the interior gap is dominated by conduction-convection. Thus, by reducing the conduction-convection here, we can significantly change the net heat transfer coefficient. On the other hand, when C_{12} is 0.79, convection and radiation across the gap play a comparable role; thus a reduction in the conductive-convective part here has a smaller effect on the net heat transfer coefficient.

Table 1.5. Heat transfer through double-glazed Kr-filled window system for variable gap separation and emissivities of surfaces. T_1 and T_2 are the equilibrium temperatures of the inner and outer surfaces, respectively. Calculations are for 1-meter-high windows. Room temperature = 21.11 °C, 70.00 °F. Exterior temperature = –2.22 °C, 28.00 °F. WV = 12.0 mph turbulent flow.

		T_1		T_2		U		Q_{LOSS}	
E1M	C_{12}	C	F	C	F	SIU	EU	SIU	EU
Window gap separation = 0.63 cm, 0.25 in.									
0.88	0.05	16.3	61.4	–1.2	29.9	1.34	0.24	31.35	9.95
0.88	0.10	15.8	60.4	–1.0	30.1	1.50	0.27	35.08	11.13
0.88	0.20	14.5	58.0	–0.8	30.5	1.76	0.31	41.05	13.03
0.88	0.50	12.1	53.9	–0.3	31.4	2.40	0.42	55.98	17.76
0.88	0.79	10.5	50.9	–0.0	32.0	2.82	0.50	65.69	20.84
Window gap separation = 1.27 cm, 0.50 in.									
0.88	0.05	16.7	62.0	–1.3	29.7	1.18	0.21	27.62	8.76
0.88	0.10	16.2	61.1	–1.2	29.9	1.34	0.24	31.35	9.95
0.88	0.20	15.0	59.0	–1.0	30.2	1.60	0.28	37.32	11.84
0.88	0.50	12.7	54.8	–0.5	31.1	2.24	0.39	52.25	16.58
0.88	0.79	11.1	52.1	–0.1	31.8	2.69	0.47	62.70	19.89
Window gap separation = 1.90 cm, 0.75 in.									
0.88	0.05	16.5	61.8	–1.3	29.7	1.18	0.21	27.62	8.76
0.88	0.10	16.0	60.8	–1.2	29.9	1.34	0.24	31.35	9.95
0.88	0.20	15.1	59.1	–0.9	30.3	1.63	0.29	38.07	12.08
0.88	0.50	12.7	54.8	–0.4	31.2	2.27	0.40	53.00	16.81
0.88	0.79	11.0	51.7	–0.1	31.8	2.69	0.47	62.70	19.89

There are not many published values of the U coefficients available in the literature for comparison to the values calculated here. One such reference is the ASHRAE *Handbook of Fundamentals*,[2] which gives a few values which can serve as "benchmarks" for comparison, although neither the parameters ASHRAE used nor the approximations made in arriving at their results are specified. Our results differ from theirs by at most 10–15%, which would tend to indicate that both the reported ASHRAE values and the much more extensive tabulations given here and in the more complete version of this report[1] are both accurate to within the inherent errors of ≃15%. The reported U values for storm windows are typically ~18% higher than those given here for the 2-in. gap double-glaze system. If a storm window is purposely vented or has a "leaky" construction, the infiltration will substantially reduce the temperature of the air gap. The situation then will be very close to that of a single pane, with free rather than forced convection at the exterior surface, which predicts a U value ~3.68 (W/m^2 °C), 0.65 (Btu/hr ft^2 °F). Storm windows can be manufactured with low infiltration—for such systems it is anticipated that the U values will be closer to the results indicated in our tables.

3. Comparisons and conclusions

We have performed calculations of the heat transfer for existing and possible future window systems. We anticipate that the absolute accuracy of the calculated coefficients are the order of 10–15%, while the relative accuracy for making comparisons of different window types should be better. In this section, we will combine our calculated heat transfer values with the solar input values determined in Section 1 B in order to case the results in a more meaningful cost-effectiveness perspective.

As we discussed previously, there are two components of energy transfer through a window—one is the heat transfer out of the window (in the winter), and the other is the solar energy input. These two factors are relatively independent for window systems which transmit relatively high percentages of the solar spectrum; hence the relative cost effectiveness here is primarily determined by the total thermal heat loss per unit area through the different window systems during the winter heating cycle. Obviously, this total heat loss depends upon the climatic factors as measured by the total number of heating degree-days and average wind velocities during the winter heating cycle.

The total loss per square foot for a given window system over the full winter heating cycle can be written as

$$Q_{\text{loss}} = 24\overline{U}d_w .$$

Here $\overline{U}$ is the average net heat transfer coefficient for the window system, d_w is the total number of heating degree-days for a specific locality, and the factor of 24 is a conversion from days to hours. We can put a cost figure on this heat loss by equating the loss per ft² to the number of gallons of heating oil consumed over the heating season. For example, an oil furnace operating at 67% efficiency supplies 8.7×10^4 Btu of space heat energy per gallon of fuel oil consumed. Present costs of fuel oil are $\simeq$ \$0.40 per gallon, which implies a cost of $\simeq$ \$4.60 per million Btu for space heating at current fuel prices.

We will take a few representative window systems and analyze the heat loss and solar energy input for the metropolitan New York City area climate. The results are illustrated in Figure 1.3. Here, the average number of degree-days per year is ~4900; and the solar influx results are taken directly from Section 1 B.

In Figure 1.3, we see that there is a bar associated with each window system which contains the symbols HT (heat transfer), N (north), E-W (east-west), and S (south). The HT value represents the heat transfer loss/ft² for the heating season computed from the average U value and the total degree-days. The differences between the HT and the N, E-W, S represent the "maximum" weather-averaged solar energy which can be transmitted through the windows for each of the individual directions. Intermediate window orientations can be obtained readily by interpolation. Obviously, if white interior curtains are used, or exterior shading exists on a window, the "maximum" solar transmission will be reduced. Windows with "black" lined curtains or, of course, with curtains open should utilize almost the "maximum" solar flux available.

To determine the cost effectiveness of one system relative to another, we first compare the relative fuel costs per ft². Then we use the interest-depreciation

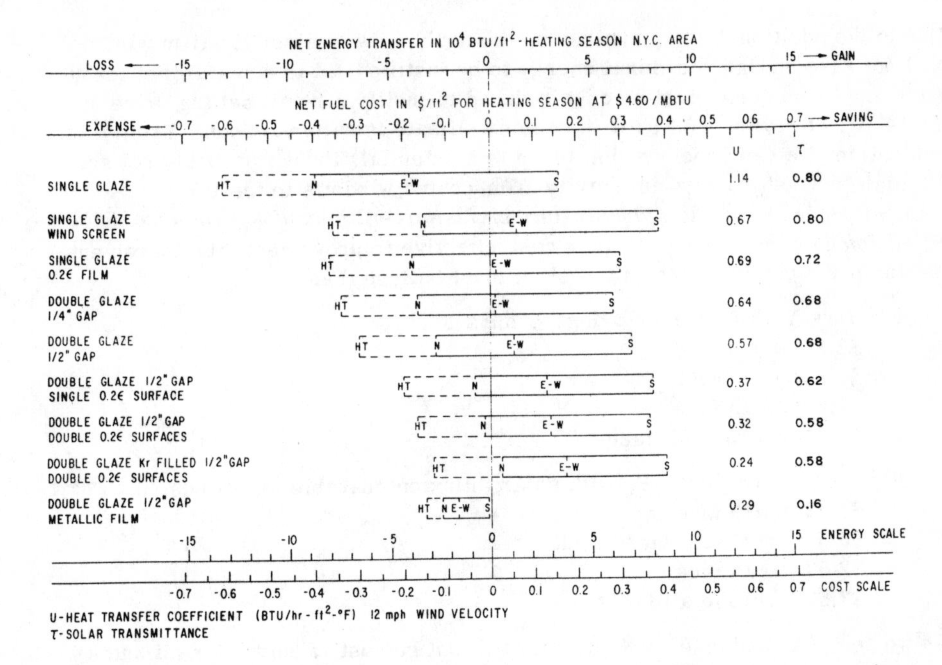

Figure 1.3. Performance chart of representative window systems in New York City area climate.

scheduling formula given in Appendix B. In our evaluation for systems which should have a long life we will use the example quoted in the appendix: A 20-year life at 10% interest and 6% fuel inflation rate implies that the initial fuel savings per year must be greater than 6.4% of the initial incremental capital expense for the improvement to be cost effective. Let us now give some examples.

Comparison of double-glaze units with and without single low-emissivity surface. To be cost effective, the incremental selling price per ft² for the system with the film must be less than (each dollar here corresponds to a fuel savings of $0.064 per ft²):

For New York City @4900 degree-days F,

$1.40 north face
$1.25 east-west face
$0.78 south face
$1.14 average all faces

For 8500 degree-days F, with solar influx comparable to latitude 45°–47° N (using the solar heat gain from Sault Ste. Marie),

$2.58 north face
$2.19 east-west face
$1.87 south face
$2.21 average all faces

Due to the additional solar exclusion, the double-low-emissivity-film window will not save enough additional energy to be justified. It is doubtful that manufacturing techniques could be cost-reduced to justify a retail selling price of $1.14/ft²; however, $2.21 for the colder climate is within reason. From this evaluation, we can conclude that there is a potential, though probably not substantial, market for low-emissivity-film-coated windows in the U.S.

Comparison of Kr-filled double-low-emissivity-film window with uncoated air-filled double-glazed unit. To be a cost-effective improvement, the incremental selling price per ft² for the Kr system must be less than:

For New York City @ 4900 degree-days F
- $2.34 north face
- $1.88 east-west face
- $1.25 south face
- $1.84 average all faces

For 8500 degree-days F, with solar influx comparable to latitude 45°–47° N
- $4.22 north face
- $3.59 east-west face
- $2.97 south face
- $3.59 average all faces

As we see, the double-film Kr system will not be cost effective for climates like or warmer than New York City, but could perhaps be cost effective for the very coldest parts of the U. S. We note, of course, that due to reduced population in the coldest climates the potential markets are not large. Hence development incentives are reduced. It is important to note that the single-film Kr-filled window will be more cost effective in these colder climates than the double-film Kr-filled window.

A very meaningful comparison to make is the ordinary single-glaze vs. the ordinary double-glaze units. This analysis should apply to well-sealed retrofitted storm windows also. Taking the half-inch-gap uncoated double-glaze system and comparing it to the standard single-pane system, we see that the incremental selling price per ft² must be less than:

For New York City
- $4.38 north face
- $3.75 east-west face
- $3.13 south face
- $3.75 average all faces

As current storm windows retail for $1.50–$2.50 per ft² installed,[22] they represent good investments in most climates. At $2.00-per-ft² storm window which is well sealed should be a cost-effective improvement for a climate with greater than 2600 degree-days F (at a fuel rate of $4.60/MBtu). For storms with high infiltration rates which cause substantial cooling of the interior pane, a U value of ≃0.65 would be more appropriate. For New York City, the incremental fuel savings would be reduced by $0.043/ft² in going from a U value of 0.57 to 0.65. The average selling price for all faces would then have to be less than $2.97 to be cost effective in the New York City climate.

The low-emissivity film on a single-glaze system is in a different category than the double-glazed system with a protected film between the glass panes. Here, the exposed surface would be subjected to mechanical abrasive degradation and the lifetime would be considerably shorter than 20 years. Let's assume an 8-year lifetime. Also, for a SnO_2-type film, both the glass and the film would have to be replaced as these films are deposited from chemical spray-pyrolysis at elevated temperatures of ≃500 °C (see Sec. 2 A). Using the interest-depreciation formula given in Appendix B with the same interest rate (10%) and fuel inflation rate (6%) but with an 8-year lifetime, we obtain that initial fuel savings must be greater than 15% of the capital costs to be cost effective. From this we see that the selling price must be less than:

For New York City
- $1.53 north face
- $1.33 east-west face
- $1.20 south face
- $1.35 average all faces

From these results we conclude that low-emissivity surfaces on single-pane systems are not practical—rather, a residential home owner should install double-glazed or storm windows.

The metallic-film–double-glazed system (solar control glass) is in a different category than the other window systems given here in that it is a double-glazed system with an interior metal film which acts both as a low-emissivity surface and as a solar reflector. These windows are used mostly in commercial office buildings to reduce the solar load on air conditioning. Certainly, these windows when compared to an ordinary uncoated double-glazed window are poorer performers during the heating cycle due to the solar exclusion. However, to be fair one must estimate the positive effects of solar control glass for saving energy in the summer. We will now perform this estimation. Taking the summer solar irradiance factors relevant to the New York City area from Section 1 B, we compute and tabulate the solar input in 10^3 Btu/ft^2 in Table 1.6.

In computing the air conditioning cost, we make the assumption that a commercial air conditioning system has a cooling efficiency of ~7 Btu/watt-hr of electrical power. If we assume an electric peak power rate of $0.03/KW-hr, and assume that all the solar input during the air conditioning season must be removed by the cooling system, each million Btu of solar input costs $4.29 in electrical power costs to remove. In Section 1 B, it was determined that the contribution of the U heat transfer is negligible for New York City area in the summer. We must remember that the results given here are for New York City area climates. In more southerly climates, the situation will undoubtedly improve somewhat for the metalized solar control glass. The net costs/ft^2 are determined as the sum of the values from the winter given in Figure 1.3 and the summer costs in Table 1.6. The comparison of ordinary double-glazed windows with the addition of a metalized film for solar control shows that the metalized solar control glass saves $0.15/year/ft^2 on a north face, costs $0.15/year/ft^2 *more* on a southerly face and saves $0.10/year/ft^2 on east and west faces. If the solar control glass does not incorporate a low emissivity metalized film, the cost situation is far worse

Table 1.6. Solar input (in 10^3 Btu/ft^2).

	North	South	East/West
Double glaze			
Summer solar input 10^3 Btu/ft^2	38	52	67
Air cond. cost \$/yr ft^2	−0.16	−0.22	−0.29
Net cost \$/yr ft^2 summer and winter	−0.29	+0.10	−0.24
Double glaze, metal film			
Summer solar input 10^3 Btu/ft^2	8.9	12.0	15.8
Air cond. cost \$/yr ft^2	−0.04	−0.05	−0.07
Net cost \$/yr ft^2 summer and winter	−0.14	−0.05	−0.14

than that indicated above. Some obvious conclusions regarding solar control glass are:

If you can get by with manual selective summer shading, do so.
Don't ever put solar control glass on southerly-facing windows.

We have discussed a number of window modifications which give energy savings in climate control. We reiterate the importance of cost-effectiveness evaluations of window innovations. The analysis used in this section should provide a good rejection filter to eliminate wasteful R&D on systems which are intrinsically too expensive. Speculative ideas such as wind screens are certainly worth investigation, as such wind screens can conceivably be retrofitted inexpensively to existing structures. For the residential home owner, the conventional uncoated storm window really works very well. In northern climates it is cost effective and adds comfort by reducing drafts and provides a warmer interior glass surface than a single pane window. Moreover, in the "triple track" models, the storm window provides the residential home owner with the convenience of a built-in screen.

Acknowledgment

The author would like to especially thank Professor S. M. Berman of Stanford University for his continued interest and inputs to this work. He would also like to thank Drs. H. B. Vakil and P. G. Kosky of General Electric Company for discussions related to this work, and Mr. R. Hosmer Norris of the General Electric Company (retired) for a critical review of the technical contents.

[1]S. D. Silverstein, General Electric Corporate Research and Development Report (to be published).
[2]C. MacPhee, ed., *Handbook of Fundamentals*, American Society of Heating, Refrigerating, and Air Conditioning Engineers (ASHRAE), New York (1972).

[3]The thermal conductivity of window glass is ≃0.78 W/m °K (0.45 Btu/hr ft^2 °F). For a 0.32 cm ($\frac{1}{8}$ in.) glass pane, the heat transfer coefficient is 244 W/m^2 °K (43 Btu/hr ft^2 °F). For a 5.36 m/sec (12 mph) wind, the temperature drop across the glass will be less than ~0.6 °C (1.1 °F). The approximation neglecting temperature gradients in the glass simplifies the computation considerably.

[4]Window glass is opaque to radiation at wavelengths longer than ~2.7 microns. Therefore all long-wavelength thermal radiation is absorbed except for the small amounts which are specularly reflected at the air-glass interface. The integration over all angles for a diffuse infrared source gives a reflectivity of $\bar{r}_g \simeq 0.12$. The hemispherical emissivity ϵ is the absorption, $1-\bar{r}_g=0.88$. If one coats glass with another material which is highly reflecting in the infrared with a reflectivity r', the laminate will now have an emissivity $\simeq 1-r'$. For example, heavily antimony doped SnO_2 will have a selective infrared reflectivity of ≃0.8 and hence an emissivity of ≃0.2. Non-oxidized metallic films will be very reflecting in the infrared, e.g., gold has an emissivity of ≃0.03.

[5]"Speeds of Winds in United States and Canada," 1974 World Almanac, p. 267, Newspaper Enterprise Association, New York.

[6]Conventional vinyl window shades cost from $0.1–$1.0/ft^2, depending upon quality and size. Source: Montgomery Ward & Company General Catalog (1975).

[7]In computing the radiation transfer between the exterior window surface and the ambient surroundings, it is useful to define an average black body temperature of the exterior. This is possible because the glass is uniformly absorbing throughout the thermal radiation spectrum; hence the window in the atmospheric emission spectrum around 10 microns between the water vapor and CO_2 emission bands does not manifest itself. Accordingly, the integral of the atmospheric spectral emission can be equated to an effective T^4 radiation relation. The emission spectrum of the horizon is that of the ambient temperature, while at higher angles there is less atmospheric (dust, CO_2, H_2O vapor) path length and the spectral emission is reduced and altered (Reference 8). For a horizontal surface, the average nighttime effective sky black body temperatures (References 9, 10) are ≃20 °C less than the ambient for a clear sky. Taking into account the ground emission, and the variation in emission with angle above the horizon, we have calculated that the 20 °C horizontal reduction corresponds to an effective reduction of 5.8 °C for a vertical face. There are both weather factors and exterior sky blocking factors (other buildings, trees) which should reduce the 5.8 °C figure on the average. Accordingly, we have assumed throughout our heat transfer calculations that the effective average exterior radiation temperature is 4 °C below the ambient air temperature.

[8]R. Kauth, "Sky Backgrounds," in *Handbook of Military Infrared Technology*, W. L. Wolfe, ed., U.S. Government Printing Office, Washington, D.C. (1965).
[9]A. Bar Cohen and C. Rambach, "Nocturnal Water Cooling by Sky Radiation in Israel," Proc. 9th IECEC Conference, p. 298 (1974).
[10]J. A. Duffie, W. A. Beckman, "Solar Energy Thermal Processes," John Wiley and Sons, New York (1974).
[11]M. Jakob, Trans. ASME 68, 189 (1946); *Heat Transfer*, John Wiley and Sons, New York (1958).
[12]W. H. McAdams, *Heat Transmission*, 3rd ed., McGraw-Hill Book Co., Inc., New York (1953).
[13]F. Kreith, *Principles of Heat Transfer*, 3rd ed., Intertext Educational Publishers, New York (1973).
[14]J. R. Welty, *Engineering Heat Transfer*, John Wiley and Sons, New York (1974).
[15]G. K. Bachelor, "Heat Transfer by Free Convection Across a Closed Cavity Between Vertical Boundaries at Different Temperatures," Quart. of Appl. Math., Vol. XII, No. 3, 209 (1954).

[16]J. G. A. de Graaf and E. F. M. Van derHeld, "The Relation Between Heat Transfer and the Convection Phenomena in Enclosed Plane Layers," Appl. Sci. Res., Vol 3A, 393 (1953).

[17]H. Tabor, "Radiation, Convection and Conduction Coefficients in Solar Collectors," Bull. Res. Council Israel, 6C, 155 (1958).

[18]E. R. G. Eckert and W. O. Carlson, "Natural Convection in an Air Layer Enclosed Between Two Vertical Plates with Different Temperatures," Int. J. Heat and Mass Transf. 2, 106 (1961).

[19]H. Kostlin, "Double Glazed Windows With Very Good Thermal Insulation," Philips Tech. Rev. 34, 242 (1974) No. 9; London Financial Times, May 30, 1974; New Scientist, p. 679, June 13, 1974; Electro-Optical Systems Design, p. 4, September 1974.

[20]I. Amdur, "Low Temperature Transport Properties of Gases. II. Neon, Argon, Krypton and Zenon," J. Chem. Phys. 16, 190 (1948).

[21]Union Carbide Corporation, Linde Division (private communication, October 1974).

[22]Including installation, typical prices for "triple-track" storm windows are between \$1.50–\$2.50 per ft^2, depending upon size and quality of window. Source: Montgomery Ward and Company retail store, Menands, New York (January 1975).

B. The positive aspects of solar heat gain through architectural windows *

The energy crisis of the mid 1970's has caused a substantial resurgence of interest in further harnessing the earth's primary source — the sun. A device of considerable interest today is the flat plate solar collector, which shows promise for residential and small commercial climate-control applications. Substantial efforts are being made to simultaneously raise the efficiency and lower the costs of such collector systems. A typical flat plate solar collector incorporates a window, quite similar to an architectural window, and an internal surface which both absorbs the solar irradiance and acts as a heat exchanger with some transferable energy storage medium.

The architectural window and the interior of a building are in a sense also solar heat collectors, albeit not optimized. If solar heat collectors can be useful in broad geographical areas for auxiliary space heating it is quite evident that a substantial amount of beneficial heat can be gained from solar transmission through architectural windows, especially in residences. Moreover, the incremental capital costs to make some windows better in conserving net energy than insulated walls are not high.

The positive aspect of the use of the solar heat gain through windows as a fuel-free energy source for auxiliary space heating is the most significant message of this section.

The glass industry has long recognized the existence of solar heat gain — but primarily as a detrimental rather than positive factor. Here we are referring to the additional solar load on air conditioning of commercial buildings. In order to reduce this air conditioning load, the glass industry has marketed various kinds of solar control glass. The solar control glass substantially reduces the transmission of the visible radiation and short-wavelength infrared radiation.

The major thrust of this section is to consider both the beneficial and the detrimental aspects of the solar heat gain. The relative weighting of these two factors depends on latitude and climate. Accordingly, we will pick representative geographic areas in our discussion. An important conclusion of our section is that for certain window systems, particularly on a southern face, the net heat gain from solar influx *can exceed the net loss* due to thermal heat transfer averaged over the whole heating season. The thermal heat transfer is that part which is proportional to the temperature difference.

---

*This section by S. M. Berman and D. E. Claridge.

(a) Potential solar heat gain for some representative geographical areas

The total energy flux from solar irradiance falling on a window can be divided into two parts. The first component is often referred to as "beam" irradiation and corresponds to direct unscattered solar radiation. The second component is diffuse irradiance which arises from a combination of atmospheric and ground scattering. Both of these components contain a significant amount of energy which can be efficiently transferred to the interior of buildings through architectural windows. The total daily irradiance as well as the relative amounts of beam and diffuse irradiance depends upon latitude, season, weather, window direction, shading, etc. For example, in winter on a north face the irradiance will be purely diffuse, while on a southern face there will be, in addition, a large component of direct irradiance.

We know that the reflectivity of diffuse irradiance differs from that of direct irradiance. Also, the reflectivity will vary with the incidence angle of the "beam" irradiance. In order to estimate the potential solar energy gain, we must first determine whether there is a significant change in the transmission factor from month to month over the year. To answer this question, we have calculated the transmission and associated potential solar heat gain (transmission $\times$ incident irradiance) through $\frac{1}{8}$-in. double-strength-sheet (DSS) glass vertical windows. The calculation, which has been done for windows facing each of the four cardinal (compass) directions, includes weather and latitude factors for each geographical area. The calculational procedure which we used is a slight modification of the approach of Liu and Jordan,[1-3] who developed a method for calculating the weather-averaged insolation on vertical surfaces from measured values of solar insolation on horizontal surfaces. Such calculations are, of course, very important for thermal and/or photovoltaic solar collectors where optimum orientation and angular inclination positioning are necessary. Similar calculations for solar collectors are illustrated extensively in the literature.[4] The results of our calculation are given in Table 1.7. It was assumed that the incident solar radiation was not reduced by external shading of any type (i.e., roof overhangs, adjacent structures, trees, etc.). The absorption of incident irradiance by the glass contributes 4% to 5% to the solar heat gain and was omitted from this calculation.

It is interesting to note that the variations of the transmittivities for the four representative geographical areas are not significant when comparing the same specific direction from area to area. The southern face shows the greatest seasonal variation. In the absence of exterior shading the small variation in transmission simplifies future analyses, in that one can choose an average seasonal transmission factor for each of the four cardinal directions and apply this factor to almost any locality in the United States without significant error. The solar heat gain factors given in Table 1.7 represent the amount gained on an average day (in Btu/ft^2 day) for the month and location specified. Such numbers are obviously of value in determining long-term energy balance but do not represent the extreme conditions necessary for designing heating and cooling systems of proper capacity. Daily averages for each month are presented for north and south facing windows. Both morning and afternoon totals are given for east facing windows. The calculation assumed insolation through east and west facing

Table 1.7. Average daily solar heat gain and transmission factor for $\frac{1}{8}$-in. DSS glass (Btu/ft^2-day).

Sault Sainte Marie, Michigan

	North		East				South	
	SHG	τ	Morn. SHG	Aft. SHG	Tot SHG	τ_{tot}	SHG	τ
January	113	0.80	220	56	276	0.77	868	0.85
February	178	0.80	413	89	502	0.80	1130	0.84
March	260	0.80	626	130	756	0.81	1179	0.81
April	350	0.78	601	173	774	0.81	846	0.77
May	451	0.76	710	206	916	0.81	752	0.74
June	495	0.76	709	218	927	0.81	670	0.72
July	490	0.76	771	214	985	0.81	725	0.72
August	401	0.77	682	187	869	0.81	800	0.75
September	285	0.80	460	142	602	0.81	811	0.79
October	199	0.80	324	99	423	0.80	814	0.83
November	118	0.80	153	59	212	0.78	498	0.85
December	92	0.80	161	46	207	0.77	658	0.86

Seattle, Wash.

	North		East				South	
	SHG	τ	Morn. SHG	Aft. SHG	Tot SHG	τ_{tot}	SHG	τ
January	87	0.80	77	43	120	0.77	297	0.85
February	142	0.80	184	71	255	0.80	509	0.84
March	236	0.80	372	118	490	0.80	752	0.81
April	335	0.79	513	166	679	0.81	762	0.78
May	421	0.77	573	198	771	0.81	675	0.75
June	455	0.77	566	208	774	0.81	607	0.74
July	452	0.77	615	207	822	0.81	656	0.74
August	388	0.77	607	183	790	0.81	760	0.76
September	277	0.80	425	139	564	0.81	774	0.80
October	175	0.80	242	88	330	0.80	623	0.83
November	107	0.80	119	53	172	0.79	385	0.84
December	75	0.80	76	37	113	0.77	291	0.85

New York, New York

	North		East				South	
	SHG	τ	Morn. SHG	Aft. SHG	Tot SHG	τ_{tot}	SHG	τ
January	140	0.80	235	70	305	0.79	734	0.84
February	201	0.80	307	100	407	0.80	788	0.83
March	280	0.80	459	140	599	0.80	837	0.79
April	351	0.80	475	175	650	0.81	667	0.76
May	440	0.77	587	204	791	0.81	593	0.72
June	499	0.76	654	219	873	0.81	556	0.69
July	479	0.76	648	214	862	0.81	584	0.70
August	401	0.77	517	189	706	0.81	636	0.74
September	312	0.80	488	156	644	0.81	782	0.78
October	235	0.80	369	118	487	0.80	853	0.82
November	158	0.80	253	79	332	0.79	720	0.84
December	123	0.80	178	61	239	0.78	661	0.85

Table 1.7. *Continued.*

Dallas–Fort Worth, Texas

	North		East				South	
	SHG	τ	Morn. SHG	Aft. SHG	Tot SHG	τ_{tot}	SHG	τ
January	205	0.80	385	103	488	0.79	1086	0.84
February	260	0.80	434	130	564	0.80	1005	0.82
March	328	0.80	621	164	785	0.81	938	0.77
April	389	0.79	599	193	792	0.81	663	0.71
May	493	0.75	750	211	961	0.81	531	0.67
June	564	0.73	949	218	1067	0.81	493	0.70
July	532	0.74	803	215	1018	0.81	504	0.69
August	423	0.75	802	198	1000	0.81	623	0.69
September	347	0.80	700	173	873	0.81	867	0.74
October	283	0.80	588	142	730	0.81	1109	0.79
November	236	0.80	466	118	584	0.80	1227	0.83
December	192	0.80	398	96	494	0.79	1179	0.84

windows to be basically anti-symmetric, i.e., morning east = afternoon west and daily totals *are identical* for east and west.

We will now consider the annual expenditure of energy for climate control for various types of window systems. First, we consider the incremental energy per unit window area consumed during the winter heating season for a given window system. This quantity, E_W, is the combination of loss due to thermal heat transfer from the interior to the exterior and gain due to solar input:

$$E_W = \overline{(\mathrm{SC})_W}\,\overline{(\mathrm{SHG})_W} - 24\,U_W d_W\,. \tag{1a}$$

For the summer, both solar input and thermal transport will cause energy consumption if air conditioning is used. Hence, the contributions are given negative (energy consuming) values:

$$E_s = -\overline{(\mathrm{SC})_s}\,\overline{(\mathrm{SHG})_s} - 24 U_s d_s\,. \tag{1b}$$

The nomenclature is a little awkward, but we have used shading coefficients $(\mathrm{SC})_{w,s}$ for winter and summer to conform to the nomenclature adopted by the glass industry. Let us explain the definitions so the meaning will be more apparent. We recall that the solar heat gain was defined as the product of the net (beam plus diffuse) solar irradiance on a vertical surface multiplied by the transmittivity of $\frac{1}{8}$-in. DSS glass. Therefore, $\overline{(\mathrm{SHG})}_{w,s}$ are the totals of these quantities over the heating and cooling season, respectively. The shading coefficient (SC) is defined as the transmittivity of a given window system relative to the transmittivity of $\frac{1}{8}$-in. DSS. The $\overline{U}_{w,s}$ are the average winter and summer heat transfer coefficients (see Section 1A for details.) The $d_{w,s}$ are the heating (cooling) degree-days. The cooling degree-days are defined by

$$d_s \equiv \sum f(\overline{T} - 75\ {}^\circ\mathrm{F})\,, \tag{2}$$

where

$$f(\overline{T} - 75\ {}^\circ\mathrm{F}) \equiv \begin{cases} \overline{T} - 75\ {}^\circ\mathrm{F} & \text{for } T > 75\ {}^\circ\mathrm{F} \\ 0 & \text{for } T \gg 75\ {}^\circ\mathrm{F}\,. \end{cases}$$

Here T is the mean daily temperature and the sum is over all days. Heating degree-days are in common use. The number of degree-days for a particular day corresponds to the difference between 65 °F and the mean outdoor temperature for a given day.

There are a number of clarifying points which we would like to make relative to the evaluation of Equations (1a) and (1b) for different localities.

The use of degree-days in the evaluation of the net winter energy consumption per unit area of a window system, (E_w), assumes that the energy supplied by an internal space heater exceeds the internal heat generation from lighting, occupying loads, machinery, etc., by an amount greater than the net loss through the window. For the E_s summer relation, the cold reservoir, which is to be kept cold, is the interior; therefore, E_s will still correspond to the incremental air conditioning load per unit window area.

In defining the winter solar heat gain as positive, we have assumed that the solar influx is all utilized to decrease the energy supplied by the building's heating system. Accordingly, the net solar flux through the windows should not exceed an amount which could cause overheating in the winter. In the usual residence with negligible internal heat load this is generally the case and thus the results are most meaningful for residential buildings. Commercial buildings with their large internal heating loads will not be able to use much of the winter solar heat gain, and hence the results are not as relevant in that case but do show the potential energy possibilities if the internal loads were to be lessened.

We have essentially divided the year into two basic seasons:

(a) The winter heating season — when it is assumed that all SHG is usable and energy saving.

(b) The summer cooling season — which corresponds to overheated periods when all SHG is undesirable and energy consuming.

There are intermediate crossover seasons in the spring and fall when neither heating nor air conditioning uses significant quantities of fuel in typical residential structures. We have assumed that the SHG has a negligible effect during these periods. The summers in both Seattle and Sault Ste. Marie fall into this category and hence for these cities we have taken the $(\mathrm{SHG})_s$ as zero.

The winter heating season was chosen to be those months for which the mean daytime temperature $t_0 < 60$ °F, as tabulated in Reference 1. This corresponds to at least 240 degree-days in a given month, as tabulated in Reference 3. The air conditioning periods were assumed to correspond to the periods indicated as overheated for New York and Dallas by Olgyay.[6] Seattle and Sault Ste. Marie are assumed to have no overheated periods $[(\mathrm{SHG})_s = d_s = 0]$. These periods are:

Winter heating period:

Sault Ste. Marie	September–May
Seattle	October–May
New York	October–April
Fort Worth	November–March

Summer cooling period:

New York	June–August
	plus afternoon ($\frac{1}{2}$ day) for September
Fort Worth	May 15–September 15
	plus afternoons April 15–May 15
	September 15–October 15

The ground reflectance was assumed to be $\rho = 0.2$ throughout the year. Snow-covered ground has a significantly higher reflectance (2) and assuming $\rho = 0.5$ for November–March in Sault Ste. Marie results in an additional 12,000 Btu/ft^2 winter solar heat gain through clear single glazing in this location.

Let us now look at the results for the winter and summer solar heat gain. In Figures 1.4 and 1.5 we show the results for the daily averaged and total seasonal winter solar heat gain per unit area as a function of the window orientation. The computations were made specifically for the cardinal directions, and we have made extrapolations for intermediate orientations. It is very interesting to note that the further north you go, as illustrated by Seattle and Sault Ste. Marie, the more closely the points conform to a straight line. As you move further south, more curvature is exhibited. We can explain this as follows. In winter months at higher latitudes the sun's angle above the horizon is small. Hence the solar radiation transverses a large and relatively constant atmospheric air mass throughout the day. Therefore, if we had a vertical plate which tracked the sun, the transmission through the plate would be relatively constant throughout the day. Then, the integrated beam SHG would be proportional to the time the surface in a given direction sees the direct solar rays. To illustrate this, we note that the north face is exposed to a diffuse background component only (which is assumed to be independent of direction). The difference in SHG for any direction and that of the north face represents the contribution to the beam irradiance. For higher latitudes, the southern face receives almost exactly twice the beam irradiance of the east-west faces. As one moves to more southerly latitudes the diurnal solar trajectory is higher and the solar irradiance reaching the surface of the earth is more sharply peaked around the noon hour. Hence, the southern face will receive more than twice the energy of east-west faces. Accordingly, more curvature is exhibited in the $(SHG)_w$ as a function of window orientation for lower latitudes.

In Figure 1.6, we illustrate the calculated values of the summer solar heat gain for the Dallas–Fort Worth and New York City areas. Based on the overheating criterion previously discussed, Seattle and Sault Ste. Marie *have no* $(SHG)_s$. Because of the high daily solar trajectory in summer, it is not nearly as easy to extrapolate between the four cardinal directions as we have done for the $(SHG)_w$.

In order to evaluate the thermal transport contributions to the winter and summer climate control energy expenditures, the degree-days d_w, d_s were obtained as follows:

d_w was assumed to be the number of degree-days given in Reference 5 for the months included in the winter heating season.

d_s is not tabulated as we have defined it. Liu and Jordan[1] have tabulated mean daytime temperatures for 80 locations in the United States. New York has maxi-

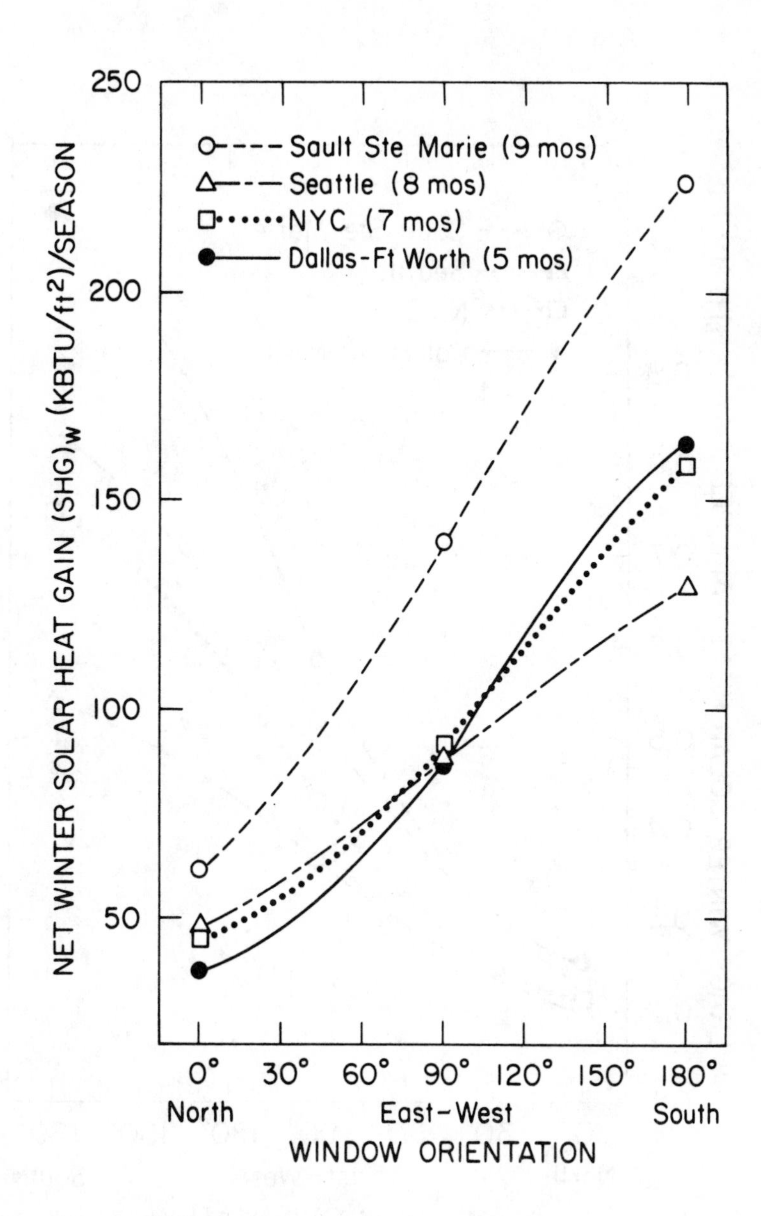

Figure 1.4. The curves represent the winter solar heat gain $(SHG)_w$ for the four cities as a function of window orientation with respect to the cardinal directions. The heating degree-days and lengths of heating season are:

Sault Ste. Marie	8646 degree-days —	9 months
Seattle	4081 degree-days —	8 months
New York City	4714 degree-days —	7 months
Dallas–Fort Worth	2241 degree-days —	5 months

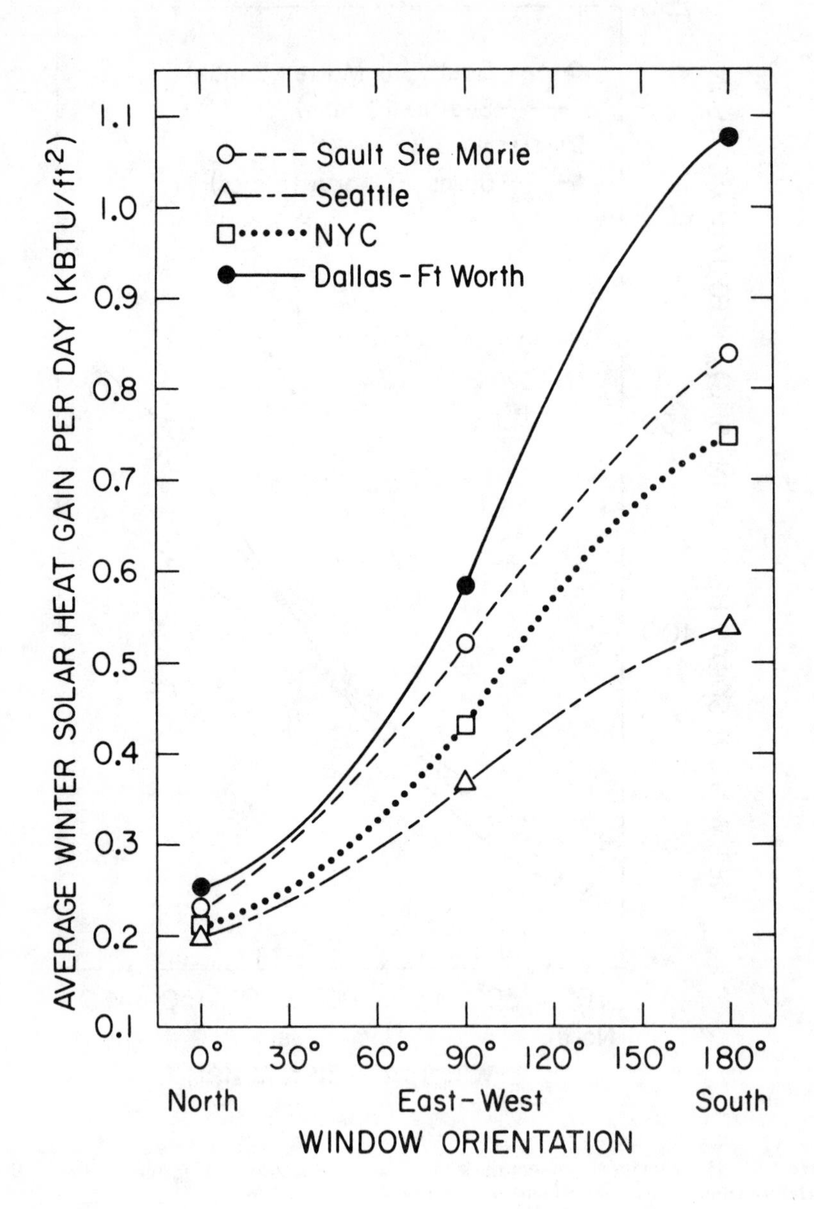

Figure 1.5. Daily average solar heat gain for four United States cities shown.

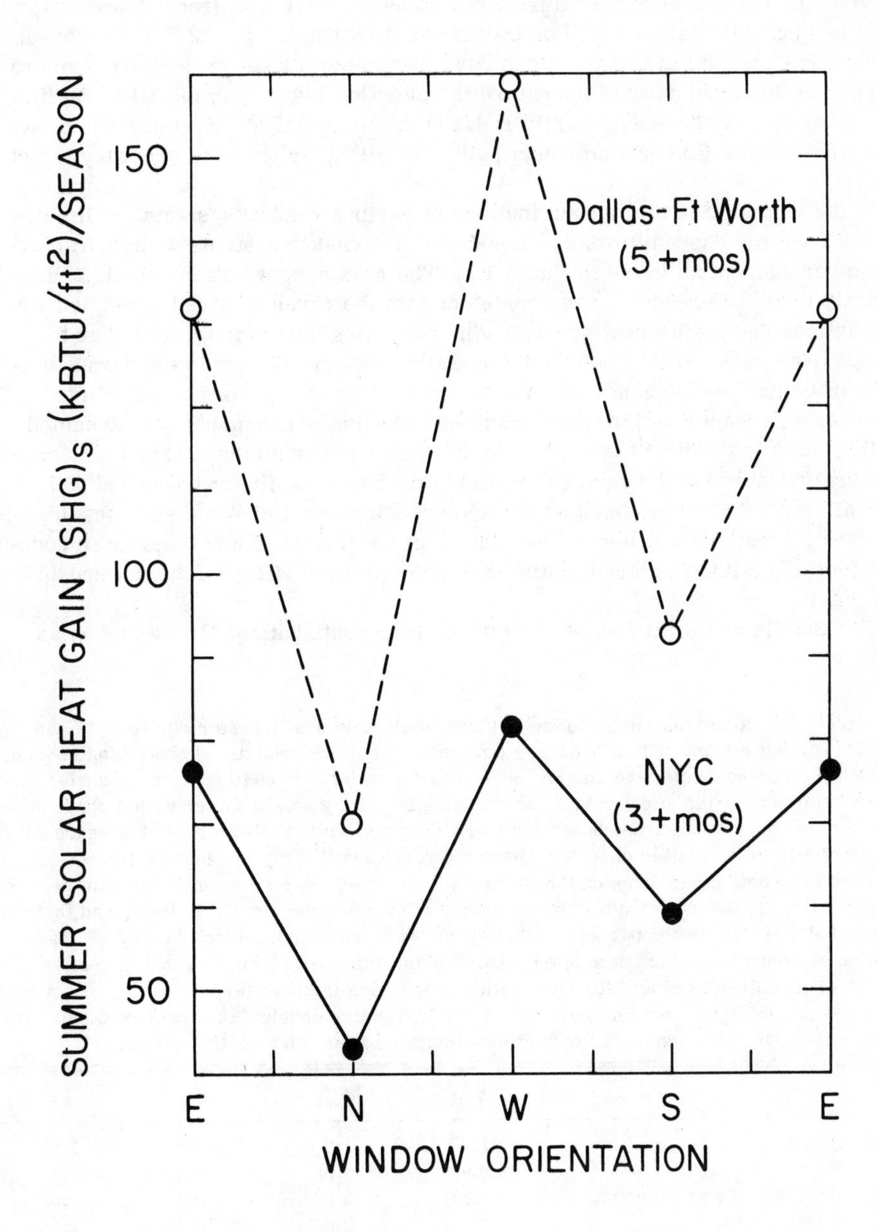

Figure 1.6. The points represent the summer solar heat gain $(SHG)_s$ for Dallas–Fort Worth and New York City areas. Both Seattle and Sault Ste. Marie have no substantial overheating periods; hence their $(SHG)_s$ are zero. The heating degree-days are zero in New York City and 1000 in Dallas–Fort Worth. The curves drawn between the calculated points are linear connections — there will be curvature.

mum mean daytime temperatures of 76.9 °F (July) and 75.3 °F (August). It seems obvious that nighttime temperatures will be below 75 °F and hence d_s for New York will be approximately zero. For Dallas–Fort Worth, we get 1275 "summer-degree-days" assuming that the mean daytime temperature = mean daily temperature. As the mean daily temperature is somewhat lower than the mean daytime temperature, we have arbitrarily reduced this to $d_s = 1000$. Even for north facing windows the $U\Delta T$ summer heat gain $< \frac{1}{3}$ $(\text{SHG})_s$, so extreme accuracy is not critical.

In the compilation of the contributions of various window systems to climate-control energy expenditures we considered the window systems with characteristic parameters as listed in Table 1.8. The U values, which were selected primarily from Reference 5, are consistent with the U values given in Section 1 A. Variations due to different average wind velocities allow for the U values to range over some 15%. The shading coefficients were chosen to be representative of typical performance on an annualized basis except for the reflective glasses and shades, where near maximum shading performance was assumed.

In Tables 1.9 and 1.10, we give the net winter and summer energy transfer per square foot through the various architectural windows illustrated in Table 1.8. Recall that we have assumed no exterior shading effects. When such shading is present, it can have either a desirable (e.g., deciduous shade tree) or an undesirable (e.g., large adjacent building in cold climate) effect on the net energy transfer.

For Seattle and Sault Ste. Marie the summer contribution (E_s) was taken as

Table 1.8. Representative window systems with their heat transfer coefficients and shading coefficients. The window systems included are defined as follows. Single: Standard $\frac{1}{8}$-in. glass. Selective single: Selective low emissivity coating on inner surface with high visible transmission. Single + shade: Single $\frac{1}{8}$-in. glass in winter with a shade added in the summer. The shade chosen here is a standard white roll-down window shade with a reflectivity in the visible of ~0.7. The same results will apply to venetian blinds. There are some new shades on the market which incorporate a thin metallic film protected by plastic. These shades offer the advantage of direct transmission of light, and their energy balance will be comparable to that of the standard shade. Selective single + shade: Same as above for selective single glass. Storm: Standard storm window. Selective double: Double-glazed unit for storm window with one interior surface coated with a high-solar-transmission, low-emissivity coating. Reflective double: Double-glazed unit with low emissivity with high solar reflectivity (metallic film solar control glass).

	SC_w	SC_s	U_w	U_s
Single	1	1	1.13	1.06
Selective single I	0.9	0.9	0.74	0.7
Aluminized Mylar coating	0.25	0.25	1.02	0.95
Single + shade	1	0.34	1.13	1.06
Selective single + shade	0.9	0.31	0.74	0.7
Storm	0.85	0.85	0.55	0.53
Selective double	0.77	0.77	0.35	0.34
Reflective double	0.2	0.2	0.35	0.34
Storm + shade	0.85	0.3	0.55	0.53

Table 1.9. Values for the winter energy expenditure (E_w) for selected window systems for Sault Ste. Marie and Seattle in units of KBtu/ft².

	Sault Ste. Marie, Mich.			Seattle, Wash.		
	N	E=W	S	N	E=W	S
Single	−173	−94	−8	−63	−23	18
Selective single	−98	−28	50	−30	7	43
Storm/double	−62	5	79	−14	21	56
Selective double	−25	35	102	2	33	65
Reflective double	−60	−45	−27	−25	−17	−9
Aluminized Mylar coating	−196	−177	−155	−88	−78	−68

zero based upon the criteria previously discussed.

Tables 1.9 and 1.10 give some very useful numbers from which one can compute the relative cost effectiveness of any window system. To do this one follows a procedure similar to that described in Section 1 A. We recall that fuel at \$0.40/gallon gave a space heating cost of \$4.60/MBtu. An electric air conditioning system with a cooling efficiency of 7 Btu/W-hr will cost ~\$3.57/MBtu at an electric power cost \$0.025/KWh. The cost/ft² in dollars is then given by $-[4.6\ (E_w) + 3.57\ (E_s)]/1000$.

Table 1.10. Winter (E_w) and summer (E_s) climate control energy expenditure for various window types (in KBtu/ft²-season). Negative values denote consumption, positive denote saving. $T = E_w + E_s$ represents total yearly energy balance.

(A) Dallas-Ft. Worth

Window type	N			E			S			W		
	E_{wN}	$+E_{sN}$	$=T_N$	E_{wE}	$+E_{sE}$	$=T_E$	E_{wS}	$+E_{sS}$	$=T_S$	E_{wW}	$+E_{sW}$	$=T_W$
Single	−24	−95	−119	+26	−156	−130	+102	−118	−16	+26	−185	−159
Selective single	−7	−80	−87	+38	−135	−97	+107	−101	+6	+38	−161	−123
Aluminized Mylar	−46	−41	−87	−33	−56	−89	−14	−46	−60	−33	−63	−96
Single + shade	−24	−49	−73	−26	−70	−44	+102	−57	+45	+26	−80	−54
Selective single + shade	−7	−39	−46	+17	−57	−40	+86	−46	+40	+17	−66	−49
Storm/double	+1	−73	−72	+44	−124	−80	+109	−92	+17	+44	−149	−105
Selective double	+9	−62	−53	+48	−109	−61	+107	−80	+27	+48	−131	−83
Reflective double	−12	−22	−34	−2	−34	−36	+14	−27	−14	−2	−40	−42
Storm + shade	+9	−34	−25	+48	−52	−4	+107	−41	+66	+48	−61	−13

(B) New York City

Window type	N			E			S			W		
	E_{wN}	$+E_{sN}$	$=T_N$	E_{wE}	$+E_{sE}$	$=T_E$	E_{wS}	$+E_{sS}$	$=T_S$	E_{wW}	$+E_{sW}$	$=T_W$
Single	−84	−43	−127	−38	−76	−114	+29	−59	−30	−38	−81	−119
Selective single	−43	−39	−82	−2	−68	−70	+58	−53	+5	−2	−73	−75
Aluminized Mylar	−105	−11	−116	−93	−19	−112	−76	−15	−91	−93	−20	−113
Single + shade	−84	−15	−99	−38	−26	−64	+29	−20	+9	−38	−28	−66
Selective single + shade	−43	−13	−56	−2	−24	−26	+58	−18	+40	−2	−25	−37
Storm/double	−25	−37	−62	+14	−65	−51	+71	−51	+21	+14	−69	−55
Selective double	−5	−33	−38	+30	−59	−29	+82	−45	+37	+30	−62	−32
Reflective double	−31	−9	−40	−22	−15	−37	−8	−12	−20	−22	−16	−38
Storm + shade	−25	−13	−38	+14	−23	−9	+71	−18	+53	+14	−24	−10
Selective double + shade	−5	−17	−22	+30	−30	0	+82	−24	+58	+30	−32	−2

Some interesting observations from these tables are:

1. For southern faces, any of the windows considered above which have high winter solar transparency and are suitably shaded in the summer will be *more effective in saving energy than the best-insulated wall*. Indeed, most of the systems have a net positive energy gain. For the better thermally insulated window systems this will be true for the east and west facing windows also. This is a *very important point* because the intuitive feeling that the most effective building design from an energy view (presumably not from an aesthetic view) would be a structure without windows *is not true*.

2. It is interesting to compare the relative performance of solar control glass (reflective double) vs. a single pane of glass with shade in the Dallas–Fort Worth area. Using the energy cost figures given above we see that the reflective double saves $0.15/ft^2 for a north facing window, costs the same for an east face, costs $0.30/ft^2 more on a southern face, and saves $0.01/ft^2 on a west face. In the more northern climates like New York City, one should compare the shaded double or the selective single plus shade with the solar control glass. Our results indicate that for residences in northern latitudes solar control glass is not a desirable solution when energy conservation is of interest. In southern latitudes there may be some possibility that solar control glass with larger shading coefficients is consistent with energy conservation. A further detailed study would be required to verify this possibility.

Since winter solar heat gain is desirable and summer solar heat gain is generally undesirable, it is clearly advantageous to use seasonal management of windows. A simple procedure would be to use solar control shades during the cooling season.

Acknowledgment

We are grateful to Dr. S. Silverstein for pointing out the simple and useful method for determining the solar heat gains for all window orientations. His critical reading and editorial contributions have substantially improved the usefulness of this section.

---

[1]Benjamin Y. H. Liu and Richard C. Jordan, "A Rational Procedure for Predicting the Long-Term Average Performance of Flat-Plate Solar Energy Collectors," Solar Energy 7, 53 (1963).

[2]Benjamin Y. H. Liu and Richard C. Jordan, "Daily Insolation on Surfaces Tilted Toward the Equator," ASHRAE Journal 53, October 1961.

[3]Benjamin Y. H. Liu and Richard C. Jordan, "The Interrelationship and Characteristic Distribution of Direct, Diffuse and Total Solar Radiation," Solar Energy 4, (3), 1 (July 1960).

[4]J. A. Duffie and W. A. Beckman, *Solar Energy Thermal Processes*, John Wiley and Sons, New York (1974).

[5]C. MacPhee, ed., *Handbook of Fundamentals*, American Society of Heating, Refrigerating and Air Conditioning Engineers, New York (1972), pp. 621–26.
[6]Aladar Olgyay and Victor Olgyay, *Solar Control and Shading Devices*, Princeton University Press, Princeton, New Jersey (1957), pp. 32, 33.

2. INFRARED-REFLECTING SELECTIVE SURFACE MATERIALS WHICH CAN BE USEFUL FOR ARCHITECTURAL AND/OR SOLAR HEAT COLLECTOR WINDOWS *

Introduction

In Section 1A of this report, the contribution of thermal radiation transport to the space heating losses through architectural windows has been determined for many different types of window systems. The results show that thermal radiation transport contributes significantly to space heating losses. These thermal radiation losses can be reduced substantially by the use of selective surface materials which can be deposited on glass and/or plastics. The "selectivity" of these materials refers to selective optical properties. These optical properties are, primarily, high infrared reflectivity and high visible transmission. For specific situations where solar load on summer air conditioning is substantial, a reduction in the visible transmission of the selective surface may be desirable.

This section will review selective surface materials. The discussion given will pertain primarily to architectural windows, but will have relevance also to solar heat collector windows. The efficiency of a solar heat collector depends upon minimizing the thermal heat losses.

We will divide this section into three parts:

(A) The first part will deal with a discussion of the wavelength selective properties of certain semiconductor and metallic films which show desirable properties for prospective window coatings. We will not attempt to make cost evaluation of materials in cases where technology can substantially reduce cost in the future. We also include a brief discussion of materials suitable for protecting a selective film from mechanical and/or chemical degradation without altering the selective optical properties of the surfaces.

(B) The second part deals with the important problem of bonding selective surface materials to plastics, without altering the selective aspects of the films. If *through further research* we can successfully develop inexpensive laminates, we will have a valuable product to accommodate the architectural window retrofit market. Also, such plastic laminates could be useful for solar heat collector windows where reduction of weight loading of collector systems would be desirable.

(C) The third part of this section deals with the basic physics of selective infrared-reflecting doped wide-band-gap semiconductors. We know that SnO_2 and In_2O_3 when doped with certain impurities have highly desirable characteristics.

---

*Edited and compiled by S. D. Berman and S. D. Silverstein from technical contributions by D. E. Claridge, R. C. Langley, L. Muldawer, S. Schnatterly, and S. D. Silverstein.

We ask the questions:

Why do these materials work?
Are there any other candidates which could perhaps be better utilized?

A. Discussion of known selective surface materials

As discussed in other sections of this report, we basically desire surface materials which when deposited on glass or transparent plastic possess properties which:

(1) Provide high infrared reflectivity in infrared region commensurate with the thermal radiation spectrum. The peak of the thermal radiation spectrum is at ~10 μ for low temperatures and and ~5 μ for high-temperature solar collectors (400 °C).

(2) Provide high visible transmission for aesthetic purposes as well as for additional winter space heating from insolation.

In climates where air conditioning is necessary and costly, special solar control surfaces which exclude insolation in the summer are desirable.

Items (2) and (3) are apparently diametrically opposed. It would take some exotic, yet to be developed, thermochromic or electrochromic type materials to fit all three desirable criteria (see discussion of electrochromic materials in Appendix A). Categories (1) and (2) are satisfied by some known wide-band-gap semiconductors, such as heavily doped SnO_2 or In_2O_3.[1-3] See Figure 2.1 for the optical properties of these wide-band-gap semiconductors. These semiconductors can be doped with impurities with a plasma wavelength in the 1.3–4 μ range. Evidently categories (1) and (2) are required for windows protecting solar heat collectors. Categories (1) and (3) are simultaneously satisfied by thin metallic films which are commonly used today for architectural window solar control glass[4,5] to reduce the solar load on air conditioning. See Figure 2.1 for the transmission and reflection properties of metallic films. As discussed in 1 A and 1 B, the low-emissivity metallic surfaces preclude the transport of thermal radiation—but at the expense of solar heating in the winter. The results obtained from the analyses performed in Section 1 were that in northern climates solar control glass was not consistent with energy conservation. In southern latitudes there may be some possibility that solar control glasses with larger shading coefficients are consistent with energy conservation and cost effectiveness. A specific evaluation would have to be made for each locality taking into account the difference between heating and cooling energy rates. A very simple, conventional alternative to solar control glass is selective summer shading by white curtains, white window shades, venetian blinds, or external building overhangs.

Table 2.1 tabulates the selective properties, film thickness, deposition method,[6] and substrate temperature at which the film was deposited for a number of materials. The solar transmittance, T_{sol}, was determined by folding the transmission $T(\lambda)$ given in the references listed in Table 2.1 with the terrestrial solar spectrum as given by Drummeter and Hass.[7] The materials have been grouped according to substrate type (glass or plastic) and film type (semiconductor or metal). Glass substrates can be heated to much higher temperatures than plastics, either during or after deposition. This often results in signifi-

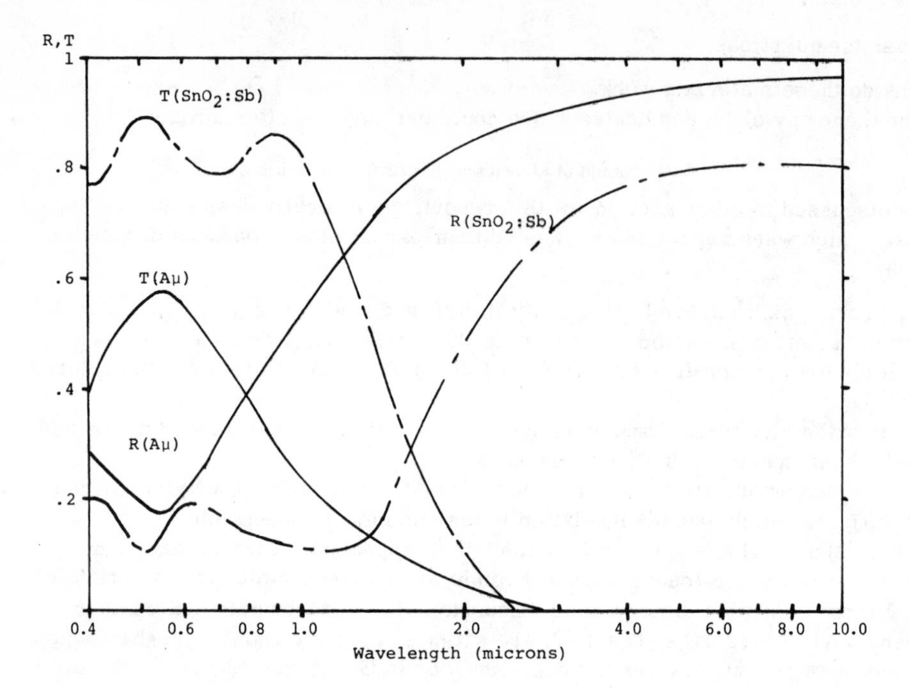

Figure 2.1. Schematics of reflectivity and transmission of thin films. Heavily doped semiconductor, $SnO_2:Sb$, at a thickness of ~3000 Å; and gold films at a thickness of ~130 Å.

cantly better optical and/or adhesive properties. As mentioned in the introduction to this section, plastic film substrates have potential for retrofit application. Metals exhibit selective optical properties at typical film thicknesses of ~100 Å, while semiconductor films typically require film thicknesses of a few thousand Å to achieve maximum selective properties.

In thin films, thickness control is important because non-uniform deposition will lead to "rainbow" (iridescent) effects due to interference. Such coloration, if excessive, would make these materials unacceptable aesthetically for the architectural window. The iridescence should be of no concern for solar collector windows.

Tin oxide doped with antimony or fluorine is a well-known selective infrared-reflecting film which can be deposited readily on glass.[1-3] The material is chemically stable and has excellent mechanical and abrasion properties. For the case of Sb as a substitutional impurity for Sn, high solubility is exhibited and free carrier concentrations in the range of 6×10^{20} carriers/cc have been obtained. The high free carrier concentration coupled with high mobility leads to infrared emissivities of $\simeq 0.2$. In fluorine doping, substitutionally for oxygen, the solubilities are lower than the antimony case and the infrared reflectivities are not as high. It is interesting to note, also, that the SnO_2 films doped with Sb

Table 2.1. Selective properties, film thickness, method of deposition, and substrate temperature at deposition for a number of materials.

Reference	Material	T_{sol}	R (10 μ)	Film thickness	Deposition	$T_{substrate}$
			Glassy substrates			
1, 2, 3	SnO_2 : Sb	0.77	0.8	3200 Å	Spray pyrolisis	460 °C
1	In_2O_2 : Sn	0.79	0.92	3100 Å	Spray pyrolisis	500 °C
3	In_2O_3 : Sn	0.71	0.92	5730 Å	Sputtered	> 500 °C
9, 10	Cd_2SnO_4 : In	~0.75-0.8	0.9	3500 Å	Sputtered	> 450 °C
12	LaB_6	0.17	0.86 (2.5 μ)	1100 Å	Sputtered-annealed	1100 °C
12	LaB_6	~0.54	~0.8	200 Å	Theory—bulk properties assumed	
	. . .	. . .	. . .	. . .	. . .	. . .
2	Ag	0.41	0.95	100 Å	. . .	. . .
2	Ag + ZnS	0.57	0.95		. . .	. . .
2	Au	0.38	0.95	130 Å	. . .	. . .
2	Au + ZnS	0.53	0.95		. . .	. . .
13	Cu	0.11*	0.92 (2.1 μ)	280 Å	Sputtered	. . .
			Plastic substrates			
15	In_2O_3 : Sn	~0.74	0.80	1.4 μ	Sputtered	. . .
16	SiO : Ag	~0.47	0.95	. . .	Evaporated	< 80 °C
16	SiO : Au	~0.51	0.83	. . .	Evaporated	< 80 °C
17	Cu_2S_{1+x}	0.37	0.74	. . .	Evaporated	80 °C
18	Cu_3PSeO_x	0.45	0.68	. . .	Evaporated	. . .
	Ag, Au, Cu: See text.					

*Solar Visible Transmission = 0.22.

have a bluish cast. This is due to a localized state caused by the impurity. The localized state has selective absorption in the red part of the visible spectrum. This reduction in transmission due to the absorption is not a problem for architectural windows—but for solar heat collectors such a reduction would be of considerable concern. The fluorine-doped system does not appear to exhibit any pronounced local state absorption.

Tin oxide coated glass is already in limited production for use in airplane windows. Here, it is used as a resistance heater for de-icing. The major barriers to architectural use are uniform deposition and cost.

Tin doped indium oxide has excellent selective properties combined with good durability.[1,8] It is somewhat more susceptible to acid attack than tin oxide, but a more serious barrier to widespread architectural use is cost and limited supply. If the total national annual consumption of indium were devoted to coating windows, it would be insufficient to coat all of the new window glass presently produced. As consumption is supply limited, widespread use of indium oxide as a window coating seems improbable. Glass coated with doped In_2O_3 is currently produced for use in liquid crystal displays, the retail cost running ~$6–$8/ft².

Cadmium stannate shows promise as a selective coating material with visible transmission and infrared reflectance properties comparable to SnO_2:Sb and In_2O_3:Sn.[9] The free carrier concentrations in cadmium stannate arise from non-stoichiometric excesses of metallic ions. Unfortunately, this method of generating free carriers is not nearly as desirable for long-term performance as impurity doping because oxygen diffusion can negate the desirable properties. The mechanical, adhesion, and corrosion properties of the cadmium stannate are quite favorable.[11] We believe the potential applicability of this material as a selective window surface would be substantially enhanced if methods could be developed to provide the free carriers by impurity doping. (Very recent results indicate that it is possible to achieve impurity doping with In and Ta.[11] The films prepared to date have not been single phase Cd_2SnO_4 but have included $CdSnO_3$ and CdO. Work is proceeding toward elimination of the CdO phase in the hope that this will result in even better solar transmission and long wave infrared reflection.)

Lanthanum hexaboride is typical of the class of metallic refractory compounds. It has excellent mechanical and chemical properties. Its bulk optical properties imply that it should have reasonable selective properties for a film 200 Å thick. Unfortunately, the optical properties of thin films prepared to date are distinctly inferior to the bulk properties, even when the film is annealed at 1100 °C.[12] The properties of the sputtered films become even worse as the thickness decreases. If good thin films can be perfected, they will be superior to metallic films.

Thin metal films offer selective properties in films approximately 100 Å thick and they do not suffer from iridescence. The properties of the metallic films are most suited to solar control glass, as the solar transmittance is usually low. The room-temperature emissivities of the metallic films are typically of order $\epsilon \leqslant 0.1$. They also require a protective cover if deposited on an exposed surface. The addition of antireflection coatings offers significant increases in solar transmission at added expense. The gold-coated solar control glass developed by Flachglas[5] utilized all of these factors to good advantage.

There are individual differences worth noting. Silver films have higher transmission toward the blue end of the spectrum and alter the color of transmitted light. The addition of a zinc sulfide antireflection coating peaks the visible transmission in a manner very similar to the eye's response curve.[2] Silver also tarnishes readily. Gold films don't tarnish, but the material cost is significant although not prohibitive. At $200/ounce, a 130-Å gold film has a material cost of ~$1.60/m^2 ($0.15/ft^2), if 100% use of the material is assumed. The transmission properties of a 280-Å copper film[13] and the bulk properties of copper lead us to expect that copper films of optimal thickness would have selective properties comparable to or slightly poorer than gold.

The best selective films deposited on plastic substrates reported to date are indium oxide doped with tin.[14,15] In_2O_3 has been successfully sputtered onto Lexan (trademark) polycarbonate polymethyl methacrylate (plexiglas) and allyl diglycol carbonate (CR-39). The best results were obtained with Lexan substrates. The film is somewhat more scratch resistant than the substrate. The films on plastics have not passed durability tests as severe as would be necessary for an exposed window coating. Desirable optical properties, adhesion, and durability are strongly correlated with a high sputtering rate. However, high sputtering rates typically lead to high levels of substrate heating. A satisfactory tradeoff has been found for plastic substrates ~3 mm thick, but problems would arise if high sputtering rates were attempted on large plastic films ~0.0025 cm thick. Promising results have also been obtained with Cd_2SnO_4 on plastic substrates.[11]

Kryzhanovskii and co-workers have investigated several semiconducting selective materials suitable for deposition on plastic substrates.[16-19] All of these materials are readily evaporated at low substrate temperatures. The SiO:Ag and SiO:Au films could be useful; they have selective properties comparable to thin metal films, good chemical stability, and reasonable mechanical properties. Their adhesive properties are not specified. The Cu_2S_{1+x} and Cu_3PSeO_x films do not appear to have application as window coatings. Their selective properties are poor and they are both mechanically fragile and chemically unstable.

Selective metallic films can be applied to thin plastic films. The Sierracin Corporation and Lockheed have applied selective thin films of gold, silver, and copper to plastic substrates, but we have no quantitative data on their optical properties. It seems safe to assume that the optical properties are not better than those obtained on glass substrates. They adhere well, but would require a protective covering just as they do on a glass substrate.

We have noted that some materials which have optical properties suitable for use as selective window coatings would not last long on an exposed window surface. A protective cover material seems to be indicated. We now consider the radiant heat transfer properties of the composite system (glass, reflective coating, protective coating). To minimize the thermal radiative heat transfer, the composite should have high reflectivity for room temperature radiation. The emissivity of any system is equal to the absorption. Accordingly, if we coat a low-emissivity metallic film with a protective material, the emissivity of the composite will remain low only if the protective film also exhibits a low infrared absorption. If we define r_p as the front-surface infrared reflection of the protective film, r_m as the reflectivity at the film metal interface, and β as the transmission of the protective film in the infrared, the net absorption of the system as a function of incident angle can be written as

$$\alpha(\theta) = 1 - r_p(\theta) \frac{[1 - r_p(\theta)]^2 \beta^2(\theta) r_m(\theta)}{1 - \beta^2(\theta) r_p(\theta) r_m(\theta)} . \tag{1}$$

Here θ is the angle of incidence relative to normal incidence of a particular ray. To get the actual emissivity of this composite structure, appropriate hemispherical averages must be made over the angles. Rather than go through a full derivation here, we will illustrate the behavior of the emissivity in the limiting cases. As $r_p \ll r_m$ we can expand the relation to obtain

$$\alpha(\theta) = 1 - r_p(\theta) - \beta^2(\theta) r_m(\theta) + \beta^2 r_m(\theta) r_p(\theta) + O(r_p^2) . \tag{2}$$

In the limit of small absorption, $\beta \to 1$, the emissivity goes to

$$\epsilon \to 1 - r_p - r_m , \tag{3}$$

which is quite close to the value for the pure metal. On the other hand, when the absorption is large, $\beta \ll 1$,

$$\epsilon \to 1 - r_p - \beta^2 r_m , \tag{4}$$

which is very close to unity. The relations given here illustrate the importance of having low-infrared-absorption protective coatings over metals. If the protective coatings have high infrared absorption, the low-emissivity selective properties of the metal are completely negated.

As an example of a metal plastic laminate currently on the market, partially transparent aluminized Mylar (trademark) polyethylene terephthalate films are available from a number of suppliers. These films are the least expensive retrofit product and the easiest to install as they can be readily cut to fit any window. These films when added to typical clear single pane of glass have solar transmission values of 14–31%. The Mylar film which protects the aluminum from abrasion is highly absorbing in the infrared.[20] This is particularly true in the region 7–15 μ, which coincides with the peak in the room-temperature black-body distribution. Thus applying one of these films to a typical single-glazed window lowers the quoted U value[21] from $U \simeq 1.06$ to $U \simeq 0.92$–0.96 instead of the $U \simeq 0.71$ as it would if the protective layer were transparent in the infrared. *This means that aluminized Mylar films will actually increase winter heating requirements because of the substantial reduction of solar energy input to the building through the windows.*

Examination of infrared transmission data reveals several plastics which should be considered as alternatives for this application. Polyethylene has very little absorption, even in a 0.1-mm thickness.[22] It does degrade rather severely from ultraviolet radiation, but the combination of window glass and the selective coating may preclude enough UV to permit reasonable life for polyethylene film. A 0.2-mm thickness of plexiglas has significant absorption from 7 to 9 μ [22] but is still much better than the 1-mil Mylar, so in a thin film it seems likely that low infrared absorption would result. It is possible to optimize the manufacture of plexiglas to reduce its infrared absorption. Plexiglas has good resistance to UV degradation. A third possibility would be the use of very thin films of Spraylon, a film similar to FEP Teflon (trademark) which has been developed by Haslim *et al.*[23] It can readily be applied in films as thin as 0.1 mil, does not require a high-temperature cure, and would probably have $\epsilon < 0.3$ if coated on a highly reflecting surface. Thus it seems probable that satisfactory protective coatings can be made from one or more plastics. *Further research investigations in this area are encouraged.*

B. Possible methods and research directions to obtain improved adherence of thin films on plastics

The window report has discussed in some length the desirable aspects of highly transparent infrared-reflecting surface materials. As costs get cheap enough, we would anticipate that the glass manufacturers will eventually manufacture and market windows using selective coatings in new window installations. However, there is a substantial need for methods of applying inexpensive, infrared-reflecting coatings to existing windows. This need could be fulfilled by a flexible plastic laminate which encompasses a surface which is an infrared reflector. Such plastic films could also have potential uses in residential solar heat collectors where additional weight load due to glass collector windows could pose problems. Here the plastic laminate would replace the glass altogether.

As discussed in Section 3A, a plastic laminate retrofit system is commercially available for reflection of solar radiation to reduce the load on air conditioning systems. There is no known plastic which is inherently a selective reflector of room-temperature thermal radiation. This necessitates that a thin film which reflects this radiative spectral region be used in conjunction with the plastic. As discussed in Section 3 A, the plastic used in the laminate system should be transparent in the infrared.

Fortunately, there is at least one commercially available plastic (polymethacrylate) which can be produced under conditions to give a flexible film having high transmission over the wavelength range of 7 to 13 microns. This property is important since it means that a thin film of an infrared-reflecting material can be sandwiched between a transparent plastic film and opaque (in the infrared) glass to give a system which reflects thermal radiation back into a heated area. When protected in this way, mechanical properties of infrared-reflecting thin films become substantially less important. However, the thin films must adhere sufficiently to the plastic to permit handling during processing (e.g., rolling and flexing) and must not delaminate due to changes in temperature or humidity after installation on glass.

It is in the area of adherence of metal and oxide thin films to plastic that research appears necessary if the goal of inexpensive, easily applied retrofit coatings for existing windows is to be realized. Some of the processes for applying infrared-reflecting films on glass use temperatures of 350 to 600 °C. For most plastics, there is an upper allowable temperature limit of about 100 °C. It is this constraint which requires variation in process methods to obtain optically useful thin films on plastic. Optical coatings have been applied to plastics by sputtering[24,25] and by evaporation. Processes which do not require vacuum appear preferable from a cost viewpoint.

An idea of how plastics might be coated with thin films of metals by variations of the mirror technique has evolved from our discussions. The concept is use of a difunctional organic molecule which has a reactive group at one end which can bond with the polymer surface. The other end of the difunctional molecule has a metal—the same one to be bonded. When applied in solution in a volatile solvent, it is expected that the difunctional molecule will orient itself with the metal side away from the plastic, thus providing a surface much more receptive to metalization than the untreated plastic. Using polymethylmethacrylate and silver as examples, the monomer and the difunctional molecule have the structures shown below in Figure 2.2.

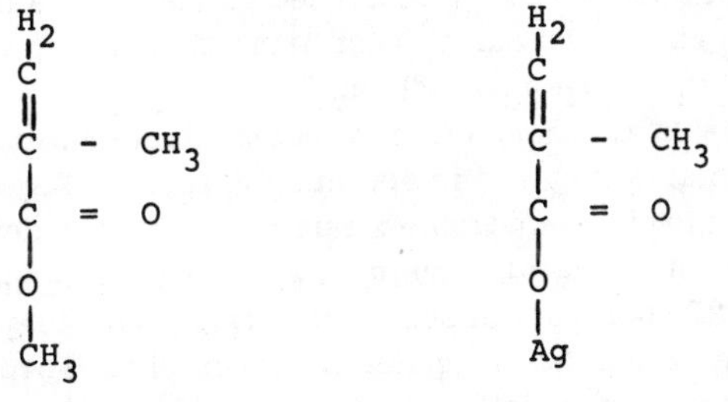

```
H2                 H2
|                  |
C                  C
||                 ||
C  -  CH3          C  -  CH3
|                  |
C  =  O            C  =  O
|                  |
O                  O
|                  |
CH3                Ag
```

Figure 2.2. Structures of monomer and difunctional organic molecule.

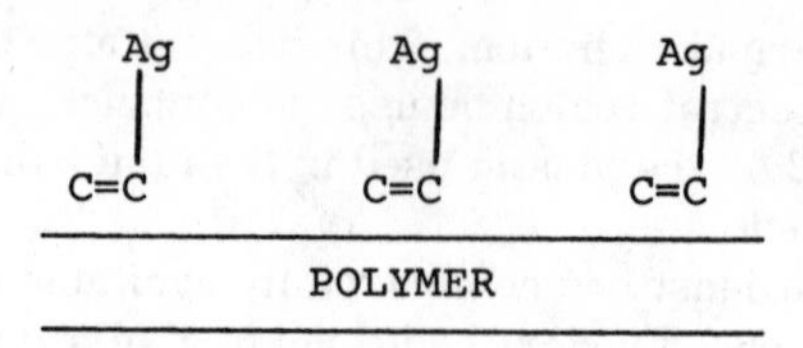

Figure 2.3. Ideal situation referred to in text.

Polymerized methylmethacrylate has a non-polar surface with some unreacted double bonds. This surface will attract the non-polar end of the difunctional molecule, i.e., the carbon-carbon double bond. Orientation of the polar, silver end of the molecule will occur during gradual evaporation of the solvent. Ideally, the situation will then be as shown in Figure 2.3.

Assuming that the oriented coating is present as a monomolecular layer, exposure to ultraviolet energy for a few seconds will cause the double bonds of the difunctional molecules to polymerize with available double bonds from the polymer and from adjacent difunctional molecules. The result will be a strongly bonded monomolecular layer.

Conceptually, the next processing step is to apply silver in a thickness of 100 to 200 Å to the coated polymer by the conventional silver mirroring technique. This process operates at about 18 °C and deposits silver by chemical reduction from a silver nitrate solution onto a sensitized surface. In our example, the oriented difunctional molecule serves as the sensitizer. The result expected is a firmly bonded infrared-reflecting silver film having 30% to 60% transmission of visible light.

The idea can be expanded to envision a silver methacrylate monolayer metalized by the mirror process with 100 to 200 Å of gold, nickel, or copper. Alternatively, approximately 2000 Å of tin might be applied by electroplating to a silver layer of minimum thickness, and then oxidized by electrochemical methods to give a tin oxide layer on the order of 3000 Å in thickness. Electroplating of tin on metallic substrates is carried out at about 90 °C, a temperature which most plastics can tolerate. However, tin oxide formed at low temperature tends to be amorphous and non-conductive. This difficulty might be overcome by electroplating platelets of crystalline, doped tin oxide on the metallic tin deposit before conversion to amorphous tin oxide. The result will be transparent, conductive platelets of tin oxide bonded by adherent amorphous tin oxide. Since refractive indices are the same, scattering should be minimal if the platelets can be deposited in an oriented fashion.[26]

The concept requires a research effort to demonstrate validity. It has strong similarity to the problems involved in obtaining adhesion of synthetic resins to glass. This field received considerable attention during the development of fiberglass-reinforced plastics for structural applications in space vehicles and aircraft. Reference 27 contains a discussion of this technology. This concept can also be considered from the viewpoint of heat of chemisorption of polar and non-polar organic groups on metals in bulk form. The following from Reference 28 gives a comparison:

	Polar	Non-polar	
Group	$-COOH$	$-CH_3$	$-CH_2$
Calories per group	8.97	1.8	1.0

In addition to this technological approach, some basic research on mechanisms of adherence of thin metallic and oxide films to plastics in general would appear to be very helpful in achieving the goal of inexpensive retrofit systems for glass windows.

C. A search for semiconductor alternatives to SnO_2:Sb and In_2O_3:Sn infrared-reflecting thin films

SnO_2 and In_2O_3 have nearly ideal optical properties for selective coatings: high visible transmission and high infrared reflectivity. Their major drawback is that they must be applied at the factory. SnO_2 is applied by vacuum sputtering or by a chemical spray-pyrolytic oxidation of a tin halide or a tin organometallic dissolved in a solvent such as alcohol. In_2O_3 is usually applied by a vacuum sputtering technique. It is desirable then to discover additional materials having comparable optical properties which might be applied less expensively at the factory, or, ideally, with a simple room-temperature process by the homeowner or building maintenance crew.

It is useful to review here the physical basis of these optical properties as an aid in selecting possible alternative materials. A metal or a heavily doped semiconductor has a Fermi energy E_F which is large compared with kT at room temperature. In the region where interband transitions are absent the optical properties of such a material can be described by the well-known Drude model in which two parameters, the electron density n and the mean free time between collisions with the lattice, τ, completely determine the properties. Qualitatively, such a material exhibits a high reflectivity for frequencies below the plasma frequency $\omega_p = (4\pi ne^2/\epsilon m_{eff})^{1/2}$, and a low reflectivity above ω_p. The physical reason for this effect is well understood. Metals, being good conductors, always redistribute the electronic charge to screen out slowly varying electric fields, which is the reason for high metallic reflectivity of visible and longer-wavelength radiation. As the fields oscillate more rapidly, there comes a frequency at which the electrons can no longer respond fast enough to screen out the oscillating electric fields. This frequency corresponds to the plasma frequency. Thus, alkali metals exhibit transparency in the ultraviolet. The wide-band-gap doped semiconductors such as the SnO_2:Sb, F systems and In_2O_3:Sn are good conductors with plasma frequencies corresponding to wavelengths as short as 1.3 μm (depending on free carrier concentrations). Accordingly, these semiconductors are transparent in the visible range as their interband absorption edges occur in the ultraviolet, and are highly reflecting in the infrared region.

The ideal semiconductor for use as a window coating must be highly doped so that the plasma frequency falls in the near infrared, between the visible range and the room-temperature black-body range. A simple calculation of the low-frequency reflectivity using the Drude model leads to

$$R_{IR}(\omega) \simeq 1 - \frac{2}{\omega_p \tau}.$$

This is strictly valid only in the range $\omega_p \gg \omega \gg 1/\tau$ but experimentally it is known to be quite accurate in the range $\omega_p > \omega \gtrsim 0$.

The absorption length in the visible range ($\omega > \omega_p$) can also be estimated easily. We assume that the semiconductor band gap is above the visible, so there is no interband absorption and the index of refraction can be taken as constant. Then a straightforward calculation again using the Drude model yields the result

$$L = \frac{1}{\alpha} = \frac{n\lambda}{2\pi}\left(\frac{\omega}{\omega_p}\right)^3, \quad \omega_p \tau \simeq \frac{n\lambda}{\pi}\left(\frac{\omega}{\omega_p}\right)^3 \frac{1}{1 - R_{IR}},$$

where

L = absorption length for $\omega > \omega_p$,

α = absorption coefficient,

n = index of refraction,

R_{IR} = infrared reflectivity,

ω = frequency of visible light.

We see that high infrared reflectivity and long absorption length in the visible automatically go together. Putting in typical numbers ($R_{IR} \simeq 0.9$, $\omega/\omega_p \simeq 3$, $n = 2$, $\lambda \simeq 0.5\ \mu m$), we find

$$L \simeq 180\lambda \simeq 90\ \mu m.$$

As practical films are usually made to thicknesses $\sim 0.3\mu \ll L$, high visible transparency is usually achieved also in the selectively infrared-reflecting films.

We now present the results of a search of the existing known elemental and binary semiconductors and insulators which was made to discover suitable new selective surface materials. The search was based on the exhaustive listing by Strehlow and Cook[29] of 1504 semiconductors and insulators whose bandgaps had been measured as of 1973. The selection criteria used are as follows:

1. *Bandgap*. In order for the material to be transparent in the visible range, the lowest direct bandgap must be greater than approximately 3 eV. If the films are thin enough, one can tolerate indirect absorption edges in the visible.

2. *Dopability*. It must be possible to produce a large concentration ($\sim 5 \times 10^{20}$/cc) of free carriers by doping the semiconductor with impurities, or other chemical means. This eliminates the "good insulators" such as alkali halides since charged impurities in these materials produce localized defects rather than free carriers. Creating a free carrier (electron or hole) in a semiconductor raises its energy by the bandgap energy. Creating a localized defect costs an energy equal to the enthalpy of formation of the defect. So, generally speaking, small-bandgap semiconductors can be doped to produce free carriers, while large-bandgap materials cannot. This condition is clearly at odds with requirement (1) and will serve to eliminate many possibilities.

Requirements (1) and (2) together amount to saying, "Find the most covalent semiconductors you can whose bandgap is above the visible range."

3. *Insoluble in water but soluble in another room-temperature solvent.* Even though a film applied to a window could be protected with a suitable plastic film, it was felt that high water solubility would not be compatible with long life. In addition, if a simple room-temperature process is the goal, then the material must be soluble in some common solvent at ordinary temperatures. Information in the *Handbook of Chemistry and Physics* was used to apply these restrictions.

4. *Cost.* Last but not least, the material must be quite inexpensive if it is to gain popularity. High-cost materials were identified using a chemical supply company catalogue.

The resulting list of 13 materials which satisfy all four restrictions is shown in Table 2.2. This is approximately 1% of the original list which we compiled. It is likely that good candidates were bypassed by the above rather crude restrictions. In addition, other requirements, such as long mean free time in the doped material, long-term chemical stability (which might favor the oxides on the list), and health and safety requirements of the film and application must ultimately be considered.

In addition to ternary and more complex compounds, which we have not considered, there is another class of simple semiconductors which should be considered but which we have not included on the list. There are some semiconductors whose lowest gap is indirect and for which the first direct gap is above the visible, for example, silicon. The absorption length in the indirect region is of the order of 1 μm at room temperature. In addition, heavy doping will screen the electron-phonon interaction, reducing further the indirect absorption strength. (The Thomas-Fermi screening length for $n = 5 \times 10^{20}$/cc is 2.5 Å.) Consequently, materials of this type are likely to be sufficiently transparent in the visible to serve as selective coatings. Such materials are easier to dope since the first bandgap is so low.

Some additional materials which have not been considered here, such as $SiTiO_3$, $BaTiO_3$, V_2O_5, WO_3, and transition metal dischalogenides, among others, should also be scrutinized. As discussed in Section 2 A, it is interesting to note that cadmium stannate ($CdSnO_4$), which falls into this category, has been shown to possess desirable selective properties.

Materials research in semiconductors has played an important role in the technological growth of our society in the last quarter century. We don't recommend that extensive research projects be devoted specifically to new window surface

Table 2.2 Candidate binary semiconductors for selective surface properties.

CaB_6	SiN	Sb_2P_3
CuCl	S	PbO
CuI	ZnO	Bi_2O_3
MnO	ZnS	
NiO	TiO_2	

materials. We do recommend, however, that researchers working in semiconductors should have a cognizance of the applicability of selective surface materials. Then, if any new materials look promising they can then either pursue development themselves or transfer development to other interested, more product-oriented parties.

---

[1]H. J. J. vonBoort and R. Groth, "Low Pressure Sodium Lamps with Indium Oxide Filter," Philips Technical Review 29, 17 (1968).

[2]R. Groth and E. Kauer, "Thermal Insulation of Sodium Lamps," Philips Technical Review 26, 105 (1965).

[3]A. Ya. Kuznetsov, A. V. Kruglova, and B. P. Kryzhanovskii, "Films of Semiconducting Tin Oxide with Increased Conductivity," Journal of Applied Chemistry (USSR) 32, 1186 (1959).

[4]For complete descriptions see *Architectural Glass Products 8.26/Pp*, PPG Industries, 1973, and *Glass for Construction 8.26/Li*, Libbey-Owens-Ford Co., 1974.

[5]Rolf Groth and Walter Reichelt, "Gold Coated Glass in the Building Industry," Gold Bulletin 7, 62 (1974) and Rolf Groth, "Heat Reflecting Window Pane," U. S. Patent No. 3,781,077.

[6]For a discussion of film preparations see Marshall Sitting, *Producing Films of Electronic Materials*, Noyes Data Corporation, Park Ridge, New Jersey (1970), and L. I. Maissel and M. H. Francombe, *Introduction to Thin Films*, Gordon and Breach Publishers, New York (1973).

[7]Louis F. Drummeter, Jr. and George Hass, "Solar Absorptance and Thermal Emittance of Evaporated Coatings," in Physics of Thin Films, Vol. 2, George Hall and Rudolph Thun, eds. (Academic, New York, 1964), p. 305.

[8]D. B. Fraser and H. D. Cook, "Highly Conductive, Transparent Films of Sputtered $In_{2-x}Sn_xO_{3-y}$," J. Electrochem. Soc., Solid State Science and Technology 119, 1368 (1972).

[9]G. Haacke, "Research on Cadmium Stannate Selective Optical Films for Solar Energy Applications," Report NSF/RANN/SE/GI-39539/PR-73/2, January 1974.

[10]G. Haacke, "Research on Cadium Stannate Selective Optical Films for Solar Energy Applications," Report NSF/RANN/SE/GI-39539/PR-73/4, January 1974.

[11]G. Haacke, private communication.

[12]K. R. Peschmann, J. T. Calow, and K. G. Knauff, "Diagnosis of the Optical Properties and Structure of Lanthanum Hexaboride Thin Films," J. Appl. Phys. 44, 2252 (1973).

[13]Howard Gillery, private communication.

[14]H. Y. B. Mar, private communication.

[15]W. T. Boord and H. Y. B. Mar, "Research on Application of Semiconducting Films for Eye Protection from IR Lasers," Quarterly Report No. 2, Contract No. F41609-74-C-0003, Honeywell, Inc., April 15, 1974.

[16]V. P. Kryzhanovskii, A. Ya. Kuznetsov, and L. A. Pafomova, "Reflection in the Long-Wavelength Region of Semiconducting Films of Silicon Monoxide Containing Silver and Gold," Optics and Spectroscopy (USSR) 15, 447 (1963).

[17]V. P. Kryzhanovskii, "Reflection of Semiconducting Copper Sulfide Layers in the Infrared Region of the Spectrum," Opt. and Spect. (USSR) 24, 135 (1968).

[18]V. P. Kryzhanovskii, "Some Optical and Electrical Properties of Thin Films of Cu_3PSe, Cu_3PSeS_x and Cu_3PSeI_x," Opt. and Spect. (USSR) 25, 343 (1968).

[19]V. P. Kryzhanovskii, B. M. Kruglov, and A. N. Kashtanov, "Electrically Conductive Coatings of Cu_3PS_x on Organic Glass," Instruments and Experimental Techniques (USSR) 11, 725 (1968).

[20]C. L. Tien, C. K. Chan, and G. R. Cunnington, "Infrared Radiation of Thin Plastic Films," Transactions of the ASME 94C (Journal of Heat Transfer), p. 41 (February 1972).

[21]"Technical Information for Reflecto-Shield Window Insulation," Material Distributors Corporation.

[22]W. L. Wolfe, ed., *Handbook of Military Infrared Technology*, (Office of Naval Research, Washington, D. C., 1965), p. 325.

[23]L. A. Haslim, S. A. Greenberg, M. McCargo, and D. A. Vance, "A Highly Stable Clear Fluorocarbon Coating for Thermal Control and Solar Cell Applications," AIAA Paper No. 74-117 presented at AIAA 12th Aerospace Sciences Meeting, Washington, D. C., January 30–February 1, 1974.

[24]R. C. Merrill *et al.*, "Continuous Sputtering of Metals on Plastic Film," J. Vac. Sci. Tech. 9, 350 (1972).

[25]Private communication, Henry Mar, Honeywell, Inc., Minneapolis, MN.

[26]Private communication, F. Howard Gillery, PPG Industries, Pittsburgh, PA.

[27]L. Holland, *Properties of Glass Surfaces*, John Wiley and Sons, New York (1964), pp. 434-447.

[28]H. F. Payne, *Organic Coating Technology*, Vol. I, pp. 17-18, John Wiley and Sons, New York (1954).

[29]Strechlow and Cook, J. Phys. Chem. Ref. Data 2, No. 1 (1973).

APPENDIX A: SOME SPECULATIVE IDEAS

During the course of the summer study a number of participants came up with some speculative ideas relating to window systems. Such ideas *should be encouraged* at all times. Such ideas should also be evaluated by potential energy-savings and cost-effectiveness analyses of the type given in Section 1 at relatively early stages in their promotion, growth, and development.

In this appendix we will mention briefly some of the speculations generated by some of the members of the group. The fact that the idea is mentioned here does not constitute an endorsement or a recommendation. In retrospect, most of these speculations appear to be impractical. Nonetheless, they are illustrative of the type of ideas which many readers will generate—and then hopefully evaluate for their potential impact *ab initio*.

1. The following are *methods to increase transmission of metal films* while maintaining high infrared reflectivity.

(a) Both S. Schnatterly and A. S. Barker pursued some theoretical studies of the effects of surface plasmons on the enhancement of the transmittance of metallic films. They concluded that controlled surface roughness would give a positive effect, but the degree of the enhancement of the selectivity of the surface was difficult to estimate.

(b) P. Baumeister suggested the use of three-layer films consisting of a metallic film sandwiched between two dielectric layers. This laminate is an interference filter with a broad pass band in the visible spectrum. The composite structure which was suggested during the study was a $TiO_2/Ag/TiO_2$ sandwich on glass.

In a recent issue of Applied Physics Letters, Fan *et al.*[1] published studies of the composite system independently suggested by Baumeister. The performance characteristics which they experimentally obtained were 84% transmission at 0.5 μm and 98–99% reflectivity at 10 μm. The 84% transmission corresponds to the peak of the pass band; the average solar transmission *will be lower*. Another similar approach using dielectric layered films on Au has been proposed by Groth and Reichelt.[2]

Potential cost estimates for such interference layers on the large areas relevant to window glass are difficult to estimate. Interference layers require relatively precise thickness of material, which usually necessitates vacuum deposition techniques. With regard to the net heat transfer through windows, little is gained from going from a surface emissivity of $\simeq$0.15 (doped semiconductor) to a surface with an emissivity of $\simeq$0.02. The reason for this, of course, is that at these low values of emissivity the net heat transfer is dominated by convection. Therefore, performance-wise, these metal-dielectric interference composites will compete directly with the doped semiconductors. As the semiconductors are themselves cost-effectively marginal, and vacuum disposition is considerably more expensive than chemical spray, we would anticipate that the metal composites will not be competitive cost-wise.

2. The study group was quite enamored with the thought of using electrochromic materials, where one could conceivably change the optical characteristics by simply throwing a switch. This prospect has been reviewed for us by F. Howard Gillery, Senior Scientist, PPG Industries Inc.

The demand for a electrochromic variable transmission window arises mainly from a concern for occupant comfort in sunny areas of buildings. It is necessary in these circumstances to control glare and decrease the general light levels. Yet in the same areas on dark days it would be advantageous to admit more light. Naturally energy savings would also occur, partly because a decrease in solar radiation would cut air conditioning costs, and partly because admission of more light on non-bright days would decrease lighting costs.

The glass industry has investigated various ways of producing a window with varying light transmission. Some of these involved mechanisms which would be under the control of the operator, such as moving polarizers, liquid in double glazed units, or suspensions of dichroic particles between conductive coated glasses. Other systems have been automatically activated, like the silver chloride phototropic glass now quite common for eyeglasses, or thermotropic glass. Up to the present time all the devices had severe disadvantages in architectural applications because of cost or other miscellaneous problems.

The relatively recent discovery of the phenomenon of electrochromism appears to have promise for such a window. The term describes a wide variety of situations where a material develops a color by the application of electric field. Most commonly the electrochromic material is located between two transparent conductive electrodes and the change in color is seen in the light passing through the electrodes. Many materials exhibit electrochromism and several different mechanisms are possible, but the one of particular interest is a solid-state device consisting of four vacuum-deposited layers on a glass substrate, namely, a transparent conductive layer (say, of tin oxide), an electrochromic layer of tungsten or molybdenum oxide, a barrier layer of silicon monoxide, and lastly a second transparent conductive layer. Such a system is described in Ref. 3.

On applying a few volts (dc) in one direction the device darkens impressively in about 30 seconds and remains in that dark state when the voltage is removed. The clear state is recovered by reversing the polarity of the voltage. The device draws a current of a few milliamps per square inch. The performance is suitable in many respects for a variable transmission window but there is one major problem, that of fatigue. After about 50 cycles of darkening and fading, the action degenerates and the device is locked into a half-darkened half-faded state.

The electrochromic mechanism is apparently not well understood. It is postulated that electrons from the conduction band of the oxide are trapped in color centers. Some of these electrons can be removed by reversing the polarity but others must be permanently trapped leading to the irreversibility of the absorbing state.

It would be an interesting problem for the basic solid-state physicist to elucidate the mechanisms involved and propose and examine new materials which would have better performance. The practical results would perhaps be a variable transmission window giving greater comfort to the building occupants and substantial fuel savings to the economy. Another probable result, if the rate of darkening and fading could be increased, is the use of the system in digital displays, a rapidly growing market at home and abroad.

Some of the existing literature is given in References 4, 5, and 6.

3. M. Labes investigated the prospects of making an infrared reflecting paint using mixtures of different encapsulated cholesteric liquid crystals. The initial idea was to attempt to utilize the relatively narrow infrared-reflection bands exhibited by liquid crystals. By taking different liquid crystals in combination, it was hoped to produce a broadband infrared reflector. It was realized that there are a number of basic factors which rule out this approach:

To produce a broadband reflector, one would need a combination of many different liquid crystals. Such a paint would not be transparent in the visible, and would, at best, be translucent.

A paint made up of many particles, each with a sharp reflection region and a broad absorption region in different parts of the infrared spectra, will have an emissivity approaching unity.

[1]"Transparent heat-mirror films of $TiO_2/Ag/TiO_2$ for solar energy collection and radiation insulation," J. C. C. Fan, F. J. Bachner, G. H. Foley, and P. M. Zavracky, Applied Phys. Letters 25, 693 (Dec. 15, 1974).

[2]R. Groth and W. Reichelt, Gold Bulletin 7, 62 (July 1974) (published by The Research Organization, Chamber of Mines of South Africa, S. Holland Street, Johannesburg 2000); R. Groth, U. S. Patent No. 2, 781, 077 (1973).

[3]S. K. Deb and R. F. Shaw, U. S. Patent 3, 521, 941.

[4]S. K. Deb, A Novel Electrophotographic System, Applied Optics, Suppl. 3, Electrophotography 192 (1969).

[5]G. A. Castellion and D. P. Spitzer, U. S. Patent 3, 712, 710.

[6]G. A. Castellion, U. S. Patent 3, 578, 843.

APPENDIX B: FORMULAS FOR COMPUTING COST EFFECTIVENESS

Throughout the report we stressed the importance of making cost-effectiveness analyses in the evaluation of existing or innovative window systems. Let us review the formulas used in this computation:

If an item is purchased and

the initial capital cost per unit area is c (dollars),
the expected lifetime is L (years),
the interest rate is r (expressed as a fraction),

the payment per year for equal yearly payments over the lifetime L will be

$$P = \frac{(r+1)^L rc}{[(r+1)^L - 1]} .$$

If f_0 represents the initial fuel savings/unit area, and the expected average fuel inflation rate is α, then the average fuel savings/year-area, $\bar{f}$, over the lifetime L will be

$$\bar{f} = \frac{(1+\alpha)^L - 1}{L\alpha} f_0 .$$

For the capital improvement to be cost effective the average savings per year must exceed the average payment,

$$\bar{f} \geq P,$$

or

$$\frac{(1+\alpha)^L - 1}{L\alpha} f_0 \geq \frac{(r+1)^L rc}{[(r+1)^L - 1]} .$$

Alternatively, this can be reexpressed such that the ratio of the *initial* fuel savings/unit area, f_0, to the initial capital costs/unit area must be greater than

$$\frac{f_0}{c} \geq \frac{(r+1)^L \alpha r L}{[(1+\alpha)^L - 1][(1+r)^L - 1]} .$$

As an example, take $L = 20$ years, $r = 10\%$, $\alpha = 6\%$; then the ratio of the initial expected fuel savings to the initial capital costs must be greater than 0.064 to be cost effective.